Engineering Hydrology

Manish Kumar Goyal
Assistant Professor
Department of Civil Engineering
Indian Institute of Technology Guwahati

PHI Learning Private Limited
Delhi-110092
2016

₹ 425.00

ENGINEERING HYDROLOGY
Manish Kumar Goyal

ISBN-978-81-203-5243-8

Published by Asoke K. Ghosh, PHI Learning Private Limited, Rimjhim House, 111, Patparganj Industrial Estate, Delhi-110092 and Printed by Syndicate Binders, A-20, Hosiery Complex, Noida, Phase-II Extension, Noida-201305 (N.C.R. Delhi).

To

My Students

Contents

Preface

The subject of Engineering Hydrology encompasses scientific study of water. It deals with the movement, distribution, and quality of water, including the hydrological cycle and water resources. Therefore, it is essential to have a good understanding of hydrology.

There are several textbooks available on the various topics related to hydrology. However, considering the present requirement of the competitive world, these books lack much needed rigour. The present book is a response to the same. This book on engineering hydrology is a step forward to serve as a text for the course in hydrology at the undergraduate level. Explanation presented in the text, together with a number of solved examples, makes this book understandable to readers. Theory is further supported with illustrations and diagrams. Numerous multiple choice questions from competitive examinations such as GATE/IES are also included at the end of almost each chapter.

Contents and coverage

The book comprises eleven chapters, designed for one-semester course. A brief overview of each chapter is as follows:

Chapter 1 deals with the basic concepts such as hydrological cycle, water availability and water balance. Chapter 2 describes about the precipitation, which includes various types and forms of precipitation, rain gauges, mean precipitation methods and presentation of rainfall data. Chapter 3 discusses hydrological abstraction such as evaporation, transpiration, interception and infiltration. Chapters 4 and 5 deal with the streamflow measurement techniques and estimation of runoff. Chapter 6 introduces flood hydrographs and unit hydrograph derivation as well as synthetic unit hydrograph and instantaneous unit hydrograph. Chapter 7 covers the various types of probability distribution and flood frequency method, including risk and reliability. Chapter 8 deals with the flood routing and flood control measures. Chapter 9 introduces the fundamentals of groundwater flow. Chapter 10 deals with the components of rainwater harvesting and technology for the same. Chapter 11 discusses the extreme events such as droughts, floods and climate change.

The matter presented here has been considered by several authors who wrote book on Hydrology and relevant subject. I do not claim originality of ideas in any part of the book. It is hoped that this book will prove to be a boon to the engineering students who are preparing for competitive examinations such as GATE/IES.

Resources for instructors

The solution manual contains complete detailed solutions to the unsolved problems given at the end of each chapter.

Manish Kumar Goyal

Acknowledgements

I appreciate the unstinted support from my students, who helped in solving various problems. I wish to record my sincere thanks to my parents, brother and my wife for patience shown throughout the preparation of the book. The support from PHI Learning, especially from Shivani Garg, is gratefully acknowledged.

I welcome suggestions and/or criticisms from the readers of this book for improvement with all humility.

Manish Kumar Goyal

CHAPTER

1

Introduction

1.1 Definition

Water is the most abundant substance on the Earth, covering over about three-fourths of the Earth's surface and it is the source of all lives on the Earth. Water is one of the few substances that can be found in all three states (i.e., gas, liquid and solid) within the Earth's atmosphere.

The term 'Hydrology' is derived from the two Greek words—'Hudor' and 'logas' meaning 'water' and 'science', respectively. Thus, hydrology is a science related to water. *Hydrology* is the branch of scientific and engineering discipline which deals with occurrence, movement and distribution of water on the Earth. The focus of hydrology is that water circulates throughout the Earth through different pathways and at different rates. The hydrology of a region is affected by its weather pattern, geography, geology and vegetation. Practical applications of hydrology are found in flood control, irrigation, hydropower generation, navigation, sediment control, water supply, drainage system and aquatic life protection.

1.2 Hydrological Cycle

Hydrological cycle, also known as *water cycle*, is the global-scale, endless recirculatory process linking water in the atmosphere, on the land and in the oceans. This cycle has no beginning. The three fundamental processes of the hydrological cycle are—(a) evaporation and evapotranspiration, (b) precipitation, and (c) runoff, as shown in Figure 1.1. The solar radiation drives the water cycle and heats water in oceans and also in other water bodies. Evaporation from the ocean, ponds, lakes, reservoirs and transpiration from vegetation such as plant leaves take place due to the heat energy provided by the solar radiation. Water vapours rise, move upwards and form clouds, resulting in droplet growth. These clouds condense and fall back, resulting in precipitation as rain, hail, sleet, mist, dew, etc. A part of the precipitation flows over the land called *runoff*. A portion of the water reaches the ground, enters the Earth's surface through infiltration, enhances soil moisture and builds up the ground water table. A portion of runoff enters rivers in valleys in the landscape, with streamflow moving water towards the oceans.

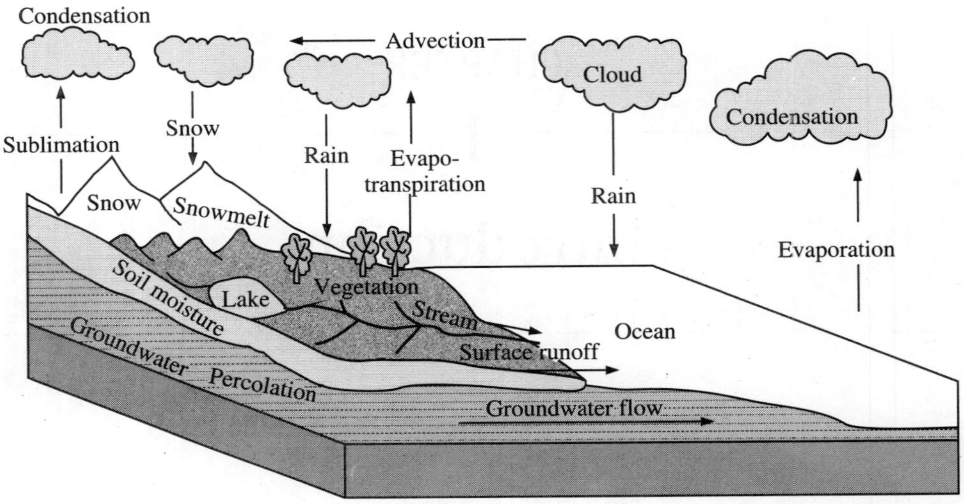

Figure 1.1 Hydrological cycle.

1.3 Key Hydrological Processes

The various processes in a hydrological cycle are as follows:

1. **Precipitation:** It is the process for all the moisture emanating from the clouds and falling on the ground.
2. **Evaporation:** Evaporation is the process by which water changes from liquid to gas or vapour.
3. **Transpiration:** Transpiration is the evaporation of water into the atmosphere from the leaves and stems of plants.
4. **Infiltration:** It is a part of precipitation which enters into the ground, and then, flows downwards.
5. **Runoff:** It is a part of precipitation that appears in surface streams, rivers, drains or sewers. This includes both surface and channel runoff.
6. **Subsurface flow:** Subsurface flow is the flow of water beneath the Earth's surface.

1.4 Water Balance Equation

The total amount of water available on the Earth is finite and conserved. The water balance equation simply incorporates principles of mass and energy continuity. This equation is a statement a mathematical description of the hydrological processes operating within a given time interval. In this way, the hydrological cycle is defined as a closed system, whereby there is no mass or energy created or lost within it. The mass of concern in this case is water. A water balance equation takes into account all of the water that goes in or out of the system. The water balance equation is given by

$$P - Q - E = \Delta S \tag{1.1}$$

where P is precipitation, Q is runoff, E is evaporation and ΔS is the change in storage. All terms in the Eq. (1.1) have the dimensions of volume.

EXAMPLE 1.1 Determine the evaporation from a reservoir of area of 25 km^2 for 1000 mm rainfall in a certain year. The average inflow and outflow are given as 2 and 2.5 cumecs [cubic metre per second (m^3/s)] in that year. Assume that there is no wet water exchange between the lake and the ground water.

Solution Given,

Area of reservoir = 25 km^2; Rainfall = 1000 mm; Average inflow = 2 cumecs; Average outflow = 2.5 cumes

Evaporation from the reservoir, (E) = ?

Since there is no net water exchange between the lake and the ground water, apply water balance equation.

Average annual outflow + Annual evaporation E – (Average annual inflow + Total annual precipitation) = 0

$$E = \frac{2 \times (3600 \times 24 \times 365)}{25 \times 10^6} + 1 - \frac{2.5 \times 3600 \times 24 \times 365)}{25 \times 10^6}$$

$$= 0.3693 \text{ m/year}$$

$$E = 369.3 \text{ mm/year}$$

EXAMPLE 1.2 Compute the average annual runoff from the catchment if the average annual precipitation in a 2000 ha catchment is 1400 mm and the corresponding value for the actual evapotranspiration is 800 mm.

Solution Given;

Average annual precipitation P = 1400 mm, catchment area A = 2000 ha, actual evapotranspiration ET = 800 mm, average annual runoff from the catchment R = ?

Assuming negligible infiltration in a catchment as well as negligible water consumption by plants, apply water balance equation.

(Actual evapotranspiration ET + Average annual runoff R)

– (Average annual precipitation P) = 0

$$ET + R - P = 0$$

$$R = P - ET$$

$$R = 1400 - 800$$

$$R = 600 \text{ mm} = 0.6 \text{ m} \times 2000 \times 10^4 \text{ m}^2 = 12 \times 10^6 \text{ m}^3.$$

Thus, average annual runoff from the catchment is 12×10^6 m^3.

1.5 World Water Distribution

Water on the Earth exists in a space called *hydrosphere*, which extends about 15 km up into the atmosphere and about 1 km down into the lithosphere, the crust of the Earth. The total quantity of water available on the earth is estimated approximately as 1386×10^6 km^3, and this may cover the Earth to an average depth of about 2.73 km (assuming the Earth as a uniform sphere of 12800 km in diameter). About 96.5% of all the water present on the Earth is in oceans as saline water. Of the remainder, about 1.7% is in polar ice, about 1.7% in ground water and only 0.1% in the surface and atmospheric water systems. An estimated distribution of water on the Earth is given in Table 1.1.

Table 1.1 Distribution of World's Water Resources

Location	*Water volume (in* km^3*)*	*Percentage of total water*
Oceans	1338000000	96.54
Ice caps, glaciers, and permanent snow	24364000	1.76
Ground water	23400000	1.69
Surface water	207610	0.015
Atmosphere	12900	0.001

1.6 Residence Time

The residence time T_r is the average duration for a water molecule to pass through a water body (i.e., subsystem of the hydrological cycle). The water contained in various subsystems or components of the hydrological cycle is continuously exchanged due to constant movement of water. It is calculated by dividing the volume of water in storage by the flow rate. This time also known as *turnover time*, provides a rough estimation in order to understand transport velocities in different components of hydrological cycle. The typical residence time of water in atmosphere is 9 days, while soil moisture generally has a residence time of one to two months. Shallow and deep ground water may have a residence time of 100 to 200 years and up to 1000 years, respectively.

EXAMPLE 1.3 The volume of atmospheric water is 12900 km^3. The evapotranspiration from land is 72000 km^3/year and that from ocean is 505000 km^3/year. Estimate the residence time (in days) of water molecules in the atmosphere.

Solution Given,

$$\text{Total flow rate} = 72000 + 505000 = 577000 \text{ km}^3/\text{year}$$

$$\text{Residence time} = \frac{\text{Volume of water}}{\text{Total flow rate}}$$

$$= \frac{12900}{577000} = 0.023 \text{ year}$$

$$= 8.2 \text{ days}$$

1.7 Catchment or River Basin

The *catchment* or *river basin* can be defined as the area of land from which water from rain and, snowmelt flow into a body of water such as river, lake, wetland or sea. The input to the catchment is the rainfall/snowmelt which gets distributed over the area of catchment. The output of a catchment is the runoff, which is concentrated to the outlet of catchment. The main features of this hydrological unit are outlet and boundary. The outlet is the point at which all surface water in the catchment lets collected. A catchment may range in size from a matter of hectares to millions of square kilometres. These hydrological units are used to carry out any water planning and management studies.

1.8 Types of Hydrological Processes

The hydrological processes are very complex in nature and involve many variables that are correlated. These processes, in general, are classified as follows:

1. **Deterministic process:** A deterministic process does not consider randomness. This process is based on exact law and chances of occurrences of variables is ignored.
2. **Stochastic process:** In this process, the chances of occurrence of variables are taken into account. This process is also known as *probabilistic process.*

In strict sense, all hydrological processes are generally stochastic in nature. These are assumed as deterministic for analysis simplicity.

1.9 Applications of Hydrology in Engineering

In hydrology, scientific knowledge and mathematical principles are applied to solve water resources-related projects. Civil engineers are concerned with the planning, designing and construction of hydraulic structures for flood control, power generation, water supply and drainage. The objective is to ensure that the structure will work for its designed lifetime, that it cannot be destroyed by action of water and the project is completed economically.

It helps us to know the required reservoir capacity to assure adequate water supply for irrigation or municipal water supply in drought condition. The magnitude of flood flows enables us for safe disposal of the excess flow. Engineering hydrology enables us to find out the relationship between a catchment's surface water and ground water resources and variation in water yield. Spillway capacity, sizes of water supply pipelines and hydraulic structures schemes, all are designed on the basis of hydrological equations.

1.10 Hydrological Data

Hydrologists are involved in monitoring, management and protection of water resources. Hydrologists use detailed data sources for the efficient planning, development and sustainable use of natural water resources, ensuring water is supplied in the most cost-effective manner. Hydrological data may be categorised as follows:

1. **Historical data:** Such data is recorded at appropriate time and it is also known as *chronological data*. Generally, rainfall, discharge, etc. are counted under historical data.
2. **Field data:** As per necessity, such data is generated and recorded at any time by conducting experiments in field. Generally, ground water depth, sediment, etc. fall under this category.
3. **Laboratory data:** This type of data is generally generated in the laboratory and is most related to hydraulic engineering.

The basic hydrological data required are as follows:

1. Hydrometeorological data such as wind velocity, humidity, etc.
2. Climatological data
3. Streamflow data
4. Ground water table data
5. Evaporation data
6. Soil moisture data
7. Cropping pattern data
8. Water quality data

In India, primarily, Indian Meteorological Department (IMD) collects the hydrometeorological data and the Central Water Commission (CWC) monitors flow in major rivers of the country. Ground water data is normally available with the Central Ground Water Board (CGWB). Department of Agriculture has the data related to evapotranspiration and infiltration characteristics of soil. The Central and State Pollution Control Boards, CWC and CGWB collect water quality data.

1.11 Annual Rainfall in India

The average annual rainfall in India is 1190 mm. The distribution of rainfall is very uneven in terms of time and space across the country. More than 75% of the annual rainfall is received in the four rainy months of June to September. The average annual rainfall varies from about 150 mm in western Rajasthan to over 2000 mm in north-east part and Western Ghats. The coefficient of variation for rainfall (C_v) varies from 15% to 70%.The coefficient of variation is least in regions of high rainfall and largest in the regions of scanty rainfall.

1.12 Characteristics of Precipitation Over India

In India, the rainfall is predominantly dictated by the monsoon climate. The precipitation in India is usually considered to occur in the following four periods (two major seasons and two transitional periods):

1.12.1 Premonsoon

During this period (from March to May), there is little rainfall in the country. Some parts of south India and Assam receive considerable amount of convective rainfall. This occurs only during the evening hours, with high intensities. Between March and June, the temperature and humidity begin to rise steadily in anticipation of the south-west monsoon.

1.12.2 The South-West Monsoon

The south-west monsoon brings heavy rains over most of the parts of the country between June and October, and 70%–80% of annual rainfall over major parts of the country. The moisture-laden air masses from the Indian ocean advance, as low-pressure areas develop over the subcontinent and release their moisture in the form of heavy rainfall, as shown in Figure 1.2. The monsoon air masses continue across the country into Bay of Bengal branch and Arabian sea branch due to topography. The Arabian Sea branch of the south-west monsoon first hits the Western Ghats of Kerala, India, and moves northwards such as Karnataka, etc. The Bay of Bengal branch of south-west monsoon flows and winds arrive at the north-east regions first. Mawsynram, situated on the southern slopes of the Khasi Hills in Meghalaya, India, is one of the wettest places on the Earth. Then, the winds turn towards the west, travelling over the Indo-Gangetic Plain covering Bihar and Uttar Pradesh. In Delhi, monsoon is received by both branches during the fourth week of June.

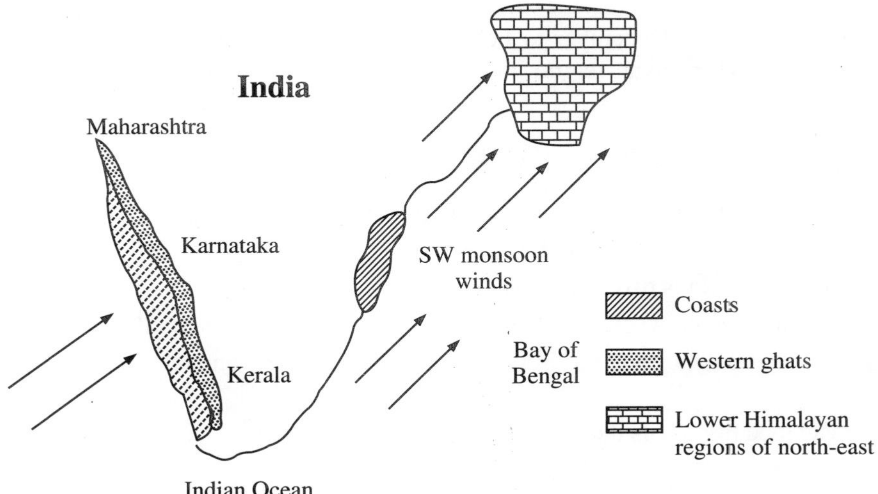

Figure 1.2 The south-west monsoon.

1.12.3 The North-East Monsoon

By ending of the south-west monsoon, the time span of this season is from October to November. During this period, rainfall is either due to tropical cyclone or due to the cold fronts displacing wet air masses generated in the Bay of Bengal, as shown in Figure 1.3.

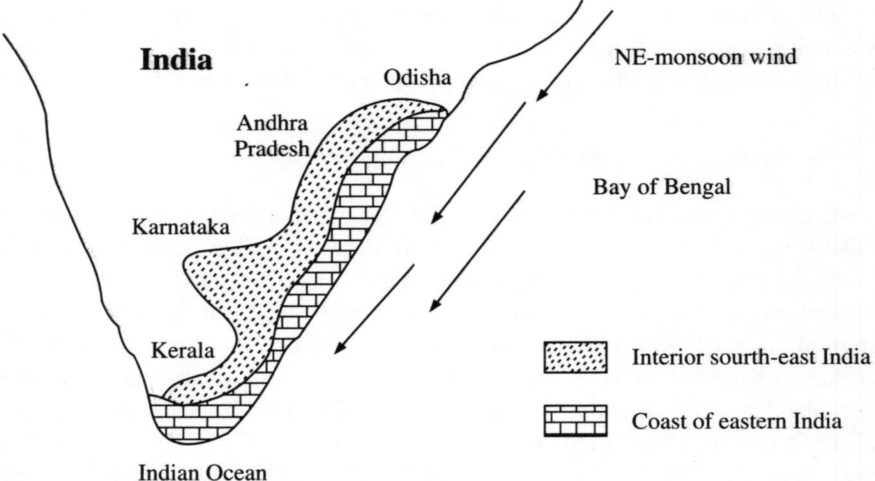

Figure 1.3 The north-east monsoon.

1.12.4 Winter Season

In the winter season (from December to February), due to western disturbances, some parts of northern India in the Himalayan region get small amounts of snowfall.

Summary

- Hydrology is the branch of science that deals with the occurrence, movement and distribution of water on the the Earth.
- Hydrological cycle is driven by solar energy.
- Of the total estimate world water, the contribution of oceans is 96.5%.
- Fresh water is only a small portion of the total water (2.5%) and is mainly stored in glaciers and ice.
- The residence time T_r is the average duration for a water molecule to pass through a water body reservoir.
- The average annual rainfall in India is 119 cm.

Objective Type Questions

1. Regional hydrological cycle is shown in the following figure:

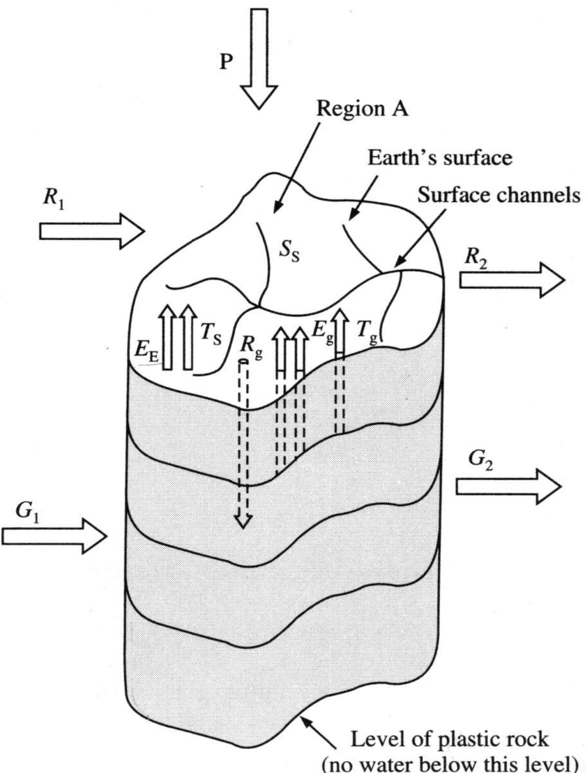

The correct hydrologic budget equation is [IES, 2002]

(a) $P + R_1 - R_2 + R_g - E_s - T_s - I = DS_s$

(b) $I + G_1 - G_2 - R_g - E_g - T_g = DS_g$

(c) $P - (R_2 - R_1) - (E_s + E_g) - (T_s + T_g) - (G_2 - G_1) = D(S_s + S_g)$

(d) $P - R - G - E - T = DS_s$

2. The quantitative statement of the balance between water gains and losses in a certain basin during a specified period of time is known as which one of the following? [IES, 2000]

1. Water budget
2. Hydrologic budget
3. Ground water budget

Select the correct answer using the codes given below:

(a) 1 only
(b) 2 only
(c) 3 only
(d) None of these

3. What is Hydrological cycle? (IES, 2009)

(a) Processes involved in the transfer of moisture from sea to land
(b) Processes involved in the transfer of moisture from sea back to sea again

(c) Processes involved in the transfer of water from snowmelt in mountains to sea
(d) Processes involved in the transfer of moisture from sea to land and back to sea again

4. Which of the following are pertinent to the realisation of hydrological cycle? (IES, 2012)
(i) Latitudinal differences in solar heating of the Earth's surface
(ii) Inclination of the Earth's axis
(iii) Uneven distribution of land and water
(iv) Coriolis effect
(a) (i), (ii) and (iii) only
(b) (i), (ii) and (iv) only
(c) (ii), (iii) and (iv) only
(d) (i), (ii), (iii) and (iv)

Answers

1. (c) **2.** (a) **3.** (d) **4.** (d)

IES Conventional Questions

PROBLEM 1.1 Following observations were made for conducting a water budget of a reservoir over a period of one month:
Average surface area = 10 km^2, mean surface inflow rate = 10 cumec, mean surface outflow rate = 15 cumec, rainfall = 10 cm, fall in the reservoir level = 1.5 m, pan evaporation = 20 cm. Assuming the pan factor as 0.7, estimate the average seepage discharge from the reservoir during the month. [IES, 2009]

Solution Given,

$$\text{Mean surface inflow rate, } I = 10 \text{ m}^3/\text{s}$$
$$\text{Mean surface outflow rate, } Q = 15 \text{ m}^3/\text{s}$$
$$\text{Rainfall, } P = 10 \text{ cm}$$
$$\text{Average surface area, } A = 10 \text{ km}^2$$
$$\text{Pan factor} = 0.7$$
$$\text{Time period, } \Delta t = \text{One month} = 30 \times 24 \times 60 \times 60 = 2.592 \times 10^6 \text{ s}$$
$$\text{Evaporation, } E = \text{Pan factor} \times \text{Pan evaporation} = 0.7 \times 20 \times 10^{-2} = 0.14 \text{ m}$$

For reservoir, the water budget equation in a time period Δt, can be written as:
Let the average seepage discharge be Q_s. Writing the water budget equation,

$$(I\,\Delta t + PA) - (Q\Delta t + EA + Q_s\Delta t) = \Delta S$$

$$\Rightarrow (10 \times 2.592 \times 10^6 + 10 \times 10^{-2} \times 10 \times 10^6) - (15 \times 2.592 \times 10^6 + 0.14 \times 10 \times 106 + Q_s \times 2.592 \times 10^6 = -1.5 \times 10^7$$

$$\Rightarrow 2.692 \times 10^7 - 4.028 \times 10^7 - Q_s \times 2.592 \times 10^6 = -1.5 \times 10^7$$

$$\Rightarrow Q_s = 0.6327 \text{ m}^3/\text{s}$$

PROBLEM 1.2 In a watershed of area 77000 hectares, the mean annual precipitation is 950 mm. About 25% of precipitation reaches the basin outlet as stream flow. Estimate the mean flow rate Q at this site in $m^3\ s^{-1}$? [IES, 2011]

Solution Given,

$$\text{Area of watershed, } A = 77000 \text{ h}$$

$$\text{Mean annual precipitation, } P = 950 \text{ mm}$$

Mean volume of precipitation over the catchment is computed by multiplying area of watershed with mean annual precipitation.

$$V = 77000 \times 10^4 \times 0.95 = 731.5 \times 10^6 \text{ m}^3$$

As stated, about 25% of precipitation reaches the basin outlet. Then, the volume of runoff that reaches outlet

$$= \frac{25}{100} \times 731.5 \times 10^6$$

$$\Rightarrow \qquad \text{Mean flow rate, } Q = \frac{\frac{25}{100} \times 731.5 \times 10^6}{365 \times 24 \times 60 \times 60} = 5.79 \text{ m}^3\text{/s}$$

Theoretical Questions

1. Define hydrology.
2. What do you mean by hydrological cycle?
3. Describe the hydrological cycle in nature with the help of a neat sketch, indicating its various phases.
4. Discuss the water balance equation.
5. What is the amount of available fresh water on the Earth and what is its distribution pattern?
6. Describe some applications of hydrology.
7. Describe briefly the sources of hydrological data in India.

Unsolved Problems

Problem 1: In a given catchment, two rainfall events were observed in close succession. After the first rainfall of 30 mm,total runoff of 15 mm was observed. Determine the runoff quantity after the second rainfall if the second rainfall was measured at about 20 mm.

Problem 2: In a given lake of surface area 150 km^2, inflow occurs from three small river channels P, Q and R and outflow occurs from river channel S. Water level in the lake was recorded 300 m and 300.15 m on 01 May and 31 August, respectively, in a given year. Use water balance method to compute the evaporation from lake surface during summer (May to August) in that year if the precipitation measured during this period is 150 mm. Average inflows and outflows are given below:

River channel	*Catchment area* (km^2)	Q_{avg} (m^3/s)
P	200	20
Q	180	25
R	160	22
S	–	35

Further Reading

Chow, V.T., Maidment, D.R. and Mays, L.W., *Applied Hydrology*, McGraw-Hill, Singapore, 1988.

Das, Ghanshyam, *Hydrology and Soil Conservation Engineering Including Watershed Management*, PHI Learning, Delhi, 2014.

Davie, Tim, *Fundamentals of Hydrology*, 2nd ed., Routledge, Taylor and Francis Group, London and New York, 2008.

Deodhar, M.J., *Elementary Engineering Hydrology*, Pearson, New Delhi, 2013.

Raghunath, H.M., *Hydrology—Principles, Analysis and Design*, New Age Publishers, New Delhi, 2014.

Subramanya, K., *Engineering Hydrology*, Tata McGraw-Hill, New Delhi, 2010.

Suresh, R., *Watershed Hydrology*, Standard Publishers Distributors, New Delhi, 2015.

Yadupathi Putty, Mysooru R., *Principles of Hydrology*, I.K. International, New Delhi, 2013.

CHAPTER

2 Precipitation

2.1 Introduction

Precipitation is the release of water from the atmosphere to reach the surface of the Earth. It is the primary input vector of the hydrological cycle. Its forms include rainfall, snowfall, and other processes by which water falls on the land surface such as hail, drizzle and sleet. In India, precipitation is mainly in the form of rainfall through there is appreciable snowfall at higher altitudes in the Himalayan range. The planning, assessment and management of water resources depend on the knowledge of the form and amount of precipitation occurring in a given region of considering over a time period of interest. The rivers of north India are perennial (i.e., sufficient quantity of water flows throughout the year), since they receive snowmelt runoff in summer, while in peninsular India, rivers receive only runoff due to rainfall and many of them have negligible flow during summer.

2.2 Types of Precipitation

Precipitation may be classified according to the conditions that generate vertical air motion/ lifting. In this respect, the three major categories of precipitation type are convective, orographic, and cyclonic.

2.2.1 Convective Precipitation

This type of precipitation is brought about by heating of the air at the interface with the ground, as shown in Figure 2.1. This heated air expands because of its lesser density. Air from cooler surroundings flows to take up its place, thus set up convective cell. The warm air continues to rise.

Dynamic cooling takes place, causing condensation and precipitation. Convective precipitation spans from light showers to storms of extremely high intensity. This type of precipitation is characterised by occurrence of hot weather conditions and it occurs in temperate zones at low latitudes. This type of precipitation is common in the equatorial and tropical areas. Convective precipitation occurs in the later afternoon hours and lasts for few hours only.

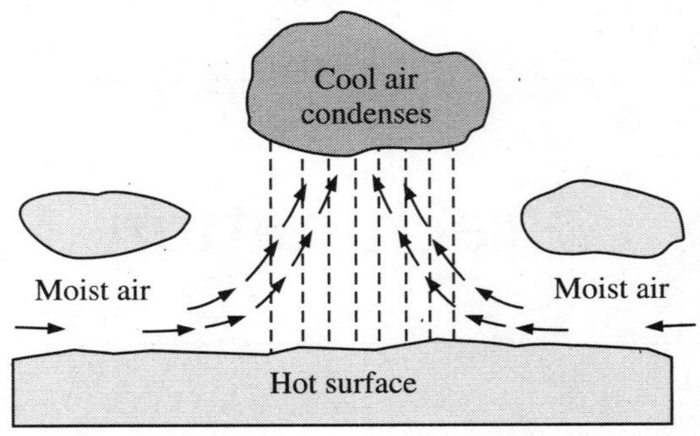

Figure 2.1 Convective precipitation.

2.2.2 Orographic Precipitation

This type of precipitation results from the mechanical lifting of moist air over mountain barriers and causes heavy precipitation on windward sides are shown in Figure 2.2. Factors that are important in this process include land elevation, local slope, orientation of land slope, and distance from the moisture source. In India, areas surrounding the lower Himalayan region and

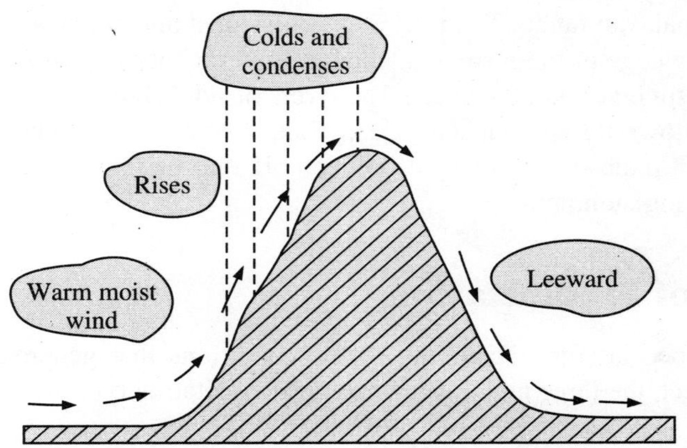

Figure 2.2 Orographic precipitation.

Western Ghats receive very heavy rainfall due to orographic lifting of south-west monsoon winds, blowing, respectively, from the Bay of Bengal and the Indian Ocean.

A comparison between convective and orographic precepitation is shown in Table 2.1.

Table 2.1 Differences between Convective and Orographic Precipitation

S. No.	*Convective precipitation*	*Orographic precipitation*
1.	Solar energy is main source of heat for the development of convective currents, thereby it occurs in the afternoon or in evening of the day.	This orographic precipitation depends on elevation and lifting of moisture-laden air mass.
2.	Convective precipitation covers small area.	It has wide head as compared to convective precipitation.

2.2.3 Cyclonic

The uneven heating of the Earth's surface by the Sun results high- and low-pressure regions. Cyclonic precipitation is associated with the lifting of air masses from high-pressure regions to low-pressure regions, as shown in Figure 2.3. This is of two types—frontal and non-frontal. *Frontal precipitation* is brought by the movement of warm air on one side over the cold air on the other side. If the air masses are moving so that warm air replaces cold air, the front is known as a *warm front*; if, on the other hand, cold air displaces warm air, the front is said to be cold.

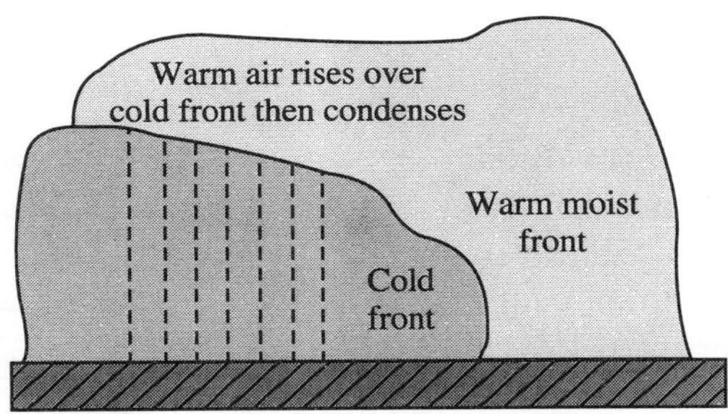

Figure 2.3 Cyclonic precipitation.

Non-frontal precipitation results from the air flows from surrounding high-pressure region to low-pressure region; lifted warm air cools down and causes precipitation.

Tropical cyclone, also known as *hurricane* or *typhoon,* is accompanied with high wind velocity and intense storm. Its diameter ranges from about 300 to 1500 km. The extra-tropical cyclone covers large diameter, which is up to 3000 km.

2.3 Forms of Precipitation

As indicated earlier a number of different forms of precipitation are listed below:

1. Rain: This includes drops of water that fall from clouds and have a diameter of at least 0.5 mm.

2. Snow: It is the precipitation in the form of ice crystals.

3. Hail: Precipitation in the form of rounded balls or irregular lumps of ice over 5 mm in diameter is called hail. Hails occur usually in violent thunderstorms.

4. Sleet: Sleet involves in the form of falling small particles of ice that are clear to translucent and are generally between 1 mm to 4 mm in diameter.

5. Drizzle: It is a fine smaller droplets of rain, with the range of diameter between 0.1 to 0.5 mm.

6. Glaze: It is the ice coating formed when freezing raindrops turn to ice on coming in contact with cold objects at the ground.

2.4 Measurement of Precipitation

Precipitation is measured by a network of rain gauges and snow gauges. Here, only rain gauges are discussed in this chapter. Generally, an open receptacle with vertical sides is used to measure rain. When installing rain gauges in the field, following points should be considered:

1. The site where rain gauge is installed be a flat ground and open space. The site should not be too close to any building or forested area.
2. The height of the fencing/protection should not exceed twice its distance from the gauge.
3. The installation of rain gauge should be avoided at steep hill side.

Rain gauges are classified into two categories—(a) non-recording rain gauge and (b) recording type rain gauge.

Non-recording Rain Gauge

This type of rain gauge does not record the data and collects only rainwater. The non-recording rain gauge extensively used in India is the Symon's rain gauge. It consists of a collector of 12.7 cm diameter connected to a funnel, as shown in Figure 2.4. In metallic container, there is a funnel that discharges the rainfall into a receiving bottle (vessel). Water collected in the bottle is measured using a suitably graduated measuring jar with 0.1 mm accuracy.

In India, rainfall is measured everyday at 08:00 AM 1ST and is recorded as the rainfall of that day. The unit of rainfall is millimetre.

During the days of heavy rainfall, measurements of rainfall are taken frequently (three to five times a day) because receiving vessel cannot hold more than 10 cm of rainfall and will overflow. Symon's non-recording rain gauge provides the total depth of rainfall

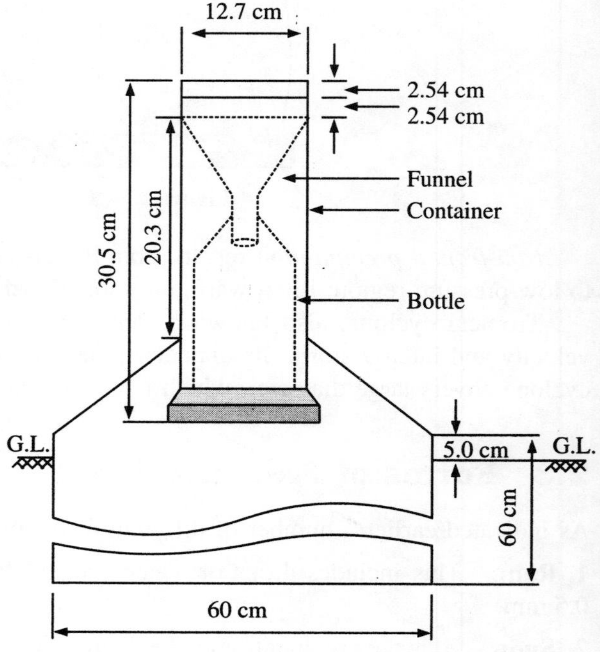

Figure 2.4 Schematic view of Symon's non-recording rain gauge.

during last 24 hours. It is to remember that this does not provide duration of rain and the variations in rainfall intensity.

Recording Type Rain Gauge

A recording or automatic rain gauge has a mechanical arrangement by which the total amount of rainfall since the time at which rainfall starts, gets automatically recorded on a graph paper. On the graph paper, time is recorded on the x-axis and the amount of rain is recorded on the y-axis. From this mass curve (a plot of cumulative rainfall versus time), depth of rainfall at any given time and intensity of rainfall during various time intervals can be computed with the help of slope of the plot.

They are three types of recording type rain gauges.

1. Tipping bucket type rain gauge: The rainwater from the funnel falls into one of a pair of small buckets as shown in Figure 2.5. One bucket will be in action at any given instant of time. When one bucket receives a rainfall of 0.25 mm (precalibrated depth of rain), it tips and brings other bucket in position. The water from the tipped bucket is collected into a tank below and the process is repeated. The tipping of bucket actuates an electrically-driven pen to trace a record on the chart mounted on a clockwork-driven drum. Such a rain gauge can be run using small battery and is useful for stations located in hilly and inaccessible areas.

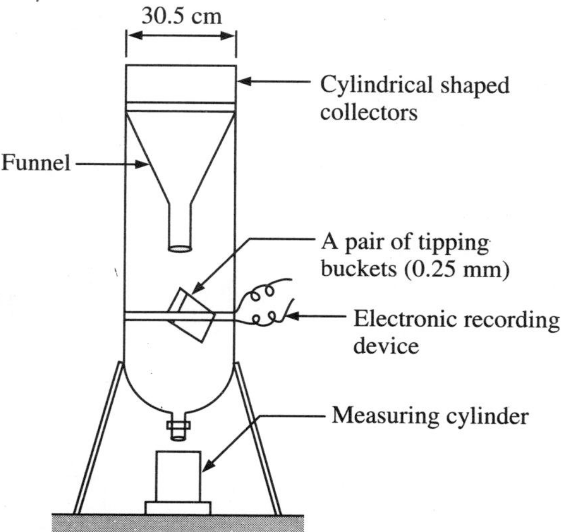

Figure 2.5 Tipping bucket type rain gauge.

2. Weighing bucket type rain gauge: In this type of rain gauges, the size of bucket is designed to collect the maximum rainfall expected to occur in a day as shown in Figure 2.6. When a certain weight of rainwater is collected in bucket, the bucket acts as a spring system, which makes the pen to move on chart. The instrument gives a plot of cumulative rainfall against time (mass curve of rainfall).

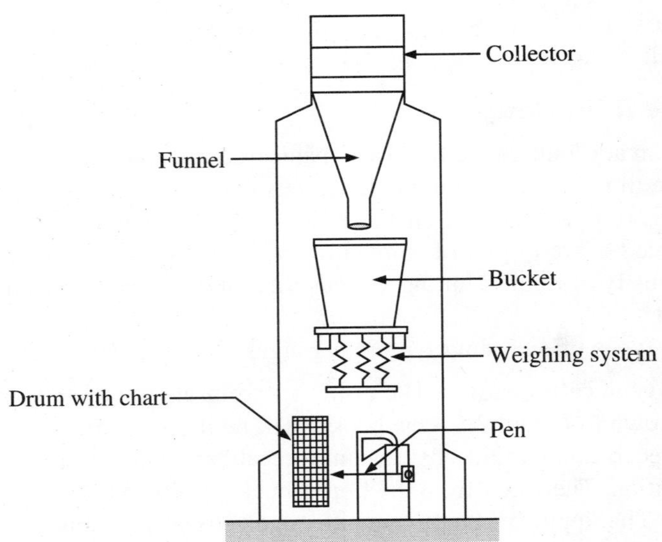

Figure 2.6 Weighing bucket type rain gauge.

3. Float type (natural syphon) rain gauge: This type of rain gauge is used by Indian Metereological Department (IMD). In this type of rain gauge, rainwater is collected into float chamber via funnel, as shown in Figure 2.7. In response to the amount of water collected, the float rises and a pen attached to the float through a lever system records the rainfall. When the float fills up, a syphon arrangement empties the float chamber.

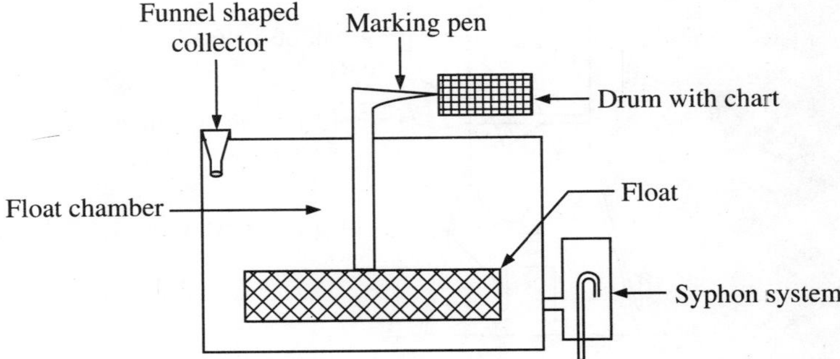

Figure 2.7 Float type rain gauge.

2.4.1 Radar Measurement of Precipitation

All of the above rain gauges measure precipitation at a point location only. However, Radio Detective and Ranging (radar) can be used to measure precipitation over a large area-spatial extent. The electromagnetic energy released and received back by radar is a measure of rainfall intensity. Indian Metereological Department (IMD) has a very well-established radar network across the country. The major limitation of this technique is that images often do not reveal rainfall-producing clouds because of overlying cloud layers.

2.5 Rain Gauge Network

The number of rain gauges and their distribution affect the nature of collected rainfall data. It is desirable to have a large number of rain gauges, to a feasible extent, in order to get a reasonably accurate estimate of the average depth of rainfall. The *rain-gauge density* is defined as the ratio of total area of catchment to the total number of rain-gauges in a catchment. McCulloch (1961) developed a relationship for computing the adequate number of rain-gauges (N) as follows:

$$N = \left(\frac{C_v}{\varepsilon}\right)^2 \tag{2.1}$$

where, C_v is the coefficient of variation and ε is the percent allowable error, generally taken as 10%.

In this regard, the World Meteorological Organisation (WMO) has made following recommendations for minimum number of rain gauges in a catchment:

1. Region I: Flat regions of temperate, mediterranean and tropical zones: Minimum one rain-gauge station for 600 to 900 km^2 is recommended. However, one rain-gauge station for 900–3000 km^2 is also acceptable.

2. Region II: Mountainous regions of temperate, mediterranean and tropical zones: Minimum one rain gauge station for 100 to 250 km^2 is recommended. However, one rain-gauge station for 250 to 1000 km^2 is also acceptable.

3. Region III: Arid and polar zones: Minimum one rain-gauge station for 1500 to 10000 km^2 is acceptable.

The Indian Standards Institute, ISI (IS: 4987-1968) has recommended the following regarding minimum number of rain-gauges.

(i) In plain regions, one rain-gauge per 520 km^2 of the catchment is recommended.
(ii) In region of average elevation, one rain-gauge per 260 to 390 km^2 of the catchment is recommended.
(iii) In predominately hilly areas with heavy rainfall, one rain-gauge per 130 km^2 of the catchment is recommended.

In India, on an average, there is one rain-gauge station for every 500 km^2 while in developing countries, it is one rain gauge station for 100 km^2.

EXAMPLE 2.1 In a catchment, the annual rainfall recorded by six rain gauge stations are given in the Table below. Determine the number of optimum stations in the catchment, for error of 10% in the estimation of mean rainfall.

Station	A	B	C	D	E	F
Rainfall (cm)	80	100.2	178.2	108.2	97.7	135.5

Solution Given, Number of statistics, $n = 6$

Allowable degree of error, $\varepsilon = 10\%$

$$\text{Mean precipitation, } \bar{x} = \frac{80 + 100.2 + 178.2 + 108.2 + 97.7 + 135.5}{6}$$

$$= 116.63 \text{ cm}$$

Standard deviation,

$$\sigma = \sqrt{\frac{(80 - 116.63)^2 + (100.2 - 116.63)^2 + (178.2 - 116.63)^2 + (108.2 - 116.63)^2 + (97.7 - 116.63)^2 + (135.5 - 116.63)^2}{6 - 1}}$$

$$= \sqrt{\frac{(-36.63)^2 + (-16.43)^2 + (61.57)^2 + (-8.43)^2 + (-18.93)^2 + (18.87)^2}{6 - 1}}$$

$$= 35.18$$

$$\text{Coefficient of variation, } C_v = \frac{\sigma}{\bar{x}} \times 100\% = \frac{35.18}{116.63} \times 100\% = 30.16\%$$

$$\text{Optimum number of stations, } N = \left(\frac{C_v}{\varepsilon}\right)^2 = \left(\frac{30.16}{10}\right)^2$$

$$= 9.1 \cong 10$$

The optimum number of station for the given catchment is 10. Therefore, four additional rain gauges are needed.

2.6 Rainfall Intensity

The *intensity of rainfall* is the rate at which rain occurs and is measured in terms of depth per unit time. Based on intensity, the rainfalls are classified into the following three categories.

1. Light rain: Its falling rate is up to 2.5 mm/h.

2. Moderate rain: Intensity ranges from 2.5 to 7.5 mm/h.

3. Heavy rain: Its falling rate is greater than 7.5 mm/h.

2.7 Estimation of Missing Data

Many a time, due to instrument failure or absence of observer, it is possible that the measurement of rainfall may consist of short breaks. Under such conditions, these missing records are computed with the help of one of the following methods:

1. Arithmetic average method: In this method, the value of missing rainfall point is calculated by taking average of the station's record. If the missing value of rainfall at station X is P_x and P_1, P_2, P_3, ..., P_m are the rainfalls at the m surrounding rain-gauges, then

$$P_x = \frac{1}{m}(P_1 + P_2 + P_3 + ... + P_m) \tag{2.2}$$

It is to be remembered that rainfall records of at least three rain-gauge stations should be there to yield good results. The limitation of this method is that variation in normal annual precipitation at the adjacent stations and missing station should be within 10%.

2. Normal ratio method: In the normal ratio method, the missing rainfall (P_x) at station X is estimated by weighing normal annual precipitation of each index station.

Let N_1, N_2, N_3, ..., N_m be the normal annual rainfalls at the m surrounding rain gauges, respectively, N_x be the normal annual rainfall at station X for which rainfall data (P_x) is missed. Then,

$$P_x = \frac{1}{m}\left(\frac{N_x}{N_1}P_1 + \frac{N_x}{N_2}P_2 + ... + \frac{N_x}{N_m}P_m\right) \tag{2.3}$$

This method is used when normal annual rainfalls at surrounding stations differ from the annual rainfall of the station at which rainfall data is missed by more than 10%.

3. Inverse distance square method: In this method, the missing rainfall is determined as a weighted average of the observed rainfall at the surrounding stations. This method provides greater weights to the points closest to the prediction location, and the weights diminish as a function of distance.

Let P_1, P_2, and P_3 are the rainfalls at the stations A, B, and C, respectively. Then, rainfall at station X for which rainfall data (P_x) is missed, is given by

$$P_x = \frac{\frac{1}{r_1^2}P_1 + \frac{1}{r_2^2}P_2 + \frac{1}{r_3^2}P_3}{\frac{1}{r_1^2} + \frac{1}{r_2^2} + \frac{1}{r_3^2}} \tag{2.4}$$

Here, r_1, r_2, and r_3 be the distances from station X whose data is to be estimated to each estimator station (A, B and C).

4. Regression method: A multiple linear regression is assumed between rainfalls at the surrounding stations and rainfall at station X for which rainfall data (P_x) is missed. This may be written as follows:

$$P_x = k_0 + k_1P_1 + k_2P_2 + k_3P_3 + ...+ k_mP_m \tag{2.5}$$

The constants k_0, k_1, k_2, ..., k_m are obtained through regression analyses of previously recorded data.

EXAMPLE 2.2 The normal annual rainfalls at the stations A, B, C and D in a catchment are 780 mm, 830 mm, 730 mm and 890 mm, respectively. Out of these four rain gauge stations, it was found that rain-gauge station D did not function for a part of month during a storm and the stations A, B, C recorded 90 mm, 80 mm and 100 mm rainfalls, respectively, produced by the storm. Calculate the missing storm rainfall at station D.

Solution Given,

Station	*Average annual rainfall* (mm)	*Normal annual rainfall* (mm)
A	90	780
B	80	830
C	100	730
D	?	890

Computing of missing storm rainfall using normal ratio method,

$$P_D = \frac{1}{m}\left(\frac{N_D}{N_A}P_A + \frac{N_D}{N_B}P_B + \frac{N_D}{N_C}P_C\right)$$

$$P_D = \frac{890}{3}\left(\frac{90}{780} + \frac{80}{830} + \frac{100}{730}\right)$$

$$P_D = 103.47 \text{ mm}$$

EXAMPLE 2.3 Consider station A to be at the origin of a rectangular coordinate system such that coordinates of four surrounding stations nearest to A in the quadrants are (20, 30) km, (–10, 15) km, (–15, –10) km, and (10, –5) km, respectively. Station X failed to record the rainfall during a storm. Estimate the missing rainfall at A if the storm rainfalls at the four surrounding stations are 75 mm, 90 mm, 70 mm, 60 mm, respectively.

Solution Let r_1 be the distance between station A and station in the first quadrant. Therefore,

$$r_1^2 = (20)^2 + (30)^2 = 1300$$

Similarly, for other stations,

$$r_2^2 = (-10)^2 + (15)^2 = 325$$

$$r_3^2 = (-15)^2 + (-10)^2 = 325$$

$$r_4^2 = (10)^2 + (-5)^2 = 125$$

Using inverse distance square method, missing rainfall of station at station A,

$$P_A = \frac{\dfrac{1}{r_1^2}P_1 + \dfrac{1}{r_2^2}P_2 + \dfrac{1}{r_3^2}P_3 + \dfrac{1}{r_4^2}P_4}{\dfrac{1}{r_1^2} + \dfrac{1}{r_2^2} + \dfrac{1}{r_3^2} + \dfrac{1}{r_4^2}}$$

$$P_A = \frac{\dfrac{1}{1300}\times 75 + \dfrac{1}{325}\times 90 + \dfrac{1}{325}\times 70 + \dfrac{1}{125}\times 60}{\dfrac{1}{1300} + \dfrac{1}{325} + \dfrac{1}{325} + \dfrac{1}{125}}$$

$$P_A = \frac{1.03}{0.014923} = 69.02 \text{ mm}$$

2.8 Mean Precipitation Over an Area

It is very important to know the areal distribution of precipitation for hydrologic analyses, water resources planning or designing the collectors for a city drain system. Usually, the rainfall over a large area is not uniform. The reliability of rainfall measured at one gauge in representing the average depth over a surrounding area is a function of (a) the distance from the gauge to the centre of the representative area, (b) the size of the area, (c) topography, (d) the nature of the rainfall of concern, etc. Generally, average depths for representative portions of the watershed are determined using arithmetic average method, Thiessen polygon method and Isohyetal method.

2.8.1 Arithmetic Average Method

The most direct approach is to use the simple arithmetic average of amounts of rainfall at the individual rain gauge stations in the area. This procedure is satisfactory if gauges are uniformly distributed and the topography is flat.

If $P_1, P_2, P_3, ..., P_m$ are the rainfalls recorded at the station 1, 2, 3, ..., m, then the average depth of rainfall $\overline{P}$ is given as

$$\overline{P} = \frac{1}{m}(P_1 + P_2 + P_3 + ... + P_m) \tag{2.6}$$

2.8.2 Thiessen Polygon Method

This method attempts to allow for non-uniform distribution of gauges by providing a weighting factor for each gauge. The area is divided into polygon subareas using the rain gauges as centres. The subareas are used as weights in estimating the watershed average depth. In the catchment, as shown in Figure 2.8, there are 12 rain gauge stations, of which seven are inside the boundary of catchment and five outside it. These rain gauges are joined together forming a network of triangles. Perpendicular bisectors are drawn to each sides of all the triangles, and then, joined together forming a polygon, one for each gauge. The area of the polygon around each rain gauge is measured.

If $P_1, P_2, P_3, ..., P_m$ are the rainfalls recorded at the station 1, 2, 3, ... m, and $A_1, A_2, A_3, ..., A_m$ are the area of polygon enclosed by them, respectively, then the average depth of rainfall $\overline{P}$ is given as

$$\overline{P} = \frac{(P_1A_1 + P_2A_2 + P_3A_3 + ... + P_mA_m)}{(A_1 + A_2 + A_3 + ... + A_m)} \tag{2.7}$$

The *advantages* of this method are as follows:

1. It makes use of data from nearby stations located outside the catchment.
2. It allocates importance of measurement according to the station spacing.

The *limitations* of this method are as follows:

1. This method does not make any allowances for orographic influences in basin. Thus, it is not suitable to compute the average areal rainfall of mountainous catchments.

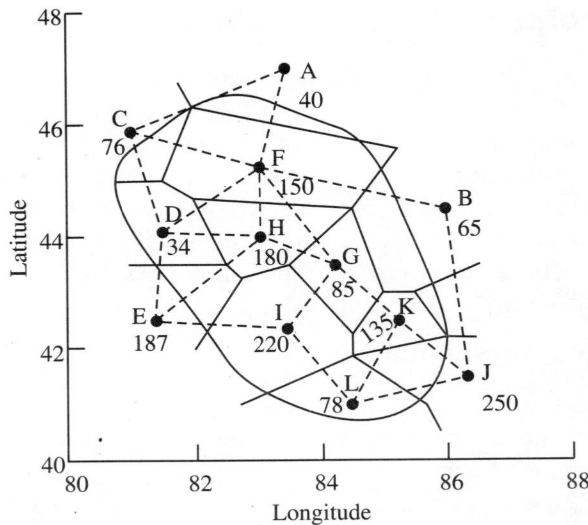

Figure 2.8 Thiessen polygon method.

2. If a new rain gauge is added to the existing network or position of rain gauge is changed in the catchment, then the network changes and new polygons are required to be sketched. Thus, in this case, a fresh computation of weights, and thus, average areal rainfall, is required.

2.8.3 Isohyetal Method

This method is based on interpolation between gauges. In this method, rain-gauge locations are plotted on a suitable map and rainfall amounts are recorded. Then, an interpolation between rain gauges is performed and rainfall amounts at selected increments are plotted. Identical depths from each interpolation are then connected to form isohyets (lines of equal rainfall depth). Once the isohyetal map is developed, the area A_j between each pair of isohyets, within the catchment, is multiplied by the average P_j of the rainfall depths of the two boundary isohyets to compute the areal average precipitation, as shown in Figure 2.9.

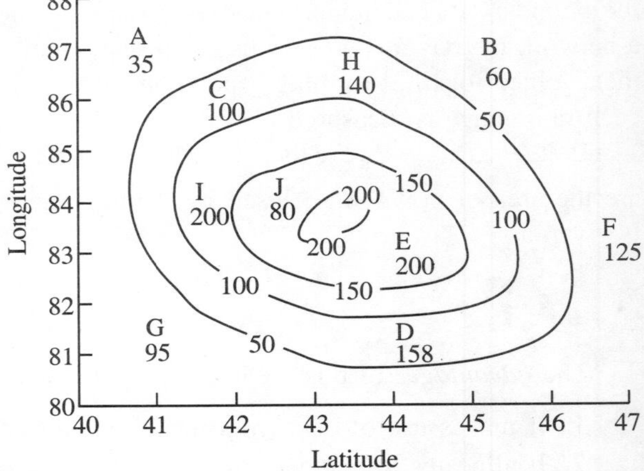

Figure 2.9 Isohyetal method.

The isohyetal method is flexible and the most accurate approach for determining the average precipitation over an area. It can show orographic effects and can be adopted for pictorial presentation. However, it needs a fairly dense network of gauges to correctly construct the isohyetal map and is time-consuming compared to the other methods.

Table 2.2 indicates the differences between Thiessen polygon method and isohyetal method.

Table 2.2 Differences between Thiessen Polygon and Isohyetal Method.

S. No.	*Thiessen polygon method*	*Isohyetal method*
1	This method does not make any allowances for orographic influences in basin.	This method is suitable to compute the average areal rainfall of mountainous catchments.
2	This method is only a mechanical and mathematical process and does not need any expertise.	This method needs special expertise in drawing of contours.
3	This method is useful in case of less number of stations.	In case of large number of stations, isohyetal method is more feasible than Thiessen polygon method.

EXAMPLE 2.4 In a river basin, Theissen weights obtained after constructing Theissen polygons for a network of 10 rain gauges are 0.16, 0.11, 0.12, 0.10, 0.09, 0.07, 0.12, 0.08, 0.06 and 0.09. Rainfalls recorded at these gauges during a storm are 112 mm, 119 mm, 130 mm, 125 mm, 140 mm, 186 mm, 177 mm, 202 mm, 145 mm and 166 mm, respectively. Determine the average depth of rainfall by Theissen mean and arithmetic mean methods.

Solution

Rain gauge station	*Theissen weightage factor*	*Station reading* (mm)	*Weighted station rainfall* (mm)
1	0.16	112	17.92
2	0.11	119	13.09
3	0.12	130	15.6
4	0.1	125	12.5
5	0.09	140	12.6
6	0.07	186	13.02
7	0.12	177	21.24
8	0.08	202	16.16
9	0.06	145	8.7
10	0.09	166	14.94

Average depth of rainfall – Thiessen mean

$$= 17.92 + 13.09 + 15.6 + 12.5 + 12.6 + 13.02 + 21.24 + 16.16 + 8.7 + 14.94$$
$$= 145.77 \text{ mm}$$

Average depth of rainfall – Arithmetic mean

$$= \frac{112 + 119 + 130 + 125 + 140 + 186 + 177 + 202 + 145 + 166}{10} = 150.20 \text{ mm}$$

EXAMPLE 2.5 Compute the average depth of rainfall due to storm for the catchment shown in the figure below. The isohyets due to a storm in a catchment are drawn and the area of the catchment bounded by the isohyets are tabulated below.

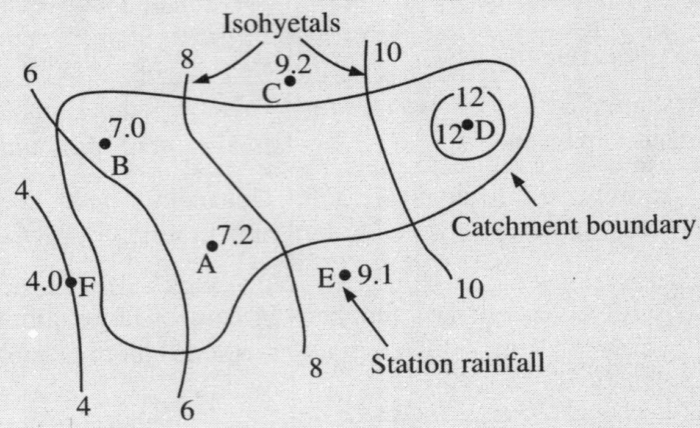

Isohyets (cm)	*Area* (km^2)
Station–10.0	15
10–8	120
8–6	80
6–5	22
4–2	10

Solution The necessary calculations are shown in the table below, for computation purpose, for the first area consisting of a station surrounded by a closed isohyets. A precipitation value of 10 cm is taken. For all other areas, the mean of two bounding isohyets are taken.

Isohyets	*Average value of P*	*Area* (km^2)	*Fraction of total area*	*Weighted P* (cm)
10	10	15	0.0607	0.6072
8–10	9	120	0.4858	4.372
6–8	7	80	0.3239	2.267
4–6	5	22	0.0890	0.445
2–4	3	10	0.0405	0.1215
Total	–	247	–	7.813

Mean precipitation, $\overline{P}$ = 7.813 cm

EXAMPLE 2.6 Following information regarding isohytes was obtained from the analysis of a storm. Determine the average depth of rainfall.

Isohyets interval (mm)	50–60	60–70	70–80	80–90	90–100	100–110
Area (km^2)	15	65	110	85	125	55

Solution Mean precipitation, $(\bar{P}) = \dfrac{\left(A_1\left(\dfrac{P_1+P_2}{2}\right)+A_2\left(\dfrac{P_2+P_3}{2}\right)+\ldots+A_{n-1}\left(\dfrac{P_{n-1}+P_n}{2}\right)\right)}{A}$

$$\bar{P} = \frac{15\left(\frac{60+50}{2}\right)+65\left(\frac{60+70}{2}\right)+110\left(\frac{70+80}{2}\right)+95\left(\frac{80+90}{2}\right)+125\left(\frac{90+100}{2}\right)+55\left(\frac{100+110}{2}\right)}{(15+65+110+95+125+55)}$$

$\bar{P} = 83.92$ mm

2.9 Frequency Analysis

The design and management of irrigation and drainage projects is based on the probability of occurrence of extreme rainfall events. With the help of a frequency analysis on historical rainfall data, the magnitude of the rainfall depths can be estimated.

Frequency of rainfall of a specified period is expressed in terms of recurrence interval. The *recurrence interval T* is defined as the value that is equal to or greater than the specified magnitude occurred once in T years.

$$T = \frac{1}{P} \tag{2.8}$$

where P is the plotting position of the event.

Using the rainfall data, plotting position of different years can be computed as follows:

$$P(\%) = \frac{m}{n+1} \times 100 \tag{2.9}$$

Here, m is the rank number of an event after arranging in descending order by its magnitude and n is total number of years of record. This formula is called *Weibull's formula*.

EXAMPLE 2.7 Determine the annual rainfall for return periods of 12 years and 60 years if the record of annual precipitation at given station covering a period of 22 years is given in the table below. What would be the probability of an annual rainfall of magnitude equal to or exceeding 105 cm occurring at given station? What is the 78% dependable annual rainfall at given station.

Year	1980	1981	1982	1983	1984	1985	1986	1987	1988	1989	1990
Annual rainfall (cm)	128	82	74	90	110	94	79	126	140	90	77

Year	1991	1992	1993	1994	1995	1996	1997	1998	1999	2000	2001
Annual rainfall (cm)	91	101	112	65	80	125	165	80	115	79	100

Solution In order to assign rank, data are arranged in descending order (as shown in table below). Then, probability is calculated using Weibull's formula.

$$N = 22$$

m (rank)	*Annual rainfall* (cm)	*Probability* $P = m/(N+1)$	*Return Period* $T = 1/P$ (years)
1	165	0.04347826	23.000
2	140	0.08695652	11.500
3	128	0.13043478	7.667
4	126	0.17391304	5.750
5	125	0.2173913	4.600
6	115	0.26086957	3.833
7	112	0.30434783	3.286
8	110	0.34782609	2.875
9	101	0.39130435	2.556
10	100	0.43478261	2.300
11	94	0.47826087	2.091
12	91	0.52173913	1.917
13	90	0.56521739	1.769
14	90	0.60869565	1.643
15	82	0.65217391	1.533
16	80	0.69565217	1.427
17	80	0.73913043	1.353
18	79	0.7826087	1.278
19	79	0.82608696	1.211
20	77	0.86956522	1.150
21	74	0.91304348	1.095
22	65	0.95652174	1.045

Annual rainfall with return periods of 12 years is obtained by interpolating between two appropriate successive values as shown below:

$$\text{Annual rainfall} = 165 + \frac{(140 - 165)}{(11.5 - 23)} \times (12 - 23) = 141.08 \text{ cm}$$

The return period of an annual rainfall of magnitude equal to or exceeding 105 cm occurring at given station can be calculated by interpolation as follows:

$$\text{Return period} = 2.875 + \frac{(2.556 - 2.875)}{(101 - 110)} \times (105 - 110) = 2.697 \text{ years}$$

Therefore, exceedence probability $= \dfrac{1}{2.697} = 0.37$

78% dependable annual rainfall at given station = Annual rainfall with probability $P = 0.78$
Therefore, return period, $T = 1/P = 1/0.78 = 1.282$ years.
Now, by interpolation,

$$78\% \text{ dependable rainfall} = 80 + \frac{(79-80)}{(1.211-1.353)} \times (1.282-1.353) = 79.5 \text{ cm}$$

2.10 Intensity-Duration Relationship

Rainfall intensity is defined as the rate of fall of rainfall per unit time, generally expressed in centimetre per hour (cm/h) or millimetre per hour (mm/h). It is generally observed that rainfall intensity is not same throughout the storm period. As the duration of storm increases, the maximum average intensity of the storm decreases. For a given place, if the observed maximum rainfall intensities for various durations such as 10 min, 20 min, ..., 1 h, 2 h etc. are plotted against the respective duration, then the resulting graph is known as *intensity-duration curve*, as shown in Figure 2.10.

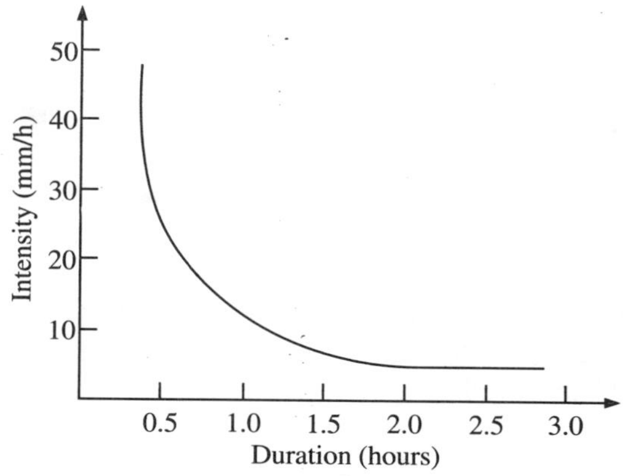

Figure 2.10 Intensity-duration curve.

2.11 Intensity-Duration-Frequency Relationship

An intensity-duration-frequency (IDF) curve is a graph with duration plotted as abscissa, intensity as ordinate and a series of curves, one for each return period, as shown in Figure 2.11. A minimum of 20 years data is desirable to develop IDF curve.

The procedure for developing the IDF curves is described below:

1. From the available rainfall data for a given station, rainfall serieses for different durations (e.g., 15 min, 30 min, 1 h, 2 h, 6 h, etc.) are developed. The rainfall intensity is computed with the help of rainfall data and duration for all durations.
2. The annual maximum intensity for all durations for each year is determined.
3. The frequency analysis of maximum intensity of rainfall for various return periods for all the selected durations are carried out.

4. Then, maximum rainfall intensity can be plotted against the return period for various durations, as shown in Figure 2.11.

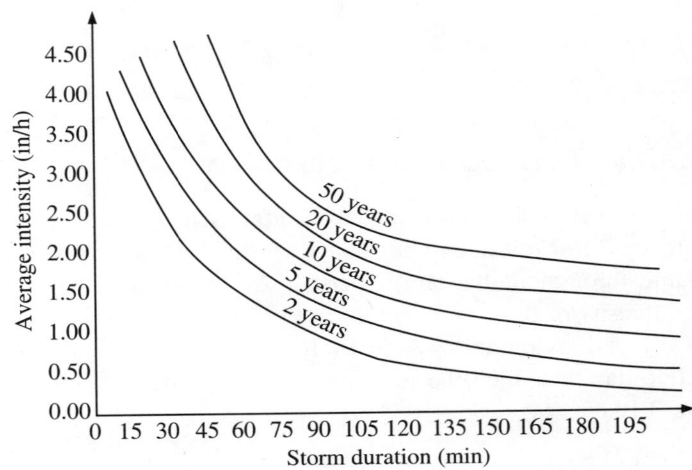

Figure 2.11 Intensity-duration frequency curve.

The relationship amongst intensity, duration and frequency of rainfall for Indian regions are given as below:

$$i = \frac{KT^a}{(t + b)^n} \tag{2.10}$$

where i is the rainfall intensity in centimetre per hour, T is the return period/recurrence interval in years, t is the storm duration in hours, and K, a, b and n are coefficients varying with geographical locations of the area.

2.12 Depth-Area-Duration Curve

The procedure for the development of depth-area-duration (DAD) curve is given below (Figure 2.12):

1. Determine 1 day, 2 days up to 5 consecutive days maximum average rainfall.
2. Plot the isohyetal maps for all maximum average rainfalls separately for each day.
3. Each isohyetal map is divided into several zones, representing the centre of main storm.
4. By using the planimeter, determine the area enclosed by each isohyet, starting from the centre of the storm of each zone.
5. Determine the incremental rainfall volume by multiplying the enclosed area between two isohyets and the mean of two adjacent values.
6. Compute the total volume of rainfall by adding the previous accumulated rainfall volume.
7. Determine the average depth of rainfall over that area by dividing the total rainfall volume with the total area of isohyetal map.

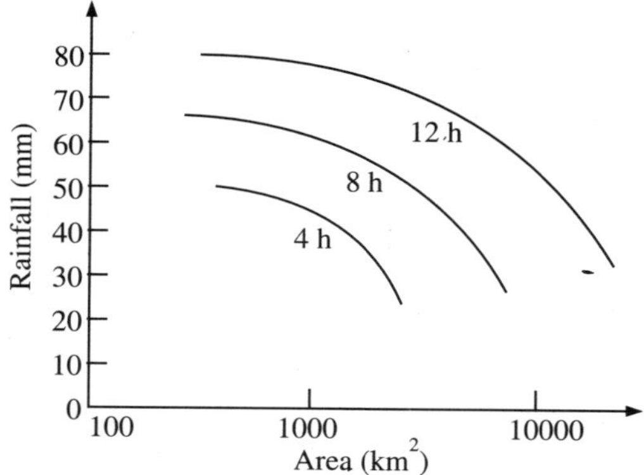

Figure 2.12 Depth-area-duration curve.

8. Compute the average rainfall depth for each zone and then combine the enclosed area by the common isohyets.
9. The highest average rainfall depth for various zones may be plotted and obtained curve is the depth-area-duration (DAD) curve for the maximum rainfall of a particular period.

2.13 Presentation of Rainfall Data

The presentation of rainfall data is given as (a) mass curve of rainfall (b) Hyetograph

2.13.1 Mass Curve of Rainfall

It is a plot of accumulated rainfall against time, plotted in chronological order, as shown in Figure 2.13. The rainfall measurements by recording type rain gauge such as float type, weighing bucket type are of this form. This type of plot is very useful in providing information on duration and magnitude of storm. The difference between any two ordinates of the mass curve is the rainfall during the corresponding duration of time.

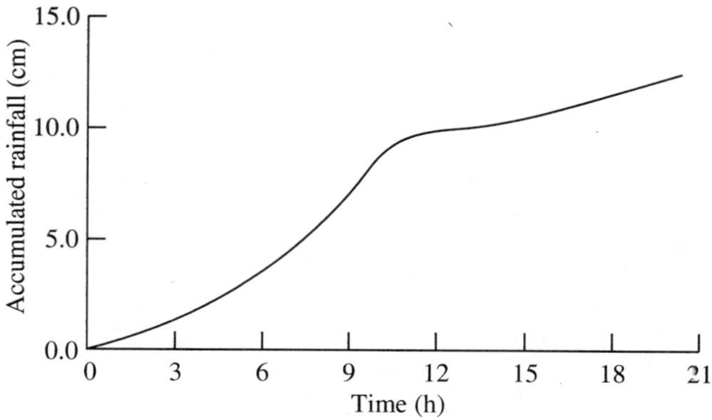

Figure 2.13 Mass curve of rainfall.

2.13.2 Hyetograph

It is a plot of rainfall intensity against time intervals, as shown in Figure 2.14. The rate at which rainfall accumulates at any given instant of time is called *rainfall intensity*. In order to develop the hyetograph, rainfall intensity data is extracted from mass curve. The total depth of rainfall received in a given time period is computed by area under the hyetograph curve. It is useful in developing design storms to predict extreme floods.

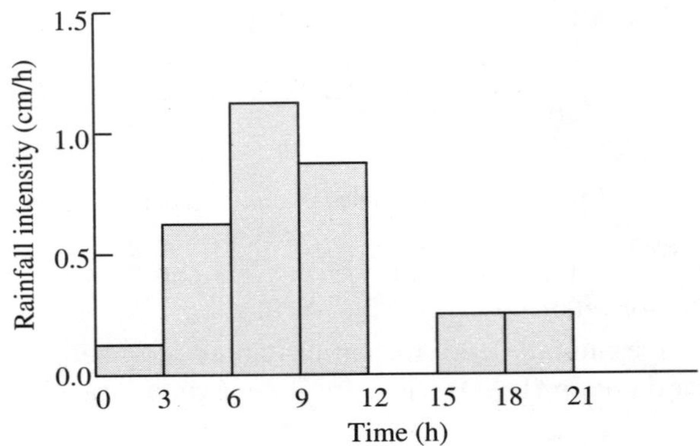

Figure 2.14 Hyetograph.

EXAMPLE 2.8 The ordinates of rainfall mass curve of a storm (in mm) that started at 9.00 AM recorded at 30 minutes interval are 0, 10, 18, 29.5, 44.5, 51, 67, 89.5, 99.5, 107, 119, 122 and 122. Plot the hyetograph for a uniform interval of 30 minutes.

Solution The necessary computations to obtain hyetograph are tabulated as below:

Time since start in minutes	*Cumulative rainfall* (mm)	*Incremental depth of rainfall* (mm)	*Intensity in* mm/h
0	0		
30	10	10	20
60	18	8	16
90	29.5	11.5	23
120	44.5	15	30
150	51	6.5	13
180	67	16	32
210	89.5	22.5	45
240	99.5	10	20
270	107	7.5	15
300	119	12	24
330	122	3	6
360	122	0	0

The rainfall hyetograph is shown below:

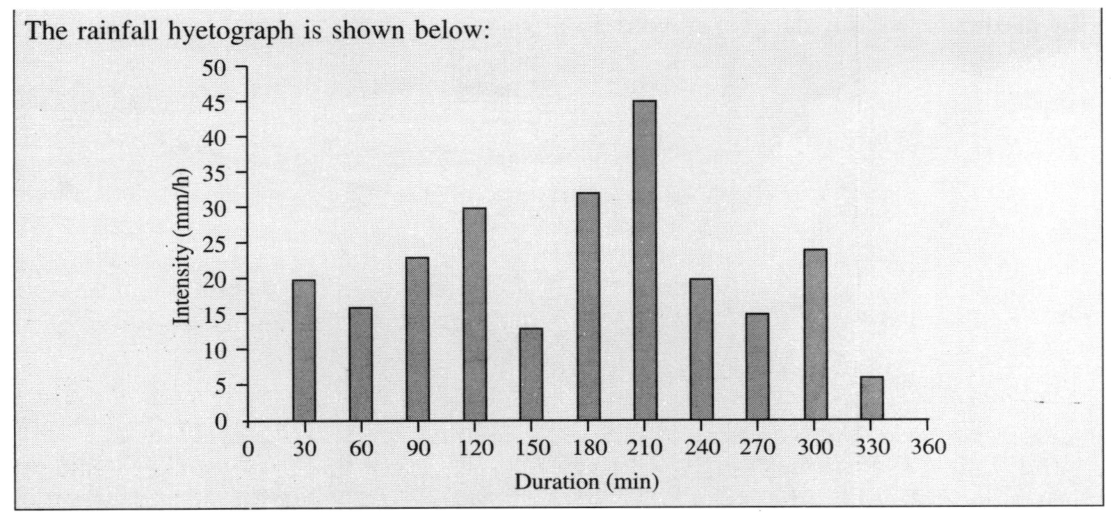

EXAMPLE 2.9 Use the rainfall mass curve ordinates given in Example 2.8 to compute the maximum rainfall intensities for the durations of 30, 60, 90, 120, 180, 240 and 360 minutes. Also, plot the intensity-duration graph.

Solution The necessary computations to obtain hyetograph are tabulated as below:

Time (min)	*Cumulative rainfall* (mm)	*Incremental depth of rainfall* (mm) *in various durations* (min)						
		30	60	90	120	180	240	360
0	0							
30	10	10						
60	18	8	18					
90	29.5	11.5	19.5	29.5				
120	44.5	15	26.5	34.5	44.5			
150	51	6.5	21.5	33	41			
180	67	16	22.5	37.5	49	67		
210	89.5	22.5	38.5	45	60	79.5		
240	99.5	10	32.5	48.5	55	81.5	99.5	
270	107	7.5	17.5	40	56	77.5	97	
300	119	12	19.5	29.5	52	74.5	101	
330	122	3	15	22.5	32.5	71	92.5	
360	122	0	3	15	22.5	55	77.5	122

The maximum intensity-duration curve is given below:

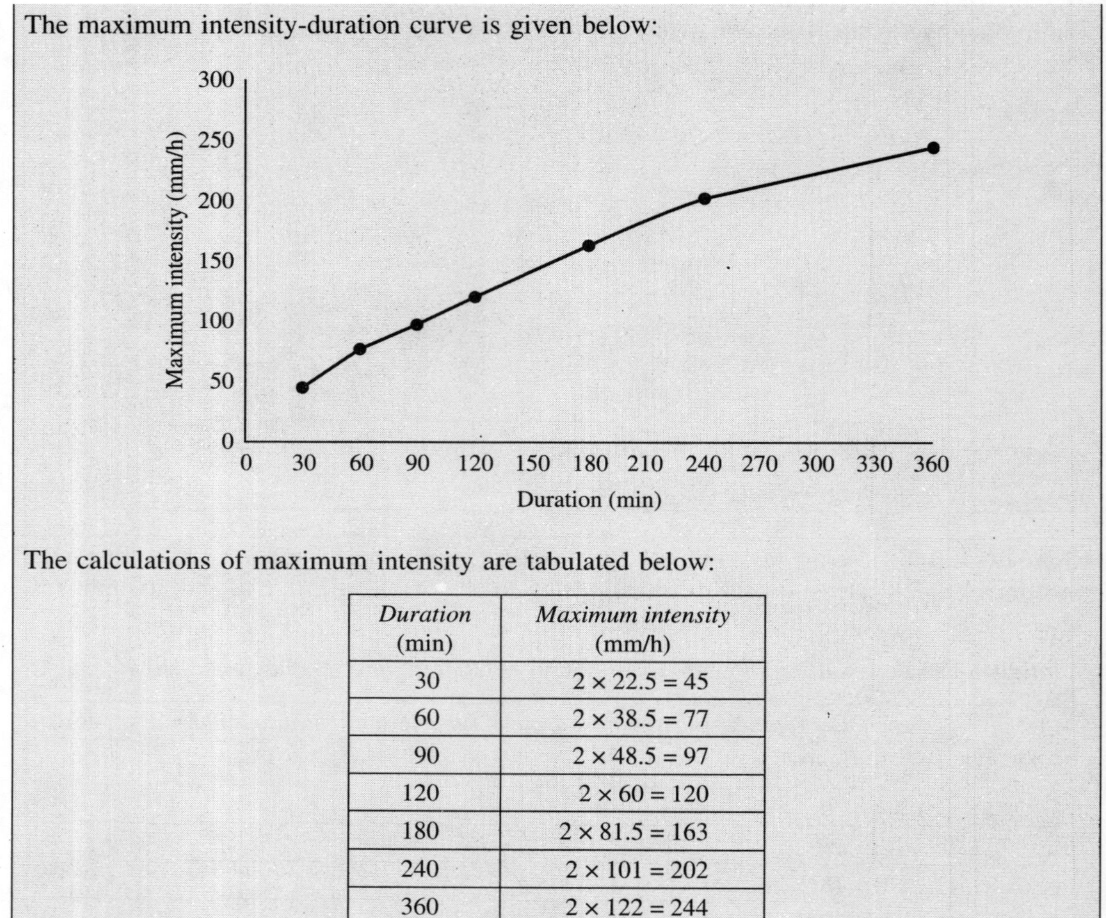

The calculations of maximum intensity are tabulated below:

Duration (min)	*Maximum intensity* (mm/h)
30	2 × 22.5 = 45
60	2 × 38.5 = 77
90	2 × 48.5 = 97
120	2 × 60 = 120
180	2 × 81.5 = 163
240	2 × 101 = 202
360	2 × 122 = 244

2.14 Consistency of Rainfall Records—Double Mass Curve

There are several reasons of inconsistency in rainfall records. This may be due to shifting of rain gauge to a new location, occurrence of observational error from a certain period, an equipment change that can affect the measurement, etc. The errors arising out of inconsistency are not acceptable for various hydrologic analyses. Test for consistency of record is carried out using double mass curve technique. The double mass curve can be used to adjust inconsistent precipitation data.

Double mass curve technique

The double mass curve is a graph of the cumulative annual values at a given station plotted against the corresponding cumulative area average rainfall obtained from the records of surrounding station, as shown in Figure 2.15. When each recorded data comes from the same parent population, they are consistent. The procedure is described below:

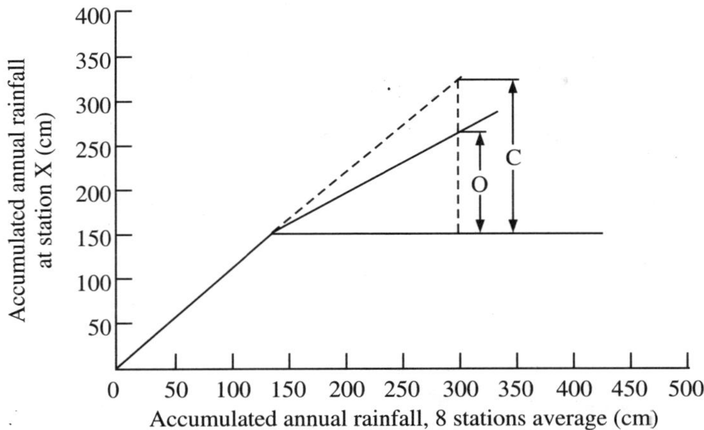

Figure 2.15 Double mass curve.

1. Select a certain number of base stations, preferably a group of 5 to 10 with reliable data of approximately same length.
2. Calculate the cumulative average rainfall of the base station after arranging in descending order.
3. Calculate the cumulative rainfall of the inconsistent station, say station X, after arranging in descending order.
4. Plot the double mass curve between cumulative rainfall of the inconsistent station on abassica and cumulative average rainfall of the base station on ordinate.
5. The corrected data for station X and the period of change in the curve is computed by using the following formula:

$$P_{CX} = P_X \frac{M_C}{M_O} = P_X \frac{C}{O}$$

where P_{CX} is corrected precipitation of station X at a particular time, P_X is recorded precipitation at station X at that particular time, M_C is corrected slope of mass curve with dotted line, M_O is original slope of mass curve, and C and O are vertical intercepts of the corrected and original mass curves.

EXAMPLE 2.10 Annual rainfall data for station S and average annual rainfall values for a group of twenty five neighbouring stations in a region are given below for a period of 30 years record.

Year	1970	1971	1972	1973	1974	1975	1976	1977	1978	1979
Annual rainfall of station S (mm)	670	573	92	460	467	690	477	430	495	500
Average annual rainfall of group (mm)	768	658	108	518	539	798	535	488	555	572

Year	1980	1981	1982	1983	1984	1985	1986	1987	1988	1989
Annual rainfall of station S (mm)	410	530	502	825	675	1240	1000	570	595	370
Average annual rainfall of group (mm)	479	598	576	948	768	1398	1136	648	642	345

Year	1990	1991	1992	1993	1994	1995	1996	1997	1998	1999
Annual rainfall of station S (mm)	632	495	384	430	560	360	680	820	425	610
Average annual rainfall of group (mm)	586	488	395	392	568	374	650	785	405	580

Test the consistency of the annual rainfall data of station S. In which year, a change in regime is indicated? Correct the record if there is any discrepancy. Estimate the mean annual precipitation at station S.

Solution The necessary computations obtained are shown in the table below.

Year	P_m	$\sum P_m$	P_{avg}	$\sum P_{avg}$	*Adjusted values of P_m* (mm)	*Finalised values of P_m* (mm)
1999	610	610	580	580	–	610
1998	425	1035	405	985	–	425
1997	820	1855	785	1770	–	820
1996	680	2535	650	2420	–	680
1995	360	2895	374	2794	–	360
1994	560	3455	568	3362	–	560
1993	430	3885	392	3754	–	430
1992	384	4269	395	4149	–	384
1991	495	4764	488	4637	–	495
1990	632	5396	586	5223	–	632
1989	370	5766	345	5568	–	370
1988	595	6361	642	6210	694.72	695
1987	570	6931	648	6858	665.53	666
1986	1000	7931	1136	7994	1167.60	1168
1985	1240	9171	1398	9392	1447.82	1448
1984	675	9846	768	10160	788.13	788
1983	825	10671	948	11108	963.27	963
1982	502	11173	576	11684	586.14	586

(*Contd.*)

Year	P_m	$\sum P_m$	P_{avg}	$\sum P_{avg}$	*Adjusted values of P_m* (mm)	*Finalised values of P_m* (mm)
1981	530	11703	598	12282	618.83	619
1980	410	12113	479	12761	478.72	479
1979	500	12613	572	13333	583.80	584
1978	495	13108	555	13888	577.96	578
1977	430	13538	488	14376	502.07	502
1976	477	14015	535	14911	556.95	557
1975	690	14705	798	15709	805.64	806
1974	467	15172	539	16248	545.27	545
1973	460	15632	518	16766	537.10	537
1972	92	15724	108	16874	107.42	107
1971	573	16297	658	17532	669.03	669
1970	670	16967	768	18300	782.29	782

Total of P_m = 18845
Mean of P_m = 628

The data is entered starting from recent year (i.e., the data is sorted in descending order of the year). Then, cumulative values of rainfall $\left(\sum P_m\right)$ of station S and twenty five neighbouring station average rainfall values $\left(\sum P_{avg}\right)$ are calculated. A graph is plotted with $\sum P_m$ on the *y*-axis and $\sum P_{avg}$ on the *x*-axis to obtain a double mass curve plot (see the figure below).

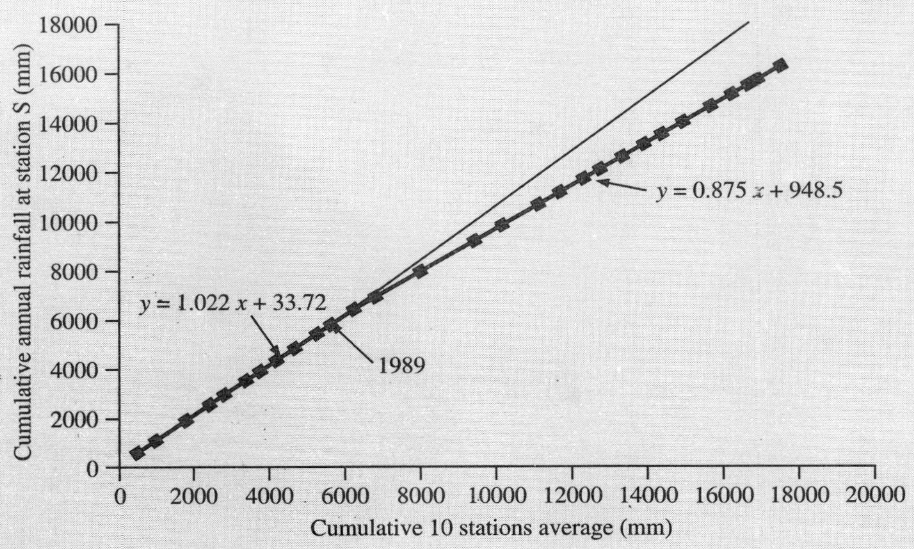

As the double mass curve is not having a uniform slope, it can be inferred that the record at station S is not consistent. Since the break in slope is observed in 1989, this represents a change in the regime of station S after 1968. The slope of the best straight line for the period 1988–70 is S_a = 0.8756. The correction ratio to bring the old records (i.e., 1970–1978) to the current (i.e., post 1989) regime is = S_c/S_a = 1.0224/0.8756 = 1.1676. Each of the pre-1989 annual rainfall value is multiplied by correction ratio of 1.1676 to get the adjusted value.

EXAMPLE 2.11 The values of annual rainfall (mm) are given for a station for 15 years from 1981 to 1995.

Year	*Rainfall* (mm)	*Year*	*Rainfall* (mm)
1981	500	1989	540
1982	550	1990	500
1983	600	1991	590
1984	580	1992	620
1985	630	1993	495
1986	570	1994	510
1987	620	1995	520
1988	480		

For the given rainfall, find the average and standard deviation. Also, plot the rainfall data as

(a) Chronological chart
(b) Bar diagram
(c) Ordinate graph.

Solution The average or mean precipitation can be given as

$$\bar{P} = \sum P/N$$

$$\bar{P} = \frac{500 + 550 + 600 + 580 + 630 + 570 + 620 + 480 + 540 + 500 + 590 + 620 + 495 + 510 + 520}{15}$$

$$\bar{P} = 553.67 \text{ mm}$$

The standard deviation can be given as

$$\sigma = \sqrt{\frac{\left(\sum (P_i - \bar{P})\right)^2}{(N-1)}} = 51.36 \text{ mm}$$

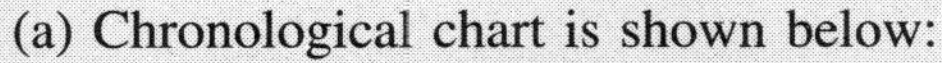

(a) Chronological chart is shown below:

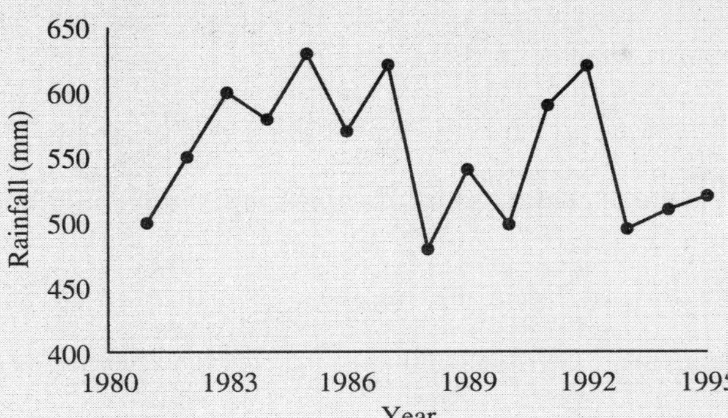

(b) Bar diagram is shown below:

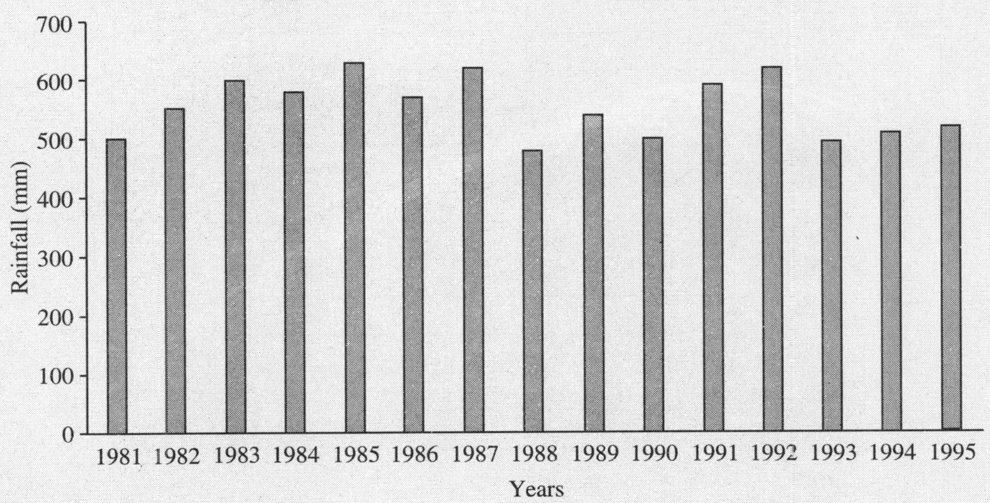

(c) Ordinate graph is shown below:

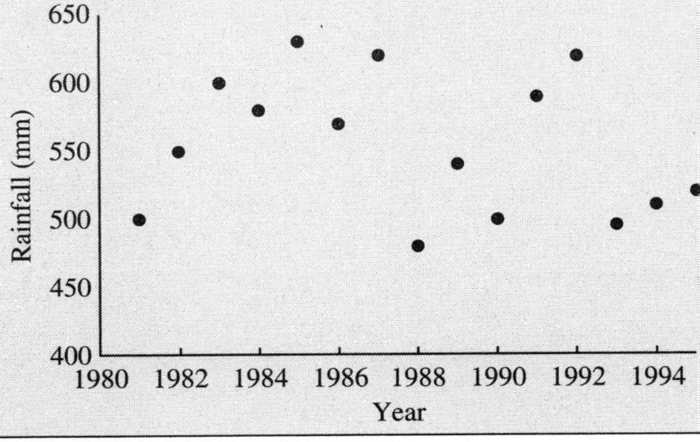

2.15 Moving Average Curve

The *moving average curve* is the plot in which the average value of precipitation of three or five consecutive time intervals is plotted at the mid-value of the time interval. The moving average values are the periodic averages of data lying in between the consecutive time intervals. This type of curve is useful in order to depict a general trend in the rainfall pattern.

EXAMPLE 2.12 Use the annual rainfall data given in Example 2.11 and construct a 3-year moving average curve.

Solution The calculations for 3-year moving mean are shown in the table below:

Year	*Rainfall* (mm)	3-*year moving mean*
1981	500	–
1982	550	550
1983	600	576.67
1984	580	603.33
1985	630	593.33
1986	570	606.67
1987	620	556.67
1988	480	546.67
1989	540	506.67
1990	500	543.33
1991	590	570
1992	620	568.33
1993	495	541.67
1994	510	508.33
1995	520	–

The following figure shows 3 years moving average.

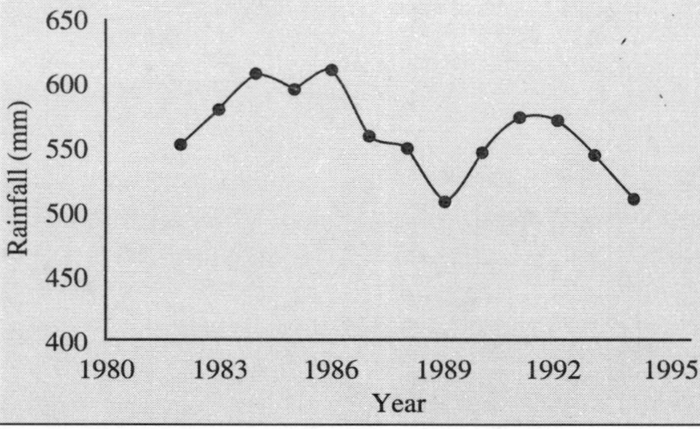

2.16 Probable Maximum Precipitation (PMP)

The greatest depth of rainfall for a given duration that is physically possible over a given size of storm area at a particular geographic region at a certain time of year is called *probable maximu precipitation* (*PMP*). The PMP is the estimated limiting value or upper value of precipitation. The significance of the PMP is to determine the probable maximum flood (PMF). The PMF is used to design dam spillways and locate essential public utilities. The occupation of the PMP is done based on the following two methods:

1. Meterological method
2. Statistical study of rainfall data

The PMP is expressed as the mean of series plus K times the standard deviation (σ) of rainfall series.

$$\text{PMP} = \overline{P} + K\sigma \tag{2.11}$$

Here, K is the frequency factor which is the function of number of years of rainfall records, return period, and statistical distribution of series.

Summary

- A cyclone is a large low-pressure region with circular wind motion, while anticyclones are the regions of high pressure, usually of large areal extent.
- Isohyet is an imaginary line on a map or chart connecting areas of equal rainfall magnitudes.
- Annual average rainfall in India is 1190 mm.
- The coefficient of variation of annual rainfall in India is from 15% to 70%.
- The rate at which the rainfall accumulates at any given instant of time is called the rainfall intensity at that time.
- A rainfall hyetograph is a plot showing the variation of rainfall intensity with time.
- A rainfall mass curve is a plot showing the cumulative depth of rainfall against time.
- Point rainfall is defined as the rainfall at single station.
- The normal rainfall is average value rainfall at a particular date, month or year over a specified 30-year period.
- The double mass curve technique is a technique to check and set right the inconsistencies in data.
- The double mass curve is a graph of the cumulative annual values at a given station plotted against the corresponding area average rainfall obtained from records of the surrounding stations.
- The PMP is the depth of precipitation which for a given area and duration can be reached, but cannot be exceeded under known meteorological conditions.

Objective Type Questions

1. Match the following: (GATE, 2003)

Group I		Group II	
P.	Rainfall intensity	1.	Isohyets
Q.	Rainfall excess	2.	Cumulative excess rainfall
R.	Rainfall averaging	3.	Hyetograph
S.	Mass curve	4.	Direct runoff hydrograph

Codes

	P	Q	R	S
(a)	1	3	2	4
(b)	3	4	1	2
(c)	1	2	4	3
(d)	3	4	2	1

2. The intensity of rainfall and time interval of a typical storm are (GATE, 2005)

Time interval (min)	*Intensity of rainfall* (mm/min)
0–10	0.7
10–20	1.1
20–30	2.2
30–40	1.5
40–50	1.2
50–60	1.3
60–70	0.9
70–80	0.4

The maximum intensity of rainfall for 20 min duration of the storm is

(a) 1.5 mm/min (b) 1.85 mm/min
(c) 2.2 mm/min (d) 3.7 mm/min

3. An accurate estimate of average rainfall in a particular catchment area can be obtained by (GATE, 1995)

(a) Arithmetic mean method (b) Isohyetal method
(c) Normal ratio method (d) Thiessen method

4. The percentage standard error of precipitation average is often expressed functionally or graphically in terms of (i) precipitating gauge network density expressed as area per gauge, and (GATE, 1996)

(a) Increases with area per gauge as well as with total area
(b) Decreases with area per gauge as well as with total area
(c) Increases with area per gauge, but decreases with total area
(d) Decrease with area per gauge, but increases with total area

5. If p is the precipitation, a is the area represented by a rain gauge, and n is the number of rain gauges in a catachment area, then the weighted mean rainfall is (IES, 1996)

(a) $\frac{\Sigma ap^3}{\Sigma a^2}$ (b) $\frac{\Sigma ap}{n}$ (c) $\frac{\Sigma ap}{\Sigma a}$ (d) $\frac{\Sigma ap^5}{\Sigma a^3}$

6. Depth-area-duration curves of precipitation are drawn as (IES, 1996)
(a) Minimising envelopes through the appropriate data points
(b) Maximising envelopes through the appropriate data points
(c) Best fit mean curves through the appropriate data points
(d) Best fit mean straight lines through the appropriate data points

7. Mean precipitation over an area is best obtained from gauged amounts by (IES, 1997)
(a) Arithmetic mean method
(b) Thiessen method
(c) Linearity interpolated isohyetal method
(d) Orographical weighted isohyetal method

8. Depth-area-duration curves would seem to resemble (IES, 1998)
(a) Arcs of circle concave upwards with duration increasing outward
(b) First quadrant limbs of hyperbolae with duration increasing outward
(c) Third quadrant limbs of hyperbolae with duration decreasing outward
(d) First quadrant limbs of hyperbolae with duration decreasing outwards

9. The following rainfall data refers to station A and B which are equidistant from station X.

	Station (A)	*Station* (X)	*Station* (B)
Long-term normal annual rainfall	200	250	300
Annual rainfall (mm) for the year 1940	140	P	270

The value of P will be (IES, 1999)
(a) 250 (b) 220 (c) 205 (d) 200

10. In a watershed, four rain gauges, I, II, III and IV are installed. The depths of normal annual rainfall at these stations are 60 cm, 75 cm and 100 cm, respectively. The rain gauge at station III went out of order during a particular year. The values of annual rainfall for that year, recorded at the remaining three stations, were 90 cm, 60 cm and 70 cm. The rainfall at station III can be considered as (IES, 2002)
(a) 60 cm (b) 70 cm (c) 80 cm (d) 120 cm

11. The moving average of annual precipitation record is carried out to determine (IES, 2003)
(a) Trend (b) Annual mean
(c) Extreme annual variation (d) Extreme seasonal variation

12. Match List-I (hydrological terms) with List-II (relationship/nature of curve) and select the correct answer using the codes given below the lists: (IES, 2003)

List-I
A. Theissen polygon B. Mass curve
C. Hyetograph D. DAD curve

List-II

1. Average depth of rainfall over an area
2. Relationship of rainfall intensity and time
3. Relationship of accumulated rainfall and time
4. Relationship of river run-off and time
5. Always a falling curve

Codes

	A	B	C	D
(a)	1	3	2	5
(b)	1	5	3	2
(c)	4	3	2	5
(d)	4	5	3	2

13. Match List-I (type of precipitation) with List-II (principal causes) and select the correct answer using the codes given below the lists: (IES, 2003)

List-I

A. Convective
B. Cyclonic
C. Frontal
D. Orographic

List-II

1. Atmospheric disturbance
2. Mountain barrier
3. Pressure difference
4. Warm and cold air masses

Codes

	A	B	C	D
(a)	1	4	5	2
(b)	4	3	5	2
(c)	1	4	2	5
(d)	4	3	2	5

14. Consider the following statements:

1. Time-area histogram method aims at developing an IUH.
2. Isochrone is a line joining equal rainfall on a map.
3. Linear reservoir having straight boundaries.
4. Linear channel is a fictitious channel in which an inflow hydrograph passes through with only translation and no attenuation.

Which of these statements are correct? (IES, 2003)

(a) 1, 2 and 3
(b) 1 and 4
(c) 2, 3 and 4
(d) 1, 2, 3 and 4

15. For which one of the following purposes is the double mass curve used? (IES, 2005)

(a) Checking on the consistency of precipitation records
(b) Prediction of annual precipitation
(c) Defining which periods of storm should be an analysed to obtain the maximum useful information from storm rainfall records
(d) For estimating the capacity of a reservoir

16. Which one of the following is not a major type of storm precipitation? (IES, 2006)
(a) Frontal storm (b) Air mass storm
(c) Orographic storm (d) Continental storm

17. Double mass curves are used (IES, 2006)
(a) To check on the consistency of precipitation records
(b) As basis for the storm rainfall analysis
(c) To determine average rainfall over an area
(d) To indicate rainfall distribution

18. What is the probable maximum precipitation (PMP)? (IES, 2007)
(a) Projected precipitation for a 100-year return period
(b) Maximum precipitation for all past recorded storms
(c) Upper limit of rainfall, which is justified climatologically
(d) Effective precipitable water

19. Which one of the following is a method of extending the length of record for a frequency curve at a station? (IES, 2007)
(a) Double mass curve method (b) Station year method
(c) Thiessen method (d) Isohyetal method

20. Match List-I (parameter) with List-II (relatable term) and select the correct answer using the codes given below the lists: (IES, 2007)

List-I
A. Rainfall intensity B. Rainfall excess
C. Rainfall averaging D. Mass curve

List-II
1. Isohyets
2. Cumulative rainfall
3. Hyetograph
4. Direct runoff hydrograph

Codes

	A	B	C	D
(a)	1	3	2	4
(b)	3	4	1	2
(c)	1	3	4	2
(d)	3	4	2	1

21. The rainfall hyetograph shows the variation of which one of the following? (IES, 2007)
(a) Cumulative depth of rainfall with time
(b) Rainfall depth with area
(c) Rainfall intensity with time
(d) Rainfall intensity with cumulative depth of rainfall

22. From the analysis of rainfall data at a particular station, it was found that a rainfall of 400 mm had a return period of 20 years. What is the probability of rainfall equal to or greater than 400 mm occurring at least once in 10 successive years? (IES, 2008)
(a) $(0.95)^{10}$ (b) $1 - (0.95)^{10}$ (c) $1 - (0.05)^{10}$ (d) $(0.05)^{10}$

23. Inconsistency of rainfall data can checked by which one of the following? (IES, 2008)
(a) Normal ratio method (b) Mass curve method
(c) Double mass curve method (d) Depth duration frequency curve

24. Consider the following with respect double mass curve: (IES, 2009)
1. Point of accumulated rainfall with respect to two chronological orders
2. Plot for estimating multiple missing rainfall data
3. Plot for checking the consistency of the rainfall data
4. Plot of accumulated annual rainfall of a station verses accumulated rainfall of a group of stations

Which of these statements are correct?
(a) 1 and 3 (b) 2 and 3 (c) 3 and 4 (d) 1 and 4

25. Generally, to estimate the PMP, $P_m = 42.16D^{0.475}$ is used (P_m is the maximum depth of precipitation, D is duration). What are the units of P_m and D in the equation? (IES, 2009)
(a) mm, s (b) cm, s (c) mm, h (d) cm, h

26. Ombrometer (pulviometer) is used to measure (IES, 2010)
(a) Soil moisture stress of a plant (b) Rainfall depth
(c) Leaf area (d) Root zone depth

27. The coefficient of variation of the rainfall for six rain gauge stations in catchments was found to be 29.54%. The optimum number of stations in the catchments for an admissible 10% error in the estimation of the mean rainfall will be (IES, 2010)
(a) 3 (b) 6 (c) 9 (d) 12

28. The rainfalls on five successive days on a catchment were 3 cm, 6 cm, 9 cm, 5 cm and 1 cm, respectively. If the ϕ index for the storm can be assumed to be 3 cm/day, the total direct runoff from the catchment due to this storm is (IES, 2012)
(a) 11 cm (b) 24 cm (c) 9 cm (d) 20 cm

29. Which of the following statements are correct?
1. A find sprinkle of precipitation of small and rather uniform water drops with all drop diameters below 0.10 mm is called drizzle.
2. The precipitation of liquid water with every drop diameter less than 0.5 mm is known as rain.
3. Precipitation in the form of balls of irregular lumps of ice each over 5 mm in diameter is called hail.
4. Dew is formed directly by condensation on the ground mainly during night when the surface has been cooled by outgoing radiation. (IES, 2013)

(a) 1 and 2 (b) 2 and 3 (c) 3 and 4 (d) 1 and 4

30. The double mass curve technique is used (IES, 2013)
(a) To find the average rainfall over a number of years
(b) To estimate the missing rainfall data
(c) To check the consistency of rain gauge records
(d) To find the minimum number of rain gauges required in basin

31. A mean annual runoff 2 m^3/s from a catchment of area 10 km^2 represents an effective rainfall of nearly (IES, 2013)
(a) 530 cm (b) 590 cm (c) 630 cm (d) 658 cm

32. Step rise in the flow-mass curve during a certain period indicates (IES, 2013)
(a) Very high evaporation losses during that period
(b) Flash floods during that period
(c) Sudden spurt in irrigation demand during that period
(d) Sudden rise in demand for water to meet hydropower generation

Directions: The following item consists of two statements, one labelled as 'Statement I' and the other labelled the 'Statement II'. You are to examine these two statements carefully and decide if the 'Statement I' and 'Statement II' are individually true and if so, whether the 'Statement II' is a correct explanation of the 'Statement I'. Select your answers to these items using the codes given below and mark your answer sheet accordingly.
(a) Both I and II are true and II is correct explanation of I
(b) Both I and II are true but II is not a correct explanation of I
(c) I is true but II is false
(d) I is false but II is true

33. **Statement (I):** Condensation of water vapour into droplets precedes the precipitation process.

Statement (II): Formation of precipitation droplets is predicted on the presence of condensation nuclei. (IES, 2013)

34. Consider the following statements:
1. The normal annual rainfall of a station is obtained as the arithmetic average of the successive annual rainfall in the last 30 years.
2. The normal rainfalls are updated by deleting the needful number of oldest years' data from the record and adding the needful number of most recent years' data to the record.
3. The standard deviation computed for the rainfall of the same 30 years of rainfall data is taken as a measure of the variability of the rainfall during the same set of years.
4. If the observed rainfall in any year is less than the current normal annual rainfall, then that year is called a dry year.

Which of the above statements are correct? (IES, 2014)
(a) 1, 2, 3 and 4 (b) 1, 2 and 3 only (c) 2 and 4 only (d) 1, 3 and 4 only

35. The double mass curve analysis is adopted to (IES, 2014)
(a) Estimate the missing rainfall data
(b) Obtain intensities of rainfall at various duration
(c) Check the consistency of data
(d) Obtain the amount of storage needed to maintain a demand pattern

36. The extreme flood that is physically possible in a region from a severe most combination, including rare combination of meteorological and hydrological factors, is designated as the (IES, 2014)
(a) Design flood (b) Standard project flood
(c) Probable maximum flood (d) Flash flood

37. A 1 h rainfall of 10 cm magnitude at a station has return period of 50 years. The probability that a 1 h of rainfall of magnitude 10 cm or more will occur of two successive years is (GATE, 2013)
(a) 0.04 (b) 0.2 (c) 0.02 (d) 0.0004

38. A isohyet is a line joining points of (GATE, 2013)
(a) Equal temperature (b) Equal humidity
(c) Equal rainfall depth (d) Equal evaporation

Answers

1. (b)	**2.** (b)	**3.** (b)	**4.** (a)	**5.** (c)	**6.** (b)	**7.** (d)	**8.** (b)	**9.** (d)	**10.** (c)
11. (a)	**12.** (a)	**13.** (b)	**14.** (b)	**15.** (a)	**16.** (d)	**17.** (a)	**18.** (c)	**19.** (b)	**20.** (b)
21. (c)	**22.** (b)	**23.** (c)	**24.** (c)	**25.** (d)	**26.** (b)	**27.** (c)	**28.** (a)	**29.** (c)	**30.** (c)
31. (c)	**32.** (b)	**33.** (b)	**34.** (b)	**35.** (c)	**36.** (c)	**37.** (a)	**38.** (c)		

Explanations

2. Maximum intensity for 20 minutes duration is for 20–40 intervals.

$$\text{Maximum intensity} = \frac{(2.2 \times 10 + 1.5 \times 10)}{10 + 10} = 1.85 \text{ mm/min}$$

9. Use arithmetic average method (since the nromal annual rainfall of stations A and B are not within the 10% of the normal annual rainfall at station X). Therefore, as per normal ratio method,

$$P = \frac{250}{2}\left(\frac{140}{200} + \frac{270}{300}\right) = 200 \text{ mm}$$

10. Using normal ratio method,

$$P_{\text{III}} = \frac{80}{3}\left(\frac{90}{60} + \frac{60}{75} + \frac{70}{100}\right) = 80 \text{ cm}$$

22.

$$R = 1 - (1 - P)^n$$

Therefore,

$$R = 1 - \left(1 - \frac{1}{T}\right)^{10} = 1 - (1 - 0.05)^{10} = 1 - (0.95)^{10}$$

27.

$$N = \left(\frac{C_v}{\varepsilon}\right)^2$$

$$N = \left(\frac{29.54}{10}\right)^2 = 8.72 \approx 9 \text{ stations}$$

28.

Day	*Rainfall* (cm)	*Runoff* (cm)
1	3	0
2	6	3
3	9	6
4	5	2
5	1	0
		$\Sigma R = 11$ cm

31. Annual runoff = $2 \times 60 \times 60 \times 24 \times 365$ m^3

Therefore, effective rainfall = $\dfrac{\text{Annual runoff}}{\text{Area of catchment}} = 630.7$ cm

37. Return period of rainfall, $T = 50$ years

Probability of occurrence once in 50 years, $P = \dfrac{1}{50} = 0.02$

Probability of occurrence in each of 2 successive years, $P = 1 - (1 - 0.02)^2 = 0.04$

IES CONVENTIONAL QUESTIONS

PROBLEM 2.1 For a drainage basin of 640 km^2, isohyetals based on a storm event yield the following data: (IES, 2002)

Isohyetal interval (cm)	*Inter-isohyetal area* (km^2)
14–12	90
12–10	140
10–08	125
08–06	140
06–04	85
04–02	40
2–0	20

Estimate the average depth of precipitation over the basin.

Solution Estimation of the average depth of precipitation over the basin is given in the table below:

Isohyetal interval (cm)	*Average value of precipitation* (cm)	*Inter-isohyetal area* (km^2)	*Fraction of total area*	*Weighted p* (cm)
Col. 1	Col. 2	Col. 3	Col. 4 $\left(\dfrac{\text{Col. 3}}{640}\right)$	(Col. 2 × Col. 4)
14–12	13	90	0.1406	1.8278
12–10	11	140	0.2187	2.4062
10–08	9	125	0.1953	1.7578
08–06	7	140	0.2187	1.5312
06–04	5	85	0.1328	0.6641
04–02	3	40	0.0625	0.1875
2–0	1	20	0.0312	0.0312
Σ		640		8.4058

Average depth precipitation over the basin is 8.4058.

PROBLEM 2.2 A watershed have five non-recording rain gauges, installed in its area. The amount of rainfall recorded for one of the years is given below:

Rain gauge station	*Annual rainfall* (cm)
I	100
II	120
III	190
IV	95
V	125

Find the required optimum number of non-recording rain gauges for this watershed. Assume an error of 10% in the estimation of mean rainfall. (IES, 2004)

Solution The total amount of annual rainfall, $\sum P = 100 + 120 + 190 + 95 + 125 = 630$ cm

Average annual rainfall, $\bar{P} = \dfrac{\sum P}{5} = \dfrac{630}{5} = 126$ cm

To compute optimum number of stations,

$$N = \left(\frac{C_v}{E}\right)^2$$

where C_v is the coefficient of variation of rainfall and E is the allowable error in percentage. Now, C_v can be calculated as

$$C_v = \frac{\sigma_{m-1} \times 100}{\bar{P}}$$

$$\sigma_{m-1} = \sqrt{\frac{\sum_{1}^{m}(P_i - \bar{P})^2}{m-1}}$$

$$\sigma_{m-1} = \sqrt{\frac{(100-126)^2 + (120-126)^2 + (190-126)^2 + (95-126)^2 + (125-126)^2}{5-1}}$$

$$\sigma_{m-1} = 37.98 \text{ cm}$$

$$C_v = \frac{37.98 \times 100}{126} = 30.14\%$$

$$N = \left(\frac{30.14}{10}\right)^2 = 9.09 \approx 10$$

About 5 more rain gauge stations are needed.

PROBLEM 2.3 What are the limitations of Thiessen polygon method when compared to isohyetal method? (IES, 2005)

Solution Please refer to text.

PROBLEM 2.4 Compute and draw the storm hyetograph and the intensity duration curve for the following storm (of a given frequency) on a drainage basin. (IES, 1998)

Duration (min)	*Accumulated precipitation* (cm)
0	–
30	5
60	7.5
90	8.5
120	9

Solution Determination of rainfall intensity and precipitation is given in the table below:

Duration (min)	*Accumulated precipitation* (cm)	*Precipitation* (cm)	*Rainfall intensity*
0	–	–	–
30	5	5	10
60	7.5	2.5	5
90	8.5	1	2
120	9	0.5	1

The hyetograph is shown below:

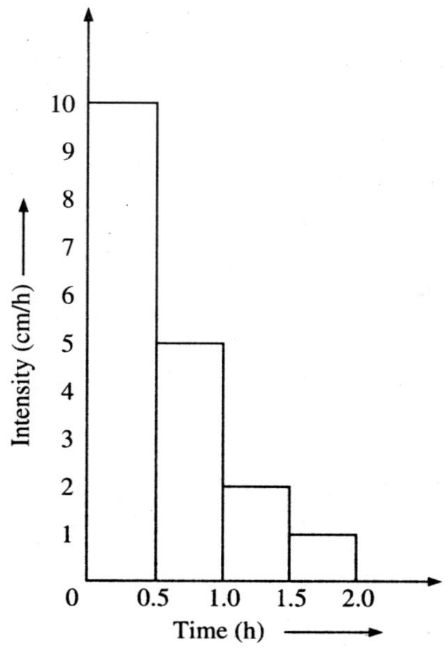

The maximum rainfall values for duration of 30 min, 60 min, 90 min, 120 min are calculated in the following table:

Duration (min)	*Accumulated precipitation* (cm)	*Rainfall in any time interval equal to*			
		30 min	60 min	90 min	120 min
0	0				
30	5	5			
60	7.5	2.5	7.5		
90	8.5	1	3.5	8.5	
120	9	0.5	1.5	4	9

Now, to draw the maximum intensity duration curve, the above maximum values of the precipitation are used to calculate their respective maximum intensities, as shown in the following table:

Duration (min) Col. 1	*Maximum rainfall* Col. 2	*Maximum rainfall intensity* (cm/h) $\text{Col. 2} \times \frac{60}{\text{Col. 1}}$
30	5	10
60	7.5	7.5
90	8.5	5.67
120	9	4.5

Using the above, a maximum rainfall intensity curve can be drawn as shown below:

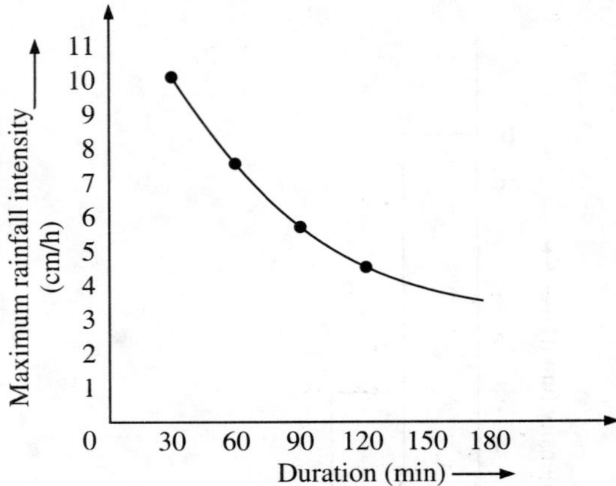

PROBLEM 2.5 There are five rain gauge sections, viz., P, Q, R, S and T. Theissen polygon network details are given in the table below. Calculate the equivalent uniform depth (EUD).

(IES, 2005)

Rain gauge	*Area* (%)	*Rainfall* (mm)
P	24	45
Q	21	57
R	37	65
S	08	67
T	10	78

Solution In order to calculate the EUD, the rainfall of each rain gauge is multiplied by the corresponding fraction of total area under the gauge station, and then, all the values for all rain gauge stations are added together.

Station	*Area*	*Fraction of total area*	*Rainfall* (mm)(%)	*Weighted rainfall* (mm) (Col. 3 × Col. 4)
Col. 1	Col. 2	Col. 3	Col. 4	Col. 5
P	24	0.24	45	10.8
Q	21	0.21	57	11.97
R	37	0.37	65	24.05
S	8	0.08	67	5.36
T	10	0.1	78	7.8
Σ	100	1		59.98 mm

Hence, the equivalent uniform depth of rainfall = 59.98 mm.

PROBLEM 2.6 The annual rainfall at station *X* and the average annual rainfall at 18 surrounding stations during 1952 to 1970 are as follows:
Annual rainfall (cm) at *X*: 30.5, 38.9, 43.7, 32.2, 27.4, 32.0, 49.3, 28.4, 24.6, 21.8, 28.2, 17.3, 22.3, 28.4, 24.1, 26.9, 20.6, 29.5 and 28.4
18 stations average annual rainfall (cm): 22.8, 35.0, 30.2, 27.4, 25.2, 28.2, 36.1, 18.4, 25.1, 23.6, 33.3, 23.4, 36.0, 31.2, 23.1, 23.4, 23.1, 33.2 and 26.4
Explain how the consistency of the record at station *X* can be verified and how to determine the year in which a change in regime has occurred. (IES, 2007)

Solution The given values of station *X* and other 18 surrounding stations are arranged in the table below:

Year	*Annual rainfall at X* (cm)	*Cumulative annual rainfall at X* (cm)	18 *station average annual rainfall* (cm)	18 *station cumulative average annual rainfall* (cm)
1970	28.4	28.4	26.4	26.4
1969	29.5	57.9	33.2	59.6
1968	20.6	78.5	23.1	82.7
1967	26.9	105.4	23.4	106.1
1966	24.1	129.5	23.1	129.2

(*Contd.*)

Year	*Annual rainfall at X* (cm)	*Cumulative annual rainfall at X* (cm)	18 *station average annual rainfall* (cm)	18 *station cumulative average annual rainfall* (cm)
1965	28.4	157.9	31.2	160.4
1964	22.3	180.2	36	196.4
1963	17.3	197.5	23.4	219.8
1962	28.2	225.7	33.3	253.1
1961	21.8	247.5	23.6	276.7
1960	24.6	272.1	25.1	301.8
1959	28.4	300.5	18.4	320.2
1958	49.3	349.8	36.1	356.3
1957	32	381.8	28.2	384.5
1956	27.4	409.2	25.2	409.7
1955	32.2	441.4	27.4	437.1
1954	43.7	485.1	30.2	467.3
1953	38.9	524	35	502.3
1952	30.5	554.5	22.8	525.1

Double mass curve is plotted as shown below:

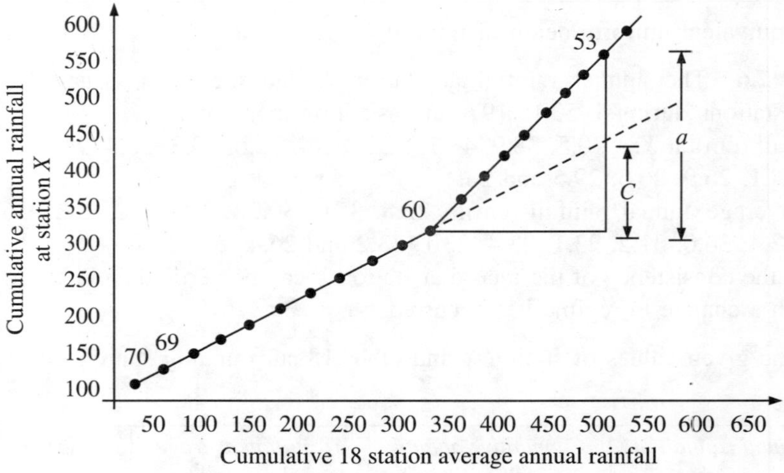

The double mass curve is not have a uniform slope. From the above double mass curve, it can be inferred that the record at a given station is not consistent. The correction ratio can be determined by the following formula:

$$\text{Correction factor} = \frac{c}{a} = \frac{173}{248} = 0.6976$$

This correction factor has to be multiplied with the previous values from 1959 to 1952 to get the corrected values as shown in the following table:

Year	*Annual rainfall at station X* (cm)	*Corrected annual rainfall at station X* (0.6976 × Col. 2)
Col. 1	Col. 2	Col. 3
1959	28.4	19.8
1958	49.3	34.4
1957	32	22.3
1956	27.4	19.1
1955	32.2	22.5
1954	43.7	30.5
1953	38.9	27.1
1952	30.5	21.3

In the following table, the corrected values of annual rainfall at station X is shown:

Year	*Corrected annual rainfall values at station X* (cm)
1952	21.3
1953	27.1
1954	30.5
1955	22.5
1956	19.1
1957	22.3
1958	34.4
1959	19.8
1960	24.6
1961	21.8
1962	28.2
1963	17.3
1964	22.3
1965	28.4
1966	24.1
1967	26.9
1968	20.6
1969	29.5
1970	28.4
Σ	469.1

From the above table, we can determine the value of average annual rainfall at station X as

$$= \frac{469.1}{19} = 24.7 \text{ cm}$$

PROBLEM 2.7 The recorded annual rainfall from rain gauge stations in a catchment and the corresponding Thiessen polygon areas are as follows:

Thiessen polygon areas (cm^2)	*Rainfall* (cm)
25	125
30	175
30	225
10	275
5	325

The scale of the map is 1:50000. Estimate the volume and the mean depth of the rainfall. Estimate the average annual discharge at the outlet if the runoff coefficient of the catchment is 0.3. (IES, 2009)

Solution The scale of map is 1:50000. It means that 1 cm on the map represents 50000 cm on the ground.

Therefore, 1 cm on map = 50,000 cm on ground = 0.5 km on the ground

Therefore, 1 cm^2 on map = 0.5 × 0.5 km^2 on the ground = 0.25 km^2 on the ground

The mean depth of rainfall is computed is the table below:

Area (cm^2)	*Area* (km^2) Col. 1 × 0.25	*Fractional total area* Col. 2/25	*Rainfall* (cm)	*Weighted P* (cm) Col. 3 × Col. 4
Col. 1	Col. 2	Col. 3	Col. 4	Col. 5
25	6.25	0.25	125	31.25
30	7.5	0.30	175	52.50
30	7.5	0.30	225	67.50
10	2.5	0.10	275	27.50
5	1.25	0.05	325	16.25
Total	25	1.00	–	195

$$\text{Mean depth of the rainfall} = 195$$

$$\text{Volume of rainfall} = 195 \times 10^{-2} \times 25 \times 10^{4} \text{ m}^3$$

$$= 487500 \text{ m}^3$$

Average annual discharge at the outlet is computed as

$$Q = \frac{\text{Runoff coefficient} \times \text{Volume of rainfall}}{365 \times 24 \times 60 \times 60}$$

$$\Rightarrow \qquad Q = \frac{0.3 \times 487500}{3153600}$$

$$\Rightarrow \qquad Q = 0.00464 \text{ m}^3\text{/s}$$

PROBLEM 2.8 (a) The areas between the isohyets are given in the table below. Obtain the equivalent uniform depth. What is the depth of flow if the coefficient of runoff is 0.4? Find the volume of runoff.

Isohyet (mm)	75	90	100	125	140	150	165	180
Area between Isohyets (km^2)	60.0	275	260	150	380	215	120	

(b) What are the limitations of the isohyetal method?
(c) What is coefficient of variation? (IES, 2011)

Solution (a) The equivalent uniform depth is calculated as

$$P = \left[\frac{60\left(\frac{75+90}{2}\right)+275\left(\frac{90+100}{2}\right)+260\left(\frac{100+125}{2}\right)+150\left(\frac{125+140}{2}\right) +380\left(\frac{140+150}{2}\right)+215\left(\frac{150+165}{2}\right)+120\left(\frac{165+180}{2}\right)}{60+275+260+150+380+215+120}\right]$$

$$P = \frac{189862.5}{1460} = 130.04 \text{ mm}$$

Coefficient of runoff is given as
c = 0.4, then

$$\text{Depth of flow} = cP = 0.4 \times 130.04 = 52.02 \text{ mm}$$

$$\begin{aligned}\text{Volume of runoff} &= 52.02 \times \text{Area of the catchment}\\ &= 52.02 \times (60 + 275 + 260 + 150 + 380 + 215 + 120) \times 10^6\\ &= 75949.2 \times 10^6 \text{ mm}^3\\ &= 75.95 \text{ m}^3\end{aligned}$$

(b) Please refer the text.
(c) Please refer the text.

PROBLEM 2.9 In a watershed, four non-recording rain gauges have been installed to record rainfall data. The annual rainfall record for one of the years is furnished below:

Location site of rain gauge station	A	B	C	D
Recorded annual rainfall (cm)	100	120	140	80

Assuming an error of 10% in the estimation of mean rainfall, find out the optimum number of non-recording and recording rain gauges for this watershed. [IES, 2012]

Solution The optimum number of rain gauge station is given as

$$N = \left(\frac{C_v}{\varepsilon}\right)^2$$

where C_v is the coefficient of variation of rainfall and ε is the allowable degree of error in percentage.

Now, $$C_v = \frac{100 \times \sigma_{m-1}}{\bar{P}}$$

where, $$\sigma_{m-1} = \sqrt{\sum_{i=1}^{m} \frac{(P_i - \bar{P})^2}{m-1}}$$

Now, $$\sum \bar{P} = 100 + 120 + 140 + 80 = 440 \text{ cm}$$

Therefore, $$\bar{P} = \frac{\sum P}{4} = \frac{440}{4} = 110 \text{ cm}$$

So, $$\sigma_{m-1} = \sqrt{\frac{(100-110)^2 + (120-110)^2 + (140-110)^2 + (80-110)^2}{4-1}}$$

$\Rightarrow$ $$\sigma_{m-1} = 25.82 \text{ cm}$$

Now, $$C_v = \frac{100 \times 25.82}{110} = 23.47$$

and $$N = \left(\frac{23.47}{10}\right)^2 = 5.51 \simeq 6 \text{ stations}$$

Therefore, in this watershed, it suggested to use 5 non-recording type rain gauge.

PROBLEM 2.10 Five rain gauge stations, namely, A, B, C, D and E are located on a circular shape basin of diameter 20 km, as shown in the figure below. Compute the mean areal rainfall over the basin using Thiessen polygon method if the rainfall at stations A, B, C, D and E are 100 cm, 90 cm, 110 cm, 120 cm, and 80 cm, respectively. (IES, 2013)

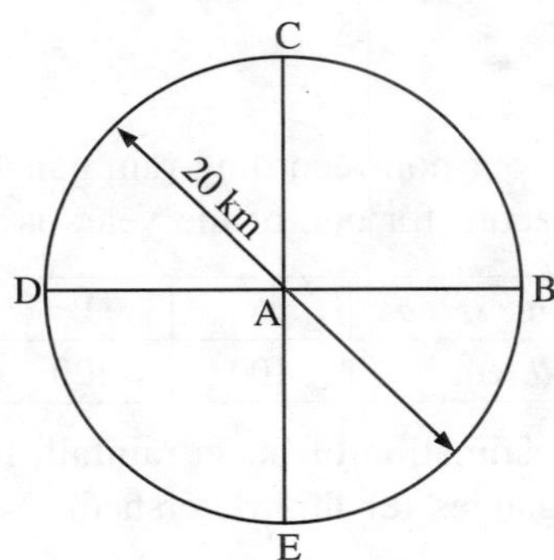

Solution The Thiessen polygons in the given basin can be drawn as follows:

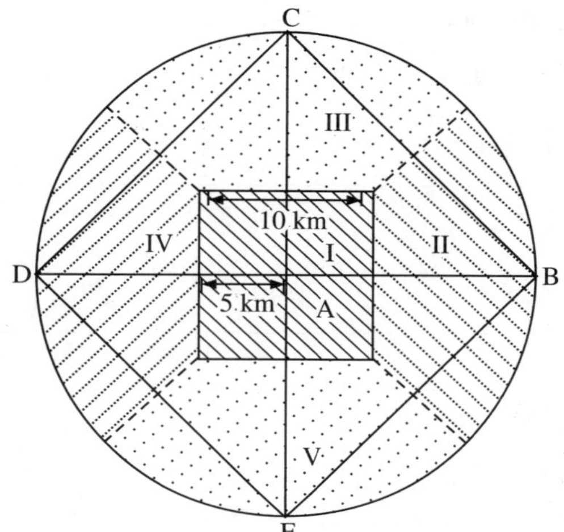

Area under influence of station A (area of region I), $A_A = 10 \times 10 = 100 \text{ km}^2$

Diameter of basin = 20 km

Areas under influence of stations B, C, D and E, i.e., A_B, A_C, A_D and A_E are equal and can be calculated as

$$A_B \text{ or } A_C \text{ or } A_D \text{ or } A_E = \frac{\pi\frac{(20)^2}{4} - 100}{4} = 53.54 \text{ km}^2$$

$$\text{Mean areal rainfall} = \frac{P_A A_A + P_B A_B + P_C A_C + P_D A_D + P_E A_E}{A_A + A_B + A_C + A_D + A_E}$$

$$= \frac{(100 \times 100) + (90 \times 53.54) + (110 \times 53.54) + (120 \times 53.54) + (80 \times 53.54)}{\frac{\pi}{4}(20)^2}$$

$$= 100 \text{ cm}$$

Therefore, mean area rainfall = 100 cm

PROBLEM 2.11 The isohyets due to a storm in a catchment were drawn (see the following figure) and the areas of the catchment bounded by the isohyets were tabulated as below:

(IES, 2014)

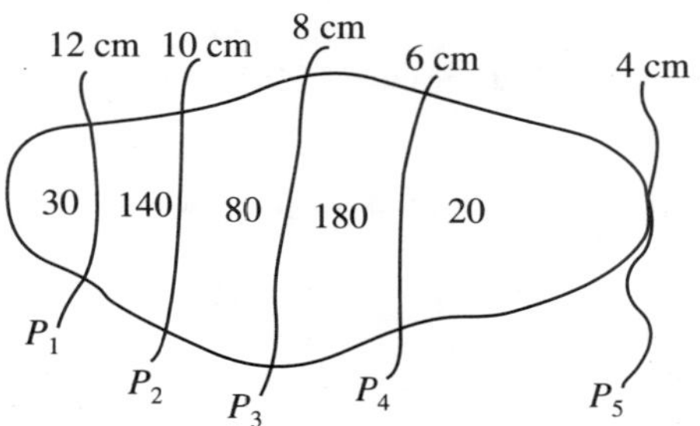

Isohyets (cm)	*Areas* (km²)
Station 12.00	30
12.0–10.0	140
10.0–8.0	80
8.0–6.0	180
6.0–4.0	20

Determine the uniform flow depth over the catchment.

Solution Since isohyets and area are given, therefore using isohyetal method

$$\text{Uniform flow depth} = \frac{A_1P_1 + \left(\frac{P_1 + P_2}{2}\right)A_2 + \left(\frac{P_2 + P_3}{2}\right)A_3 + \left(\frac{P_3 + P_4}{2}\right)A_4 + \left(\frac{P_4 + P_5}{2}\right)A_5}{A_1 + A_2 + A_3 + A_4 + A_5}$$

$$= \frac{30 \times 12 + \left(\frac{10+12}{3}\right)140 + \left(\frac{10+8}{2}\right)80 + \left(\frac{8+6}{2}\right)180 + \left(\frac{6+4}{2}\right)20}{30 + 140 + 80 + 180 + 20}$$

$$= 8.84 \text{ cm.}$$

Theoretical Questions

1. Explain different types of precipitation.
2. Define rain gauge. Describe the different methods of measurement of rainfall.
3. What are the advantages and disadvantages of recording type rain gauges.
4. Write short notes on rain gauge network.
5. What are the methods of estimating missing data of rainfall?
6. Explain different methods of determining the average rainfall over an area.
7. What do you understand by frequency analysis?
8. Write short notes on (a) intensity-duration-frequency curve (b) depth-area-duration curve.
9. What is the consistency of rainfall? Explain the procedure for checking a rainfall data for consistency.
10. What do you understand by probable maximum precipitation (PMP)?

Unsolved Problems

Problem 1: The average annual rainfalls (mm) at 5 existing rain gauge stations in a basin are 95, 75, 110, 88 and 68. If the average depth of rainfall over the basin is to be estimated within 15% error, estimate the additional number of gauges required.

Problem 2: In a catchment, the normal annual rainfall at stations P, Q, R and S in a basin are 79.5 cm, 66.4 cm, 74.2 cm and 91.1 cm, respectively. In the year 1990, station S was inoperative and station P, Q and R recorded annual precipitations of 90.1 cm, 70.2 cm and 81.7 cm, respectively. Estimate the rainfall at station S in that year.

Problem 3: The following table show the recorded rainfall at 4 rain gauge stations in a catchment for a period of 10 years:

Year	A	B	C	D
2001	120	90	86	?
2002	90	85	95	88
2003	80	115	99	96
2004	95	124	112	99
2005	75	98	107	120
2006	115	88	88	85
2007	125	85	130	115
2008	99	112	77	102
2009	85	75	89	60
2010	110	115	93	75

Estimate the missing rainfall at station D in the year 2001 using

(a) Arithmetic mean method
(b) Normal-ratio method

Problem 4: A circle of approximate 100 km diameter, encloses a river catchment, in which four rainfall stations are situated inside the catchment and one station is outside in its neighbourhood. The position coordinates of 5 rain gauge stations 1, 2, 3, 4 and 5 located are (30, 80), (70, 100), (100, 140), (130, 100) and (100, 70), respectively. The coordinate of the centre of catchment is (100,100).If the annual rainfall recorded at these rain gauges are 83 mm, 133.3 mm, 93.2 mm, 144.4 mm, and 100.2 mm, respectively, compute the average annual precipitation by the Thiessen mean method.

Problem 5: In a catchment, rainfall recorded by four rain gauges A, B, C, and D during a storm are 10 cm, 12 cm, 15 cm, and 13 cm, respectively. Assuming catchment is in square shape with side 10 km, position of four corners of catchment are (0, 0), (10, 0), (10, 10), and (0,10), respectively. The rain-gauge positions (A, B, C and D) are (2.5, 2.5), (7.5, 2.5), (7.5, 7.5), (2.5, 7.5), respectively. Compute the average depths of rainfall over the catchment by arithmetic mean and Thiessen polygon methods.

Problem 6: Analysis of data on a maximum one-day rainfall depth at Madras indicated that a depth of 270 mm had a return period of 50 years. Determine the probability of a one-day rainfall depth equal to or greater than 270 mm at Madras occurring

(a) Once in 25 successive years
(b) Two times in 18 successive years, and
(c) At least once in 25 successive years.

Problem 7: The mass curve of rainfall in a storm of total duration 360 min is given below:

(a) Draw the hyetograph of the storm at 40 min time step.
(b) Plot the maximum intensity-duration curve for this storm.
(c) Plot the maximum depth-duration curve for the storm.

Time since start (min)	0	40	80	120	160	200	240	280	320	360
Cumulative rainfall (mm)	0	8	20	24	40	45	52	54	56	60

Problem 8: Represent the data of below Table for station S as a bar diagram with time in chronological order:

Year	*Annual rainfall* (mm)	*Year*	*Annual rainfall* (mm)
1950	670	1965	1240
1951	573	1966	1000
1952	92	1967	570
1953	460	1968	595
1954	467	1969	370
1955	690	1970	632
1956	477	1971	495
1957	430	1972	384
1958	495	1973	430
1959	500	1974	560
1960	410	1975	360
1961	530	1976	680
1962	502	1977	820
1963	825	1978	425
1964	675	1979	610

(a) Identify those years in which the annual rainfall is (i) less than 20% of mean, (ii) more than the mean,

(b) Plot the 3-year moving mean of the annual time series.

Further Reading

Ram Babu, Tejwani, K.K., Agrawal, M.C., and Bhusan, L.S., 'Rainfall intensity-duration-return period equations and nomographs of India', *CSWCRTI*, ICAR, Dehradun, India, 1979.

Das, Ghanshyam, *Hydrology and Soil Conservation Engineering Including Watershed Management*, PHI Learning, Delhi, 2014.

Raghunath, H.M., *Hydrology—Principles, Analysis and Design*, New Age Publishers, New Delhi, 2014.

Suresh, R., *Watershed Hydrology*, Standard Publishers Distributors, New Delhi, 2015.

Deodhar, M.J., *Elementary Engineering Hydrology*, Pearson, New Delhi, 2013.

Yadupathi Putty, Mysooru, R., *Principles of Hydrology*, I.K. International, New Delhi, 2013.

CHAPTER

3

Abstraction Losses

3.1 Introduction

Abstraction is defined as the amount of rainfall that does not turn into runoff. Rainfall abstractions include interception of rainfall by vegetation, evaporation from the surface, transpiration from plants, small depressions on the surface and infiltration to the soil. If these abstractions (losses) are deducted from the rainfall, the surface runoff can be obtained.

3.2 Evaporation

Evaporation is a process by which a liquid on a free surface is transferred into gaseous state below the boiling point through the transfer of heat energy. Evaporation occurs from lakes, rivers, land surfaces, soils, glaciers and seas.

3.2.1 Dalton's Law of Evaporation

In atmosphere, rate of evaporation depends on the difference of saturation vapour pressure and actual vapour pressure. Thus, John Dalton (1802) suggested the formula

$$E = C(e_w - e_a) \tag{3.1}$$

where E is evaporation rate, C is constant, e_w is saturated vapour pressure at water temperature, and e_a is actual vapour pressure in the air.

Saturation vapour pressure of water surface temperature can be taken from Table 3.1

3.2.2 Factors Influencing Evaporation

The factors which influence the rate of evaporation can be listed below:

Table 3.1 Saturation Vapour Pressure of Water

Temperature (°C)	*Saturation vapour pressure e_s*		*Slope A of plot between Col. 1 and Col. 2*
	(mmHg)	(mbar)	mm/°C
Col. 1	Col. 2	Col. 3	Col. 4
0.0	4.58	6.11	0.30
5.0	6.54	8.72	0.45
7.5	7.78	10.37	0.54
10.0	9.21	12.28	0.60
12.5	10.87	14.49	0.71
15.0	12.79	17.05	0.80
17.5	15.00	20.00	0.98
20.0	17.54	23.38	1.05
22.5	20.44	27.95	1.24
25.0	23.76	31.67	1.40
27.5	27.54	36.71	1.61
30.0	31.81	42.42	1.85
32.5	36.68	48.89	2.07
35.0	42.81	57.07	2.35
37.5	48.36	64.46	2.62
40.0	55.32	73.14	2.95
42.5	62.18	84.23	3.25
45.0	71.2	94.91	3.66

1. Vapour pressure: Evaporation rate varies directly with the difference of vapour pressure between air and water. Higher the differences between saturation vapour pressure e_s and the prevailing vapour pressure e, the higher the rate of evaporation.

2. Temperature: As the air temperature is increased, its capacity to hold moisture also increases. An increase in the temperature of water helps the water molecules in escaping from liquid to a gaseous state, thereby increasing evaporation. However, it is to note that evaporation increases with the increase in temperature, even though a close relation between evaporation and temperature cannot be established.

3. Wind speed: When the wind velocity is high, turbulence gets set up in the air. The wind that blows over the water body helps in removing vapour particles. Thus, the rate of evaporation is accelerated. Hence, wind speed is an important factor influencing the rate of evaporation.

4. Atmospheric pressure: Other factors remaining same, decrease in atmospheric pressure or increase in altitude increases evaporation. At lower pressures, the evaporation rate will be higher.

5. Other factors: Other factors include the exposed surface area of water, soluble salts, etc. Larger areas of evaporating surface increase the rate of evaporation. The rate of evaporation of fresh water is always greater than that of salt water. Under identical conditions, it has been noted that evaporation from seawater is about 2–3% less than that from fresh water.

EXAMPLE 3.1 Compute the daily rate of evaporation for the catchment area (80 km^2) of an irrigation tank for a month. In a given month, the uniform precipitation over the catchment was recorded to be 110 mm. The constant water spread during this month was 2.5 km^2 and the irrigation canal discharges at a uniform rate of 1.5 cumecs. Assume 55% of the precipitation reaches the tank and seepage losses are 54% of the evaporation losses.

Solution Given, water spread = 2.5 km^2, uniform precipitation over catchment = 110 mm, percentage of precipitation that reaches the tank = 55%, discharge of irrigation canal = 1.5 cumecs, seepage loss = 54 % of evaporation loss.

$$\text{Runoff that reaches the tank} = 0.55 \times 80 \times 10^6 \times 110 \times 10^{-3}$$
$$= 4840000 \text{ m}^3 \text{ (Inflow)}$$

Let evaporation losses be e.

$$\text{Total loss} = \text{Evaporation Losses + Seepage Losses}$$
$$= e + 0.54e = 1.54e$$
$$\text{Water discharging in the canal} = 1.5 \times 31 \times 24 \times 60 \times 60 = 4017600 \text{ m}^3$$

We know that

$$\text{Inflow} - \text{Losses} = \text{Outflow}$$
$$4840000 - 1.54e = 4017600$$
$$e = 534025.974 \text{ m}^3$$
$$\text{Evaporation losses (depth)} = e/\text{Water spread}$$
$$= 534025.974/2.5 \times 10^6$$
$$= 0.21361 \text{ m}$$
$$\text{or } 213.61 \text{ mm (for 31 days)}$$
$$\text{Evaporation losses (depth) per day} = 213.61/31 = 6.89 \text{ mm/day.}$$

3.2.3 Measurement of Evaporation

Measurement of evaporation is done by using evaporimetres in the field. These evaporimetres are large pan in which water is kept and exposed to the atmosphere and the depth of loss of water is measured using a point gauge at regular intervals. The commonly used evaporimetres are the US Weather Bureau Class A Pan and the ISI Standard Pan (IS 5973-1970).

Class A Evaporation Pan

The US Weather Bureau Class A Pan is a standard pan of 1210 mm diameter and depth 255 mm, as shown in Figure 3.1. The Class A Evaporation Pan is normally installed on a wooden platform of 150 cm height on the ground level, away from bushes, trees and other obstacles. The pan is made of galvanised iron sheet and is painted white.

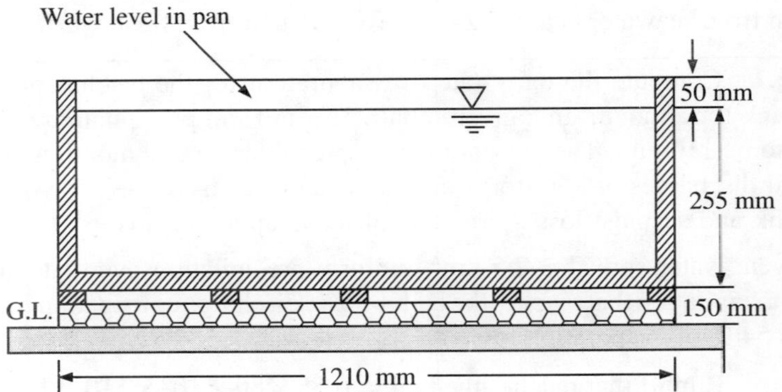

Figure 3.1 US Class A Evaporation Pan.

The pan is filled with a known quantity of water and the depth of water in the pan is maintained at 20 cm. To measure evaporation, a hook gauge is used. The amount of evaporation is computed on daily basis as the difference between the observed water levels.

ISI Standard Pan

The ISI Standard Pan consists of a circular vessel of 1220 mm diameter and 255 mm depth, as shown in Figure 3.2. This pan has a stilling well for measuring the depth of water. A fixed gauge housed in a stilling well indicates the level of water. At the top of the pan, a hexagonal wire netting of galvanised iron is done to cover the whole surface in order to protect the water in the pan from birds. The pan is placed over a square wooden platform of 1220 mm width and 100 mm height to enable circulation of air underneath the pan.

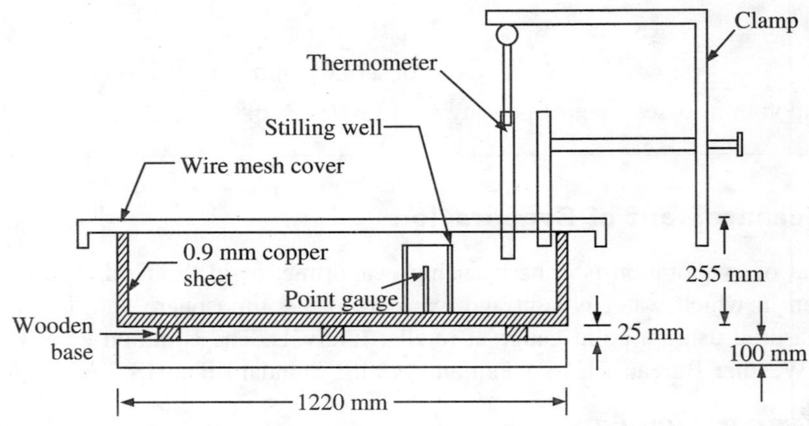

Figure 3.2 ISI Standard Pan.

Limitation of pan and pan coefficient: Since the heat transfer and heat storage characteristics of a pan is different from that of a reservoir/lake and pan contains a very little volume of water as compared to the reservoirs, therefore, normally, the pan will overestimate evaporation in comparison to nearby lakes. In an evaporation pan, the height of the rim also affects the wind action over the surface. The evaporation pan data have to be corrected to obtain actual evaporation from lakes/reservoirs by multiplying a coefficient called pan coefficient. Pan coefficient C_p is defined as the rate of evaporation from reservoir to the rate of evaporation from pan.

$$C_p = \frac{\text{Rate of evaporation from reservoir}}{\text{Rate of evaporation from pan}} \tag{3.2}$$

The value of evaporation coefficient C_p is always less than or equal to 1.0 and it is having an average value of 0.7. The values of pan coefficient for various types of pans are given in Table 3.2.

Table 3.2 Values of Pan Coefficients for Various Types of Pan

S. No.	*Types of pan*	*Average value*	*Range*
1	Class A Land Pan	0.7	0.60–0.80
2	ISI Pan (Modified Class A)	0.8	0.65–1.10
3	Colorado Sunken Pan	0.78	0.75–0.86
4	USGS Floating Pan	0.8	0.70–0.82

EXAMPLE 3.2 Determine the rate of evaporation from an evaporation pan of 1.9 m diameter. The drop in water level in evaporation pan during 24 h is 15 mm. The precipitation recorded during this period is 25 mm.

Solution Given, diameter of evaporation pan = 1.9 m, time period = 24 h, drop in water level = 15 mm, and precipitation depth = 25 mm.

$$\text{Area of evaporation pan} = A = \frac{1}{4}\pi d^2 = 0.25 \times \pi \times (1.9)^2 = 2.835 \text{ m}^2$$

$$\text{Outflow} = \text{Evaporation from the pan} = 2.835 \times 15 \times 10^{-3} = 0.04253 \text{ m}^3$$

$$\text{Inflow} = \text{Precipitation into the pan} = \frac{\pi}{4} \times (1.9)^2 \times 25 \times 10^{-3} = 0.07088 \text{ m}^3$$

$$\text{Storage} = \text{Inflow} - \text{Outflow}$$

$$= 0.07088 - 0.04253 = 0.02835 \text{ m}^3$$

$$\text{Rate of evaporation} = \frac{\text{Inflow} - \text{Outflow}}{\text{Area}}$$

$$= \frac{0.02835}{\text{Area}} = 0.01 \text{ m/day}$$

EXAMPLE 3.3 It was found that the rate of evaporation from a 1.5 m pan having a pan coefficient of 0.9, was 1.5 mm/m^2/h. Compute the total evaporation from a reservoir in a week, having a water spread of 3 ha.

Solution Given, diameter of evaporation pan = 1.5 m, pan coefficient = 0.9, rate of evaporation = 1.5 mm/m^2/h, time period = 7 days and water spread = 3 ha

$$\text{Area of evaporation pan} = \frac{1}{4}\pi d^2 = 0.25 \times \pi \times (1.5)^2 = 1.7671 \text{ m}^2$$

$$\text{Evaporation from pan} = 1.7671 \times 1.5 = 2.6507 \text{ mm/h}$$

$$\text{Total evaporation from reservoir} = 0.90 \times 3 \times 10^4 \times 2.6507 \times 10^{-3} \times 7 \times 24$$

$$= 12023.575 \text{ m}^3$$

3.2.4 Empirical Equations for Evaporation Estimation

A number of empirical equations are available for the estimation of evaporation. Some commonly used expressions for estimating evaporation are as follows:

1. Meyer's formula:

$$E = K_m(e_w - e_a)\left(1 + \frac{u_a}{16}\right) \tag{3.3}$$

where, E is evaporation (mm), e_w is saturated vapour pressure at the water surface temperature (mm Hg), e_a is prevailing vapour pressure (mm Hg), u_a is wind velocity at 9 m above the ground level and K_m is parameter accounting other factor such as size of water body. It is generally taken as 0.36 for large lakes and as 0.50 for small water bodies.

2. Fitzgerald's formula:

$$E = (0.4 + 1.24u_0)(e_w - e_a) \tag{3.4}$$

Here, u_0 is wind velocity at the land surface.

3. Rohwer's formula:

$$E = 0.771\,(1.465 - 0.000732p_a)(0.44 + 0.0733u_0)(e_w - e_0) \tag{3.5}$$

Here, p_a is the mean barometric reading (mm Hg).

3.2.5 Analytical Methods of Evaporation Estimation

The analytical methods available to estimate evaporation are that water budget, the energy budget and the mass transfer method. It is to mention that instrumentation required for energy budget and mass transfer methods are expensive, therefore water budget method is more common.

Water budget method

This is the simplest analytical method to determine evaporation from ponded water such as lakes, etc., but is seldom produces reliable results. The water budget equation can be written as

$$E = I + P - Q - Q_{gw} + \Delta S \tag{3.6}$$

where, E is daily evaporation, I is daily inflow, P is daily precipitation, Q is daily outflow, Q_{gw} is daily inflow to groundwater and ΔS is increase in lake storage in a day.

Energy budget method

In this method, the available energy for evaporation is determined by considering incoming, outgoing and stored energy in water body at one time. This method illustrates an application of continuity equation written in terms of energy. The energy budget equation may be presented in the following form:

$$Q_n = Q_h + Q_e + Q_g + Q_s + Q_i \tag{3.7}$$

where Q_n is net heat energy received by water surface, Q_h is sensible heat transfer, Q_e is evaporation energy, Q_g heat flux into the ground, Q_s is heat stored in water body, and Q_i is advected energy.

In order to compute Q_h, Bowen's ratio β is used, which is defined as below:

$$\beta = \frac{Q_h}{Q_e} = 6.1 \times 10^4 \rho_a \frac{(T_w - T_a)}{(e_w - e_a)} \tag{3.8}$$

where, ρ_a is atomspheric pressure (mm of mercury), e_w is saturated vapour pressure (mm of mercury), e_a is actual vapour pressure (mm of mercury), T_w is temperature of water surface (°C), and T_a is temperature of air (°C).

The limitation of this method is that it requires extensive instrumentation.

Mass Transfer Method

This method is primarily based on the concept of the turbulent transfer of water vapour from evaporating surface to atmosphere. Here, evaporation is driven by vapour pressure of gradient and wind speed.

$$E = (a + bu)(e_s - e_a)$$

Here, u is wind speed at some level above surface a, b are empirical constants.

Thornthwaite and Holzman (1939) provided the following equation to compute water vapour transferred to atmosphere from a lake surface.

$$E = \frac{0.000119\,(e_1 - e_2)(u_2 - u_1)}{p \times \left[\ln\left(\frac{h_2}{h_1}\right)\right]^2} \tag{3.10}$$

Here, u_1 and u_2 are wind velocities (m/s) at heights h_1 and h_2, respectively, and e_1 and e_2 are vapour pressures (Pa) at heights h_1 and h_2, respectively and p is air pressure.

EXAMPLE 3.4 For air temperature of 25°C, relative humidity of 98%, air pressure of 101.3×10^3 Pa and wind speed of 2 m/s measured at 2 m above water surface, calculate the evaporation loss from the water surface. Assume saturation height h_1 as 0.025 m.

Solution Given, $u_1 = 0$, $u_2 = 2$ m/s,

Vapour pressure e_1 $(= e_s)$ = 31.67 m bar. for air temperature of 25°C at the level close to water surface (using Table 3.1)

$$e_2 = \text{Relative humidity} \times e_s = 0.98 \times 3167 = 3104 \text{ Pa}$$

$$E = \frac{0.000119\,(e_1 - e_2)(u_2 - u_1)}{p \times \left[\ln\left(\frac{h_2}{h_1}\right)\right]^2}$$

$$E = \frac{0.000119(3167 - 3104)(2 - 0)}{101.3 \times 10^3 \times \left[\ln\left(\frac{2}{0.025}\right)\right]^2}$$

$$E = \frac{0.01499}{1945178.5}$$

$$E = 7.706 \times 10^{-9} \text{ m/s}$$

$$E = 7.706 \times 10^{-9} \times 1000 \times 24 \times 3600 \text{ mm/day} = 0.67 \text{ mm/day}$$

3.2.6 Measures to Reduce Evaporation

The losses due to evaporation can be significant at a given location. Therefore, it is necessary to reduce the evaporation losses. The various methods recommended to reduce evaporation from water surface are as follows:

1. Storing water in covered reservoirs: Water losses through evaporation can be reduced by using covered reservoir (no or less exposure to solar radiation). Generally, either permanent cover or temporary cover can be adopted, wherever feasible.

2. Storage reservoirs with less surface area: Surface area of storage reservoir may be reduced since water lost depends on the exposed surface area.

3. Applying chemical films: By spreading chemicals such as cetyl alcohol over the reservoir surface, a layer is formed on the surface which does not allow water molecules to escape from it. It is found that it reduces evaporation about 20 to 50% by preventing the water molecules to escape from the its surface in small size lakes (about 1000 ha).

4. Other methods: Other methods can also be used to reduce evaporation such as by growing tall trees on the windward side of the reservoirs to act as wind breakers and conveying water in closed conduits rather than open channels from one place to other place.

3.3 Transpiration

Transpiration is a process by which water leaves the plants and enters the atmosphere. Factors affecting the transpiration process include temperature, relative humidity, wind movement, soil moisture availability and plant type. It is observed that only about 1% of water sucked by the roots is retained by the plants.

As temperature goes up, plants transpire more rapidly because solar light stimulates the opening of the stomata where water is released to the atmosphere. When soil moisture is lacking, a plant cannot continue to transpire rapidly. Some plants transpire less, especially which grow in arid regions.

The major difference between evaporation and transpiration is that transpiration is especially confined in daylight hours while evaporation continues all through day and night.

3.4 Evapotranspiration

Evapotranspiration of crops is the total lost water due to evaporation from the soil and transpiration by the plants. The term *consumptive use* is also used to denote this loss by evapotranspiration. It is usually expressed as a depth over the area.

Potential evapotranspiration (*PET*) is the quantity of water that is removed from the surface through the processes of evaporation and transpiration when unlimited water covers the soil. The value of potential evapotranspiration is computed at a local climate station on a reference surface. This is known as *reference evapotranspiration* (*RET*). The PET can be calculated by multiplying reference evapotranspiration by crop coefficient. *Actual evapotranspiration* (*AET*) is the quantity of water that is actually removed from a surface due to the processes of evaporation and transpiration. Thus, the PET is the maximum value of the AET. The AET depends on available soil water and vegetation type.

In Figure 3.3, a relation between AET/PET and available soil moisture for various types of soil has been shown. It can be seen that for the same AET/PET ratio, clayey soil has less moisture than sandy soil.

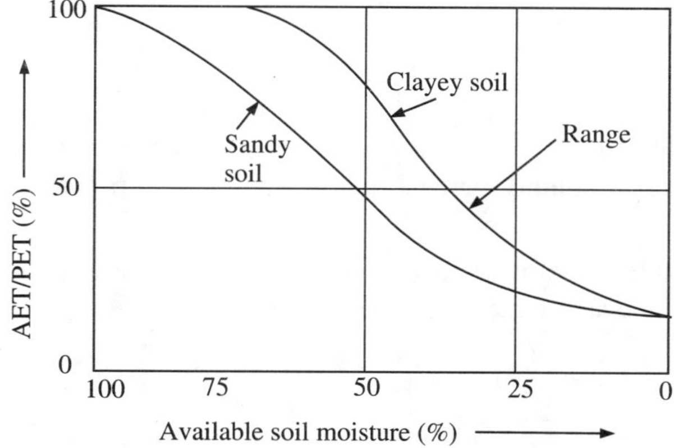

Figure 3.3 Available moisture for sandy and clayey soil.

3.4.1 Estimation of Evapotranspiration

Empirical methods

Empirical formulae are developed based on climatological data available. These methods are not universally applicable.

1. Blaney–Criddle method: This is one of oldest theoretical method which assumes that the PET is related to mean monthly temperature and daylight hours, as shown in Table 3.3. It should be noted, however, that this method is not very accurate. This method is developed in FPS units, as follows:

Table 3.3 Mean Monthly Values of Possible Sunshine Hours (N)

	North latitude (degree)					
Month	0°	10°	20°	30°	40°	50°
January	12.1	11.6	11.1	10.4	9.6	8.6
February	12.1	11.8	11.5	11.1	10.7	10.1
March	12.1	12.1	12.0	12.0	11.9	11.8
April	12.1	12.4	12.6	12.9	13.2	13.8
May	12.1	12.6	13.1	13.7	14.4	15.4
June	12.1	12.7	13.3	14.1	15.0	16.4
July	12.1	12.6	13.2	13.9	14.7	16.0
August	12.1	12.4	12.8	13.2	13.8	14.5
September	12.1	12.9	12.3	12.4	12.5	12.7
October	12.1	11.9	11.7	11.5	11.2	10.8
November	12.1	11.7	11.2	10.6	10.0	9.4
December	12.1	11.5	10.9	10.2	9.1	8.1

$$E_T = \frac{kTP}{100} \tag{3.11}$$

where E_T is PET monthly (inch), k is monthly consumptive use coefficient, T is mean monthly temperature, (degree), P is monthly percent annual daytime hours, (depend on latitude of the place).

2. Thronthwaite method: Thornthwaite developed the following formula from data of eastern USA to estimate potential evapotranspiration on per month basis.

$$E_T = 1.6L_a\left(\frac{10T_m}{I}\right)^a \tag{3.12}$$

where E_T is monthly PET (cm), L_a is adjustment for the number of hours of daylight and days in the month related to latitutde of the place, T_m is mean monthly air temperature (°C), I is heat index $\left[= \sum_{i=1}^{12} I_i, \text{ where } i = \left(\frac{T_m}{5}\right)^{1.514}\right]$ and a is empirical constant $[= 6.75 \times 10^{-7})I^3 - (7.71 \times 10^{-5})I^2 + (1.792 \times 10^2)I + 0.49239]$

The PET computed by Eq. (3.12) is reasonably accurate, but has been found more applicable in North American conditions.

Hargreave equation

Hargreave developed the following formula on the basis of data collected from grass lysimeter as follows:

$$ET_g = 0.0135(T + 17.79)R_s \tag{3.13}$$

where, ET_g is reference crop potential consumptive use, T is mean daily temperature, (°C) and R_s is incident solar radiation [$= 0.10 R_{50}\sqrt{S}$, where S is percent possible sunshine hour and R_{50} is the clear day solar radiation in Langley per day (Ly/day).

Lysimeters and Field plots

Lysimeters: The lysimeters can be used to measure potential evapotranspiration from a vegetated surface when water supply is ultimated. These lysimeters are watertight tanks filled with soil in which crops are grown under natural conditions and set in a field, as shown in Figure 3.4. The evapotranspiration is measured by using water balance equation of all parameters. Lysimeters' studies are time-consuming and expensive. The difficulty in operating lysimeters include include that soil sample is distributed and non convenient in water with snow cover and freezing temperatures.

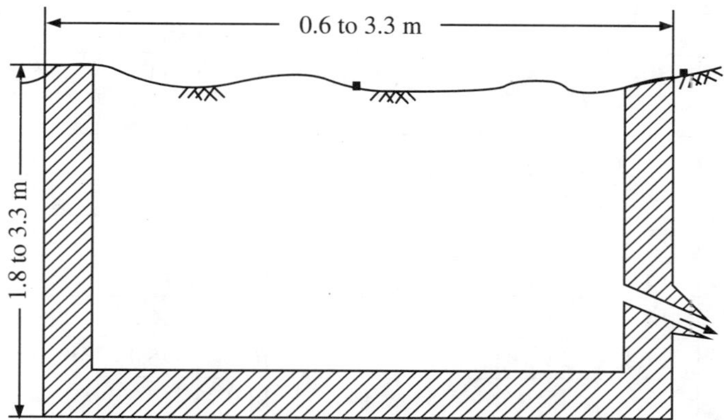

Figure 3.4 Lysimeter.

Field plots: In field plots, evapotranspiration can be measured as follows:

Evapotranspiration = Rainfall + Irrigated water – Runoff – Increase in soil moisture storage

3.4.2 Penman Equation for Evapotranspiration

The PET can be calculated using Penman equation, which is based on a combination of energy balance and mass transfer approach from open water surface. This method is relatively tedious, but is more accurate.

$$\text{PET} = \frac{\Delta Q_n + \gamma E_a}{\Delta + \gamma} \tag{3.14}$$

where PET is daily potential evapotranspiration in (mm/day), D is slope of saturation vapour pressure versus temperature curve (dE_a/dT) at the mean air temperature T_a (mmHg/°C), Q_n is net radiation (in mm of evaporable water per day), γ is psychrometric constant or the ratio of specific heat of air to the latent heat of evaporation of water (0.49) for 0°C (mmHg) and E_a is parameter including wind velocity and saturation deficit.

The net radiation Q_a is given as

$$Q_n = Q_a(1-r)\left(a + b\frac{n}{N}\right) - \sigma T^4\left(0.56 - 0.092\sqrt{e_a}\left(0.10 + 0.90\frac{n}{N}\right)\right) \tag{3.15}$$

where Q_a is mean monthly extraterrestrial incident radiation (mm of water per day, as shown in Table 3.4), r is reflection coefficient of evaporating surface (0.06 for open water surface), a is constant depending on latitude (generally 0.18), b is constant with average value of 0.55, $\frac{n}{N}$ is ratio between actual and possible hours of bright sunshine, σ is Stefan-Boltzman constant, T is mean air temperature in Kelvin and e_a is actual mean vapour pressure in the air (mmHg).

Table 3.4 Mean Monthly Solar Radiation Incident on Earth's Outer Space (Extraterrestrial Radiation in mm of Evaporable Water per day) (After Criddle 1958)

	North latitude in (degree)									
Month	0°	10°	20°	30°	40°	50°	60°	70°	80°	90°
January	14.5	12.8	10.8	8.5	6.0	3.6	1.3	–	–	–
February	15.0	13.9	12.3	10.5	8.3	5.9	3.5	1.1	–	–
March	15.2	14.8	13.9	12.7	11.0	9.1	6.8	4.3	1.8	–
April	14.7	15.2	15.2	14.8	13.9	12.7	11.1	9.1	7.8	7.9
May	13.9	15.0	15.7	16.0	15.9	15.4	14.6	13.6	14.6	14.9
June	13.4	14.8	15.8	16.5	16.7	16.7	16.5	17.0	17.8	18.1
July	13.5	14.8	15.7	16.2	16.3	16.1	15.7	15.8	16.5	16.8
August	14.2	15.0	15.3	15.3	14.8	13.9	12.7	11.4	10.6	11.2
September	14.9	14.9	14.4	13.5	12.2	10.5	8.5	6.8	4.0	2.6
October	15.0	14.1	12.9	11.3	9.3	7.1	4.7	2.4	0.2	–
November	14.6	13.1	11.2	9.1	6.7	4.3	1.9	0.1	–	–
December	14.3	12.4	10.3	7.9	5.5	3.0	0.9	–	–	–

The parameter E_a is computed as

$$E_a = 0.35(e_w - e_a)(1 + 0.00625u_2)$$

where u_2 is mean wind speed at 2 m height from ground (km/day) and e_w is saturation vapour pressure at mean air temperature (mmHg).

EXAMPLE 3.5 Using Penman's formula, calculate the consumptive use of rice for the month of February. Take the following data:

Wind velocity measured at 2 m height = 30 km/day
Elevation of the area = 220 m
Relative humidity for February = 50%
Latitude = 22°N
Mean observed sunshine hours = 7.2 h
Mean monthly temperature = 16°C

Solution From Table 3.1, for temperature of 16°C, e_s = 13.67 mm/Hg
Slope of the saturated vapour pressure versus temperature curve A = 0.86 mm/°C
From Table 3.4, Q_a = 11.94 mm of water per day for 22°N latitude
From Table 3.3, N = 11.42 h for the latitude of 22°N
Since mean observed sunshine hours is 7.2 h,

Therefore, $$\frac{n}{N} = \frac{7.2}{11.42} = 0.63$$

Vapour pressure in air, $e_a = e_s \times R_H = 13.67 \times 0.50 = 6.88$ mmHg
Drying power of air, $E_a = E_a = 0.35(e_w - e_a)(1 + 0.00625u_2)$
$= 0.35(1 + 0.00625 \times 30)(13.67 - 6.88) = 2.82$ mm/day
The reflection coefficient for close crop like paddy is assumed as 0.20.
Take, $\gamma = 0.49$, $\sigma = 2.01 \times 10^{-9}$ mm/day, $T_a = 273 + 16 = 289$ K

Therefore, $$Q_n = Q_a(1-r)\left(0.29\cos\phi + 0.55\frac{n}{N}\right) - \sigma T_a^4(0.56 - 0.092\sqrt{e_a})\left(0.10 + 0.9\frac{n}{N}\right)$$

or, $$Q_n = 11.94\left(1 - 0.2)(0.29\cos 22° + 0.55 \times \frac{7.2}{11.42}\right)$$

$$-2.01 \times 10^{-9} \times 289^4(0.56 - 0.092\sqrt{6.88}) \times \left(0.1 + 0.9 \times \frac{7.2}{11.42}\right)$$

$= 11.94 \times 0.8 \times (0.269 + 0.347) - 14.02(0.56 - 0.24)(0.1 + 0.568)$
$= 5.88 - 3.00 = 2.88$ mm of water per day

$$\text{PET} = \frac{(AQ_n + \gamma E_a)}{(A + \gamma)}$$

$$\text{PET} = \frac{(0.86 \times 2.88 + 0.49 \times 2.82)}{(0.86 + 0.49)}$$

$$\text{PET} = \frac{3.858}{1.35} = 2.86 \text{ mm/day}$$

$$\text{PET} = 2.86 \times 28 \times \frac{1}{10} \text{ cm/month} = 8.00 \text{ cm for February}$$

Consumptive use of rice for February is 8.00 cm = 80 mm of water

3.5 Interception

Interception is defined as the amount of gross precipitation that is captured by vegetation or surface cover, and subsequently, evaporated back to the atmosphere. Interception loss is high in the beginning of storms and gradually decreases. Intercepted water does not reach the ground surface. The factors which affect interception include the type of storm, plant, season and prevailing wind. A precipitation in the form of drizzle is completely absorbed by vegetative foliage. A plant with wide and dense canopy cause higher interception loss. The relationships between rainfall and interception is shown in Figure 3.5.

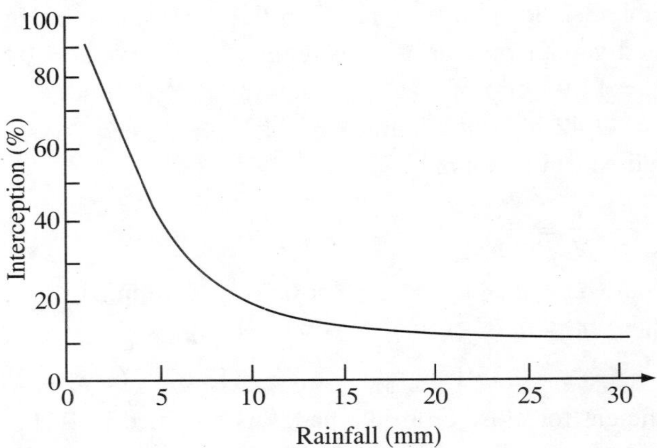

Figure 3.5 Typical interception loss curve.

3.6 Infiltration

The term *infiltration*, first used by Horton (1935), is defined as the flow of water into the ground through soil surface layer. It replenishes the soil moisture deficiency. *Percolation* is the movement of water under gravity and percolation starts after infiltration.

Infiltration capacity

The maximum rate at which the soil in any given condition at a given time is capable of absorbing water is called *infiltration capacity* f_p. The initial antecedent presence of moisture in the soil affects the infiltration capacity considerably. The actual rate of infiltration f can be expressed as

$$f = f_p \quad \text{when} \quad i \ge f_p$$
$$f = i \quad \text{when} \quad i \ge f_p$$

where i is intensity of rainfall.

The infiltration capacity of soil depends on a number of factors—(a) the soil type (texture, structure, hydrodynamic characteristics), (b) condition of soil surface, (c) crop rotations, (d) tillage operations, (e) variation in temperature of soil, and (f) soil moisture.

The relationship between infiltration capacity and time is shown in Figure 3.6, which is a plot of instantaneous rate of water entering into the soil. A loose, permeable, sandy soil has large infiltration capacity than a tight, clayey soil. At the beginning of rainfall, the intake rate is higher, and then reduces with time.

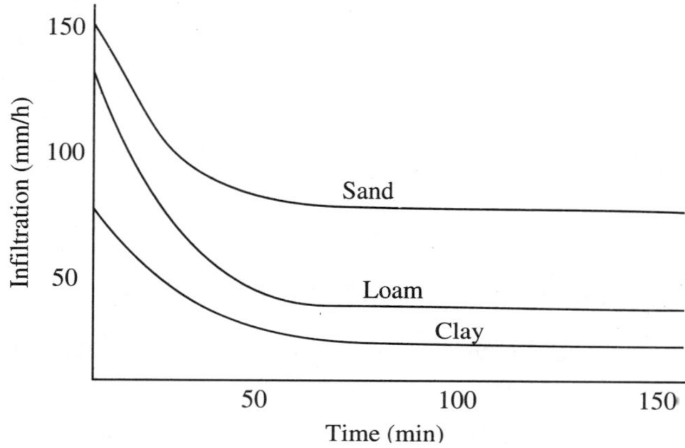

Figure 3.6 Variation of infiltration capacity.

3.6.1 Methods of Determining Infiltration

The following methods are most commonly used for determining infiltration of soil at a given location:

1. Infiltrations method
2. Measurement in pits and ponds
3. Artificial rainfall simulator
4. Hydrograph analysis

Infiltrometers method

These are two commonly used methods for the measurement of infiltration by cylinder infiltrometers.

1. Single cylinder infiltrometer: In the single cylinder infiltrometer, only one cylinder is used. The cylinder, 30 cm in diameter and 60 cm long, is made of metal and open at both ends, as shown in Figure 3.7. The metal cylinder is driven into the ground with the help of hammer up to a depth of 50 cm, leaving about 10 cm above the ground. Water is then flooded into the top part to maintain a constant water level of 7 cm. A pointer gauge may be used to measure water accurately. As the infiltration begins,water level is maintained at a constant level by adding measured water at successive intervals till the constant rate of infiltration is achieved. Knowing the quantity of water added, during different time intervals, a plot between infiltration capacity versus time is obtained. Normally, 2–3 hours of duration is required to attain constant infiltration rate.

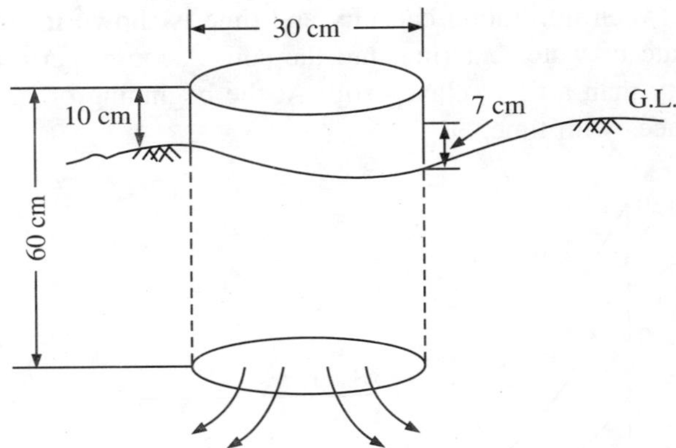

Figure 3.7 Single cylinder infiltrometer.

The major limitation of this method is that uncontrolled lateral movement of water from the cylinder bottom introduces errors.

2. Double cylinder infiltrometer: In this type of infiltrometer, two concentric cylinders are driven into the ground up to a depth of 15 cm leaving 10 cm above the ground level. Water is applied to both cylinders and the same level of water is maintained in both cylinders by adding measured quantity of water at successive intervals. To prevent any sealing of the ground surface inside the inner cylinder,a jute matting is recommended to be placed while adding water. The measurements of inner cylinder are used to determine infiltration rate, as shown in Figure 3.8. The purpose of outer cylinder is to prevent sideflows of water.

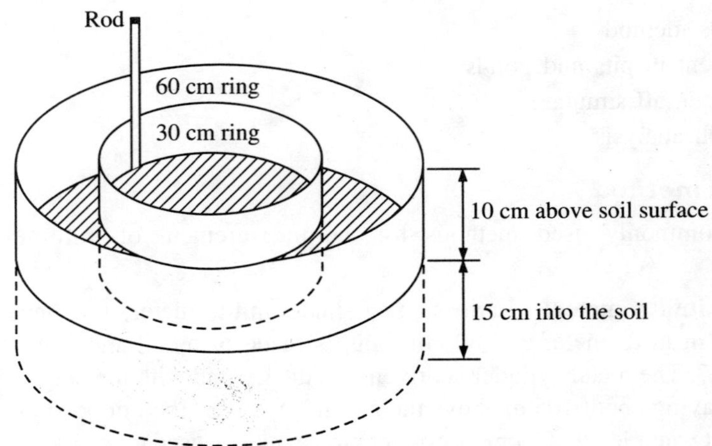

Figure 3.8 Double cylinder infiltrometer.

The limitations of using infiltrometers are as follows:

1. The soil is disturbed while cylinders are driven into the ground.
2. Lateral flow of infiltrated water occurs.
3. Effect of slope of ground is not accounted.

Measurement in pits and ponds

Depression in the water level in the pits and ponds is observed. Infiltration rate of soil may be computed with the use of water level in ponds by deducting loss due to evaporation.

Rainfall simulator

In a small area of land of about 0.1 to 50 m^2, especially designed nozzles are used to apply artificial showers at a uniform rate. Experiments are carried out in controlled conditions with various combinations of intensities and durations. The resulting surface runoff is measured in each case. Then, infiltration rate of the soil is determined.

Hydrograph analyses

If rainfall record and runoff hydrograph for a given storm is available, then infiltration capacity may be determined using water balance equation.

3.6.2 Infiltration Equations

Horton's equation

Horton (1939) developed the following equation for computing infiltration rate at any time period as an exponential decay:

$$f = f_c + (f_0 - f_c)e^{-kt} \tag{3.16}$$

where f is infiltration rate at any given time, f_0 is initial infiltration rate at time $t = 0$, f_c is constant infiltration rate at time $t = t_c$, k = Horton's decay coefficient depends on the soil and vegetation, and t is time from beginning of the storm

When $t = 0$, $e^{-kt} = 1$ then $f = f_0$.
When $t = \infty$, $e^{-kt} = 0$, then $f = f_c$.

The limitation of Horton's equation is that it is difficult to find the variation in the three parameters f_0, f_c and k based on the soil characteristics and antecedent moisture content and is not popular.

Philip's equation

Philip (1957) develop the following expression for computing the cumulative infiltration depth:

$$F_p = st^{1/2} + Kt \tag{3.17}$$

where F_p is cumulative infiltration, s is soil property known as sorptivity, K is Darcy's hydraulic conductivity.

By differentiating Eq. (3.17), the infiltration rate f can be given by

$$f = \frac{1}{2}st^{-1/2} + K \tag{3.18}$$

Kostiakov's equation

Kostiakov (1932) developed the following expression to determine the cumulative infiltration capacity:

$$F_p = kt^n \tag{3.19}$$

where k and n ($0 < n > 1$), are constant.

Lewis and Milne equation

The expression is given as follows:

$$F_p = at + b(1 - e^{-nt}) \tag{3.20}$$

where a, b and n are constants ($0 > n > 1$).

Green-Ampt equation

Green and Ampt (1991) proposed the following equation for computing infiltration rate based on Darcy's law:

$$f = k\left(1 + \frac{\eta S_c}{F_p}\right) \tag{3.21}$$

where η is porosity of soil, S_c is capillary suction at the wetting front, k = Darcy's hydraulic conductivity of soil, and F_p is cumulative infiltration.

3.6.3 Infiltration Indices

In hydrological processes, estimation of runoff from large catchment is made by using constant value of infiltration rate for duration of storm, although in actual practice, the infiltration varies with time. This average value of infiltration rate is called *infiltration index*. The following three types of infiltration indices are in common use:

ϕ Index

The *ϕ index* is defined as the rate of rainfall above which the rainfall volume is equal to the runoff volume, as shown in Figure 3.9. In this index, all losses arising out of infiltration,

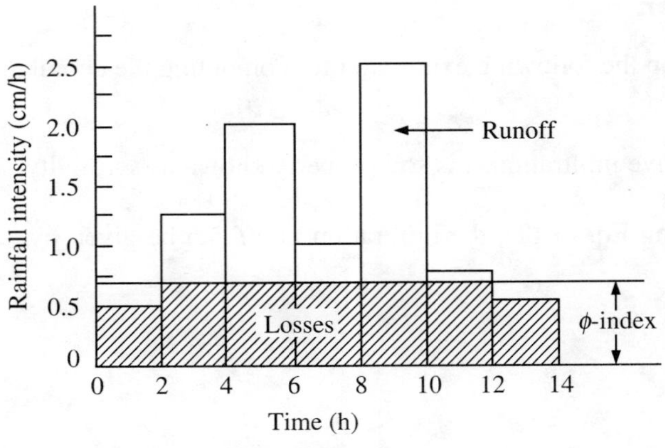

Figure 3.9 ϕ-index.

interception and depression storage are accounted. This index can be derived from the rainfall hyetograph with the knowledge of the resulting runoff volume.

$$\phi\text{-index} = \frac{\text{Basin recharge (total intake of water into the soil)}}{\text{Duration of rainfall}} \tag{3.22}$$

W-index

The *W*-index is the average infiltration rate when the time rainfall intensity exceeds the infiltration capacity rate.

$$W = \frac{P - R - I_a}{t_r} \tag{3.23}$$

where P is the total rainfall, R is the total runoff, I_a represents initial losses, and t_r is duration of storm during which $i > f_p$.

In the *W*-index, initial losses are separated from the total abstraction, thus, *W*-index is essentially equal to ϕ-index minus the average value of initial losses ($W < \varphi$). Since it is difficult to find the correct values of the initial abstraction, the *W*-index method is not popular.

f_{avg}-index

In this method, an average infiltration loss is assumed throughout the storm for the period ($i > f$).

EXAMPLE 3.6 The quantity of water added to a double ring infiltrometer of 1.5 m diameter at 30 min interval to keep the water level constant is as follows:

Time (min)	0	30	60	90	120	150	180
Quantity of water added (1)	0	15	14.2	12.2	11.8	11	11

Compute the rate of infiltration for every 30 min.

Solution Area of ring = $\pi r^2 = \pi \times 1.5^2 = 1.767\ \text{m}^2$

The calculations are shown in the table below:

Time (min)	*Quantity of water added* (1)	Q_{in} *area* (m²)	*Depth* (*m*) = $\frac{Q}{A}$	*Intensity* (mm/h)
0	0	1.767	0	0
30	15	1.767	0.008489	16.97
60	14.2	1.767	0.008036	16.07
90	12.2	1.767	0.006904	13.80
120	11.8	1.767	0.006678	13.35
150	11	1.767	0.006225	12.45
180	11	1.767	0.006225	12.45

EXAMPLE 3.7 The value of k in the Horton's equation for infiltration is 3. The maximum and the minimum rates of infiltration are 3 cm/h and 0.7 cm/h. Plot the infiltration rate curve.

Solution Given, maximum infiltration = 3 cm/h, minimum infiltration = 0.7 cm/h, k = 3

$$f_t = f_c + (f_0 - f_c)\,e^{-kt}$$

$$f_c = \text{Minimum infiltration} = 0.7 \text{ cm/h}$$

$$f_0 = \text{Maximum infiltration} = \text{Initial infiltration} = 3 \text{ cm/h}$$

$$f_0 - f_c = 2.3 \text{ cm/h}$$

$$f(t) = 0.7 + 2.3e^{-3t}$$

The computation of infiltration rate is shown in the table below:

Time (h)	*Infiltration rate*
0	3.0000
1	0.81
2	0.71
3	0.70
4	0.70
5	0.70
6	0.70
7	0.70

The infiltration rate curve is shown in below:

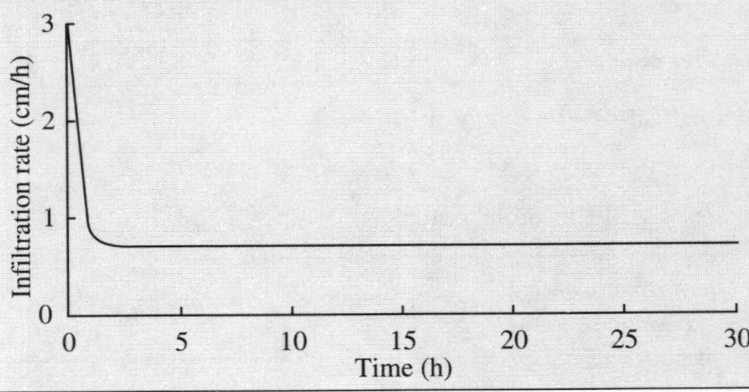

EXAMPLE 3.8 The storm over a catchment of 60 km^2 was having the following intensity:

(a) 45 min/h for 1 h

(b) 80 mm/h for 2 h

(c) 37 mm/h for 1 h

The catchment area had infiltration rate as follows:

(a) 25% area, ϕ = 15 mm/h

(b) 55% area, ϕ = 29 mm/h

(c) Balance impervious

Determine the runoff due to storm.

Solution Given, total area of catchment = 60 km^2, net runoff for different duration of rainfall

(a) $P_{1\text{ h}} = 0.25(45 - 15) + 0.55(45–20) + 0.2 \times 45 = 30.25$ mm

(b) $P_{2\text{ h}} = 0.25(80 - 15) + 0.55(80 - 20) + 0.2 \times 80 = 62.25$ mm for 1 h

So, for 2 h, i.e., $65.25 \times 2 = 130.5$ mm

(c) $P_{1\text{ h}} = 0.25(37 - 15) + 0.55(37 - 20) + 0.2 \times 37 = 22.25$ mm

The total net runoff = 30.25 + 130.5 + 22.25 = 183.25 mm

$$\text{Runoff} = 183 \times 10^{-3} \times 60 \times 10^6 = 10980000 \text{ m}^3$$

Summary

- The rainfall which is not available for surface runoff is defined as abstraction(loss).
- vaporation is a process in which a liquid changes to the gaseous state at the free surface below the boiling point through the transfer of heat energy.
- Evaporation is likely to continue till the saturation vapour pressure at water temperature e_w and actual vapour pressure (e_a) are equilibrated.
- Evaporation from seawater is about 2% to 3% less in comparison to the fresh water under identical conditions.
- Lake/reservoir evaporation can be computed by multiplying pan coefficient with pan evaporation.
- Transpiration is a process in which plans suck water through their roots and leave out the excess amount of water after building their tissues into the atmosphere.
- Evapotranspiration (consumptive use) is the water lost due to combination of evaporation and transpiration.
- Actual evapotranspiration (AET) is always less than potential evapotranspiration (PET).
- Penman's equation is based on sound theoretical reasoning and obtained by a combination of energy balance and mass transfer approach. Lysimeter is a device which is used to compute evapotranspiration by maintaining a water budget.
- Interception is defined as the amount of gross precipitation that is captured by vegetation or surface cover, and subsequently, evaporated back to the atmosphere.
- Infiltration is defined as the downward movement of water through the land surface into the soil.
- The maximum rate at which the soil in any given condition at a given time is capable of absorbing water is called infiltration capacity f_p.
- The ϕ-index is defined as that rate of rainfall above which the rainfall volume of equal to the runoff volume.

Objective Type Questions

1. Match List-I with List-II and select the correct answer using the codes given below the lists: (IES, 2001)

List-I

A. Anemometer B. Rain simulator
C. Lysimeter D. Hygrometer

List-II

1. Humidity 2. Evaporation
3. Infiltration 4. Wind speed

Codes

	A	B	C	D
(a)	4	3	1	2
(b)	3	4	1	2
(c)	4	3	2	1
(d)	3	4	2	1

2. A 3 h storm on a small drainage basin produced rainfall intensities of 3.5 cm/h, 4.2 cm/h and 2.9 cm/h in successive hours. If the surface runoff due to the storm is 3 cm, than the value of ϕ-index will be (IES, 2001)
(a) 2.212 cm/h (b) 2.331 cm/h (c) 2.412 cm/h (d) 2.533 cm/h

3. The Penman's evapo-transpiration equation is based on (IES, 2002)
(a) Water budget method (b) Energy balance method
(c) Mass transfer method (d) Energy balance and mass transfer approach

4. The plan area of a reservoir is 1 km^2. The water level in the reservoir is observed to decline by 20 cm in a certain period. During this period the reservoir receives a surface inflow of 10 ha-m, and 20 ha-m are abstracted from the reservoir for irrigation and power. The pan evaporation and rainfall recorded during the same period at a nearby meteorological station are 12 cm and 3 cm, respectively. The calibrated pan factor is 0.7. The seepage loss from the reservoir during this period in hectare-meter is (GATE, 2003)
(a) 0.0 (b) 1.0 (c) 2.4 (d) 4.6

5. Which one of the following defines aridity index (AI)? (IES, 2004)

(a) $AI = \dfrac{(PET - AET)}{PET} \times 100$ (b) $AI = \dfrac{PET}{AET} \times 100$

(c) $AI = \dfrac{AET}{PET} \times 100$ (d) $AI = \dfrac{(AET - PET)}{AET} \times 100$

(where AET is actual evapotranspiration and PET is potential evapotranspiration)

6. The rainfalls during three successive 2 h periods are 0.5, 2.8 and 1.6 cm. The surface runoff resulting from this storm is 3.2 cm. The ϕ-index value of the storm is (GATE, 2004)
(a) 0.20 cm/h (b) 0.28 cm/h (c) 0.30 cm/h (d) 0.80 cm/h

7. The area between the two isohyetes 45 cm and 55 cm is 100 km^2, and that between 55 cm and 65 cm is 150 km^2. What is the average depth of annual precipitation over the basin of 250 km^2? (IES, 2005)
(a) 50 cm (b) 52 cm (c) 56 cm (d) 60 cm

8. Penman's equation is based on (IES, 2006)

(a) Energy budgeting only
(b) Energy budgeting and water budgeting
(c) Energy budgeting and mass transfer
(d) Water budgeting and mass transfer

9. Match List-I with List-II and select the correct answer using the code given below the lists:

List-I (IES, 2006)

A. Location B. Stability
C. Variation of discharge D. Plan form

List-II

1. Perennial 2. Degrading
3. Tidal 4. Braided

Codes

	A	B	C	D
(a)	4	3	2	1
(b)	3	1	2	4
(c)	4	1	2	3
(d)	3	2	1	4

10. A standard, ground-based evaporation pan, corresponding to Indian Standards, is installed at the banks of a surface reservoir. The water surface area on a particular day is 100 ha. The recorded evaporation loss from the pan, on a certain day, is nearly 4.0 cm. What is the anticipated evaporation loss from the reservoir for that day? (IES, 2007)

(a) $(1.8 \text{ to } 2) \times 10^4 \text{ m}^3/\text{day}$ (b) $(2.5 \text{ to } 2.75) \times 10^4 \text{ m}^3/\text{day}$
(c) $(3.0 \text{ to } 3.25) \times 10^4 \text{ m}^3/\text{day}$ (d) $(3.8 \text{ to } 4.05) \times 10^4 \text{ m}^3/\text{day}$

11. The total observed runoff volume during a 4 h storm with a uniform intensity of 2.8 cm/h is $25.2 \times 10^6 \text{ m}^3$ from a basin of 280 km^2 area. What is the average infiltration rate for the basin? (IES, 2007)

(a) 3.6 mm/h (b) 4.8 mm/h (c) 5.2 mm/h (d) 5.5 mm/h

12. Consider the following chemical emulsions:

1. Methyl alcohol 2. Cetyl alcohol
3. Stearyl alcohol 4. Kerosene

Which of the above chemical emulsions is/are used to minimise the loss of water through the process of evaporation? (IES, 2009)

(a) 1 only (b) 1 and 4 (c) 2 and 4 (d) 2 and 3

13. During a 3 h storm event, it was observed that all abstractions other than infiltration were negligible. The rainfall was idealised as three 1 h storms of intensity 10 mm/h, 20 mm/h and 10 mm/h, respectively, and the infiltration was idealised as a Horton curve, $f = 6.8 + 8.7 \exp(-t)$ (f in millimetre per hour and t in hour). What is the effective rainfall? (GATE, 2006)

(a) 10.00 min (b) 11.33 mm (c) 12.43 (d) 13.63 mm

14. A unconfined aquifer of porosity 30%, permeability 35 m/day and specific yield of 0.20 has an area of 100 km^2. If the water table falls by 0.25 m during a drought, the volume of water lost from storage, in million cubic metres, is (IES, 2010)

(a) 2.0 (b) 5.0 (c) 1.0 (e) 4.0

15. Consider the following statements:

1. Hver the oceans, there is more evaporation than precipitation.
2. On lands, it is more precipitation than evapotranspiration.

Which of these statements are correct? (IES, 2012)

(a) Both 1 and 2 (b) Neither 1 nor 2
(c) 1 only (d) 2 only

16. The ratio of actual evapotranspiration to potential evapotranspiration is in the range of (GATE, 2012)

(a) 0.0 to 0.4 (b) 0.6 to 0.9 (c) 0.0 to 1.0 (d) 1.00 to 2.0

17. Storm-I of duration 5 h gives a direct runoff of 4 cm and has an average intensity of 2 cm/h. Storm-II of 8 h duration gives a runoff of 8.4 cm. (GATE, 2013)

(Assume ϕ-index is same for both the storms.)

(i) The value of ϕ-index is

(a) 1.2 (b) 1.6 (c) 1 (d) 1.4

(ii) Intensity of Storm-II in centimetre per hour

(a) 2 (b) 1.5 (c) 1.75 (d) 2.25

18. The average surface area of a reservoir in the month of June is 20 km^2. In the same month, the average rate of inflow is 10 m^3/s, outflow rate is 15 m^3/s, monthly rainfall is 10 cm, monthly seepage loss is 1.8 cm and the storage change is 16 million m^3. The evaporation (in centimetre) in that month is (GATE, 2015)

(a) 46.8 (b) 136.0 (c) 13.6 (d) 23.4

Answers

1. (c) **2.** (d) **3.** (d) **4.** (d) **5.** (a) **6.** (c) **7.** (c) **8.** (c) **9.** (d) **10.** (c)
11. (d) **12.** (d) **13.** (d) **14.** (b) **15.** (a) **16.** (c) **17.** (i) (a) (ii) (d) **18.** (b)

Explanations

4. Evaporation losses, $E = 1\times 10^6 \times \dfrac{12}{100} \times 0.7 = 84$ ha-m

Amount of rainfall, $P = 1\times 10^6 \times \dfrac{3}{100} = 3$ ha-m

Change in storage, $\Delta S = 1\times 10^6 \times \dfrac{20}{100} = -20$ ha-m

$$(I + P) - (E + Q + \text{Seepage}) = \Delta S$$
$$(10 + 3) - (8.4 + 20 + \text{Seepage}) = -20$$
$$13 - 28.4 - \text{Seepage} = -20$$
$$\text{Seepage} = 4.6 \text{ ha-m}$$

6. Total rainfall during the storm = 0.5 + 2.8 + 1.6 = 4.9 cm

$$\text{Excess rainfall} = \text{Total rainfall} - \text{Surface runoff}$$
$$= 4.9 - 3.2 = 1.7 \text{ cm}$$

Duration of excess rainfall, $t_e = 2 \times 3 = 6$ h

$$\phi\text{-index} = \frac{\text{Excess rainfall}}{\text{Duration of excess rain fall}} = \frac{1.7}{6} = 0.28 \text{ cm/h}$$

As the value of ϕ-index obtained is more than $\frac{0.5}{2}$ = 0.25 cm/h, so the excess rainfall duration will be reduced by 2 h.

Hence, $t_e = 4$ h

Now, excess rainfall = 2.8 + 1.6 – 3.2 = 1.2 cm

$$\phi\text{-index} = \frac{1.2}{4} = 0.3 \text{ cm/h}$$

7. Given, the area between the two isohyetes 45 cm and 55 cm is 100 km^2 and that between 55 cm and 65 cm is 150 km^2. Average depth of annual precipitation.

$$\frac{\left(\frac{45+55}{2} \times 100 + \frac{55+65}{2} \times 150\right)}{100+150} = 56 \text{ cm}$$

10. The total volume of evaporation = Depth of evaporation × Water surface area

$$= 0.8 \times 4 \times 10^{-2} \times 100 \times 10^4 \text{ m}^3 \text{ per day}$$
$$= 3.2 \times 10^4 \text{ m}^3\text{/day}$$

11. The average infiltration rate is also called *W*-index. We know that

$$W = \frac{P - R - I_a}{t_r}$$

Therefore,

$$P = 2.8 \times 10 \times 4 = 112 \text{ mm}$$

$$R = \frac{25.2 \times 10^6 \times 10^3}{280 \times 10^6} = 90 \text{ mm}$$

$$t_r = 4 \text{ h}$$

$$I_a = 0$$

Therefore,

$$W = \frac{112 - 90}{4} = 5.5 \text{ mm/h}$$

13. The calculation of *f* is shown below in the table:

Time	0	1	2	3
F (mm/h)	15.5	10	8	7.2

The superposition of hyetograph and infiltration curve clears that the first hour does not contribute to rainfall excess.

For $t = 1$ to $t = 3$, the rainfall excess is

$$r_e = 20 + 10 - \int_1^3 (6.8 + 8.7e^{-t})\, d_t$$

$$r_e = 30 - \left[6.8 \times (3-1) + 8.7 \int_1^3 (e^{-t}) d_t\right]$$

$$r_e = 30 - 13.6 + 8.7 \left[\frac{1}{e^3} - \frac{1}{e^1}\right] = 13.63 \text{ mm}$$

14. Volume of water lost during drought = S_y × Area × Drop in water table

$$= 0.2 \times 100 \times 10^6 \times 0.25$$

$$= 5 \times 10^6 \text{ m}^3$$

17. Let intensity of Storm-II be P cm/h.

$$1.2 = \left(\frac{P \times 8 - 8.4}{8}\right)$$

$$P = 2.25 \text{ cm/h}$$

18.

$$\Delta S = Q_t + Q_R - Q_O - Q_S - Q_E$$

$$16 \times 10^6 = (10 \times 86400 \times 30) + (0.1 \times 20 \times 10^6) - (15 \times 86400 \times 30) - (1.8 \times 10^{-2} \times 20 \times 10^6) - Q_E$$

$$Q_E = \frac{27320000}{20 \times 10^6} = 136.6 \text{ cm}$$

IES Conventional Questions

PROBLEM 3.1 The following is the set of observed data for successive 15 min periods of 105 min storm in a catchment.
If the value of ϕ-index is 3 cm/h, estimate the net runoff, the total rainfall and the value of W-index.

Duration (min)	15	30	45	60	75	90	105
Rainfall (cm/h)	2.0	2.0	8.0	7.0	1.25	1.25	4.5

(IES, 1996)

Solution

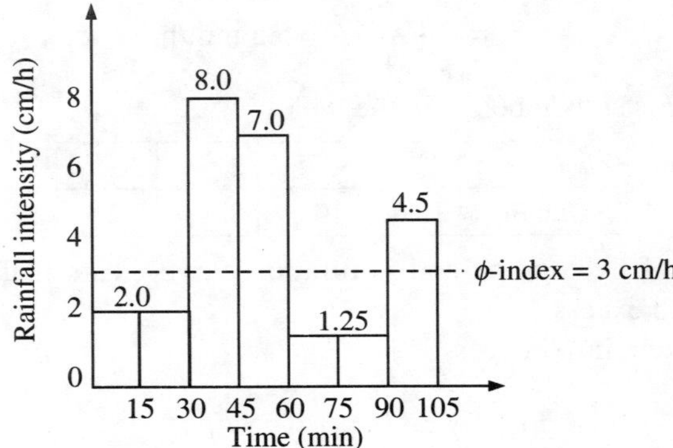

The variation of rainfall intensity with time using given data is shown in the following hyetograph: Total infiltration is shown by the dotted area and total surface runoff is shown by the hatched area. Now, the total amount of runoff can be calculated as

$$\text{Runoff} = (8-3)\times\frac{15}{60}+(7-3)\times\frac{15}{60}+(4.5-3)\times\frac{15}{60}$$

$$= 1.25 + 1.0 + 0.375 = 2.625 \text{ cm}$$

From the above hyetograph, the total amount of precipitation can be calculated as

$$\text{Total precipitation} = 2\times\frac{30}{60}+8\times\frac{15}{60}+7\times\frac{15}{60}+1.25\times\frac{30}{60}+4.5\times\frac{15}{60}$$

$$= 6.5 \text{ cm}$$

Now, W-index can be calculated by the following formula:

$$W\text{-index} = \frac{\text{Precipitation} - \text{Runoff}}{\text{Duration of rainfall (h)}}$$

$$= \frac{6.5-2.625}{105/60} = 2.21 \text{ cm/h}$$

PROBLEM 3.2 What are the factors affecting evaporation from water surfaces? Describe briefly a method of estimating the evaporation from weather data. (IES, 1998)

Solution Please see the text.

PROBLEM 3.3 Determine the amount of evapotranspiration from an area if the total rainfall precipitated during a storm is 10.0 mm. Given the antecedent moisture at the root in the soil is 5.00 mm, the loss of water due to seepage is 2.5 mm, losses due to percolation are 2.00 mm, surface runoff is 3.00 mm and the moisture retained in the soil is 1.00 mm. (IES, 2012)

Solution Given, total rainfall, P = 10 mm, antecedent moisture at root in the soil = 5 mm, loss of water due to seepage = 2.5 mm, loss due to percolation = 2 mm, surface runoff = 3 mm and moisture retained in the soil = 1 mm.

Using hydrologic water budget equation,

$$P\text{-}R\text{-}G\text{-}E\text{-}T = \Delta S$$

Therefore, the water budget equation can be written as

Precipitation + antecedent moisture at root in the soil if any – Runoff – groundwater loss – seepage loss – evapotranspiration = moisture retained in soil

$$10 + 5 - 2.5 - 2 - 3 - T = 1$$

$$\Rightarrow \quad 7.5 - T = 1$$

$$\Rightarrow \quad T = 6.5 \text{ mm}$$

Therefore, the amount of evapotranspiration = 6.5 mm.

PROBLEM 3.4 Rainfalls over a basin in three consecutive hours are 4 cm, 5 cm and 3 cm, respectively. Estimate the surface runoff from the basin, assuming negligible surface retention and evaporation losses. The infiltration loss can be estimated using the following Horton's equation:

$$f = 1.2 + 4.2e^{-2.5t}$$

Here, f is infiltration (cm/h) and t is time (h) from start of rainfall. (IES, 2013)

Solution Rainfall intensity for each hour can be calculated as

$$\text{Rainfall intensity for 1st hour} = \frac{4}{1} = 4 \text{ cm/h}$$

$$\text{Rainfall intensity for 2nd hour} = \frac{5}{1} = 5 \text{ cm/h}$$

$$\text{Rainfall intensity for 3rd hour} = \frac{3}{1} = 3 \text{ cm/h}$$

Horton's equation is given as

$$f = 1.2 + 4.2e^{-2.5t}$$

We know that Horton's infiltration equation is

$$f = f_c + (f_0 - f_c)e^{-kt}$$

where f_0 is initial infiltration rate and f_c is final infiltration rate.

Now, comparing the given Horton's equation,

$$f_c = 1.2, \quad f_0 = 5.4$$

$$f = 1.2 + (5.4 - 1.2e^{-2.5t})$$

A figure is plotted between time and rate of infiltration, as shown below. Runoff may be computed by subtracting infiltration from total precipitation.

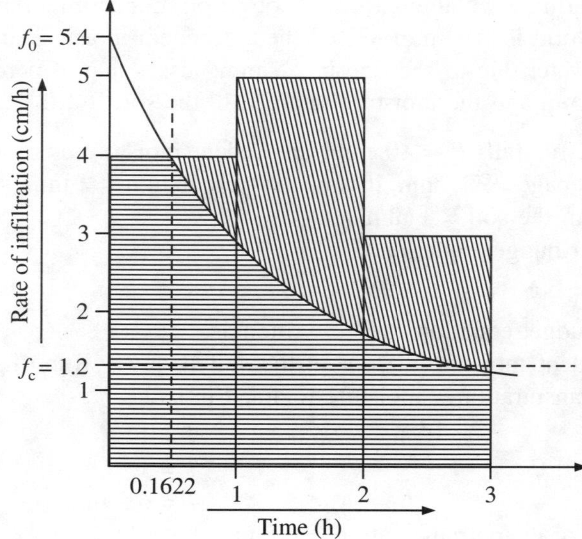

In this plot, the dotted portion shows total infiltration. We have to calculate the time when the Horton's filtration curve cuts the hyetograph. Assuming that the curve cuts the hyetograph before t = 1 h.

Therefore, $f = 4$ cm/h

$$1.2 + 4.2e^{-2.5t} = 4$$

$\Rightarrow$ $t = 0.1622$ h

Therefore, till $t = 0.1622$, there will be no runoff.

$$\text{Total infiltration} = \text{Area of dotted portion}$$

$$= 4(0.1622) + \int_{0.1622}^{3} [1.2 + 4.2e^{-2.5t}]\, dt$$

$$= 0.6488 + \int_{0.1622}^{3} 1.2\, dt + 4.2 \int_{0.1622}^{3} e^{-2.5t}\, dt$$

$$= 0.6488 + 3.4054 + 4.2(0.2664) = 5.1732 \text{ cm}$$

$$\text{Runoff} = \text{Total precipitation} - \text{Infiltration}$$

$$= (12 - 5.1732) \text{ cm} = 6.8268 \text{ cm}$$

PROBLEM 3.5 What is the difference between ϕ-index and W-index? (IES, 2014)

Solution Please refer text.

PROBLEM 3.6 A storm with a 15.0 cm precipitation produces a direct runoff of 8.7 cm. The time distribution of the storm is as follows:

Time from start (h)	1	2	3	4	5	6	7	8
Incremental rainfall in each hour (cm)	0.6	1.35	2.25	3.45	2.7	2.4	1.5	0.75

Estimate the ϕ-index of the storm. (IES, 2014)

Solution Precipitation = 15 cm

Direct runoff = 8.7 cm

Calculation of ϕ-index using trial:

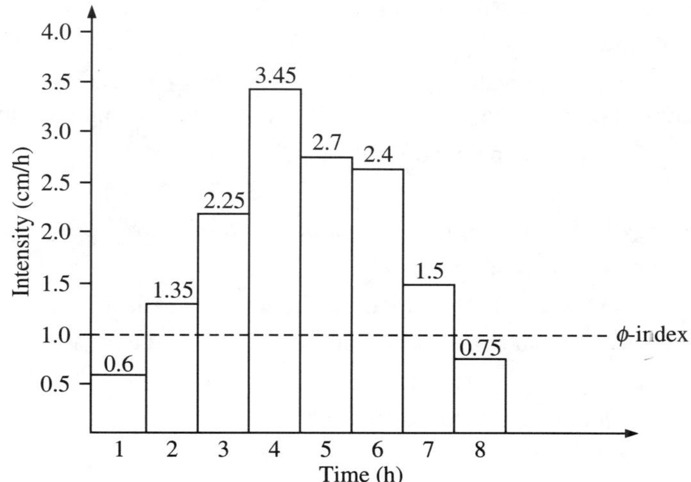

Trial 1:

Let us assume ϕ-index = 0.65 cm/h

$$\text{Runoff} = 15 - (0.65 \times 8) = 9.8 \text{ cm} > 8.7 \text{ cm}$$

Therefore, ϕ-index > 0.65 cm/h

Trial 2:

Let us assume ϕ-index = 0.70 cm/h

$$\text{Runoff} = 15 - (0.70 \times 8) = 9.4 \text{ cm} > 8.7 \text{ cm}$$

Therefore, ϕ-index > 0.7 cm/h

Trial 3:

Let us assume ϕ-index = 0.75 cm/h

$$\text{Runoff} = 15 - 0.6 - (0.75 \times 7) = 9.15 \text{ cm} > 8.7 \text{ cm}$$

Therefore, ϕ-index > 0.75 cm/h

Trial 4:

Let us assume ϕ-index = 1.35 cm/h

$$\text{Runoff} = 15 - 0.6 - 0.75 - (6 \times 13.5) = 5.55 \text{ cm} > 8.7 \text{ cm}$$

Therefore, ϕ-index lies between 0.75 cm/h and 1.35 cm/h

$$\text{Runoff} = 15 - 0.6 - 0.75 - 6\phi$$

$$\Rightarrow \quad 15 - 0.6 - 0.75 - 6\phi = 8.7$$

$$\Rightarrow \quad \phi = 0.825 \text{ cm/h}$$

Therefore, value of ϕ-index = 0.825 cm/h.

Theoretical Questions

1. Discuss briefly the various abstractions from precipitation.
2. Explain Dalton's law of evaporation.
3. Define evaporation and explain various factors affecting it. Explain the importance of evaporation control of reservoir and possible methods of achieving the same.
4. What are the limitation of using evaporation pan?
5. Define pan coefficient.
6. Explain the energy budget method of estimation of evaporation from a reservoir.
7. Write a short note on transpiration.
8. Explain the procedure to compute evapotranspiration using Blaney–Criddle method.
9. Differentiate between actual evapotranspiration and potential evapotranspiration.
10. What do you understand by interception?
11. Define infiltration and explain factors affecting rate.
12. What are the methods available to measure infiltration?

13. Write short notes on

(a) Horton's equation of infiltration
(b) Disadvantage of cylinder infiltrometer.

14. Distinguish between

(a) Infiltration rate and infiltration capacity
(b) Single and double cylinder infiltrometer

15. Define various infiltration indices.

Unsolved Problems

Problem 1: The surface are of a reservoir (m^2) is given by $A = 110y^2$, where y is the depth of water (m) in the reservoir. In one week, water depth in the reservoir has reduced from 11 m to 8 m. Determine the average hourly rate of evaporation. Assume seepage loss to be 40% of the evaporation loss. Determine the average daily rate of evaporation.

Problem 2: A trapezoidal channel of bed width 6 m and side slopes 1:1.2 carries water at a depth of 3 m. The rate of evaporation observed was 0.5 mm/m^2/h. Find the daily loss due to evaporation from the canal in a length of 15 km in hectare-metre.
Given data:

Bed width of channel = 6 m
Side slopes = 1:1.2
Flow depth = 3 m
Rate of evaporation = 0.5 mm/m^2/h
Length of channel = 15 km

Problem 3: Observations taken on a 1.5 m diameter land pan for 12 h were as follows:

(a) Quantity of water taken out of the pan = 6 l
(b) Precipitation during this time period = 25 mm

If there was no change in the water level in the pan, find the rate of evaporation.

Problem 4: The monthly evaporation (mm) observed on a pan from January to December 2011 is as follows:

100, 115, 130, 132, 134, 144, 149, 146, 141, 129, 120, 116

The water spreads of reservoir in January 2007 and December 2007 were 2.7 km^2 and 2.83 km^2, respectively. Assuming a pan coefficient of 0.9, find the total loss of water due to evaporation in 2007 from the reservoir. Neglect all other losses.

Problem 5: The rate of precipitation (mm/h) observed over a catchment of 40 km^2 for successive 40 min is as follows:

18, 22, 26, 39, 30, 14, 6

If the value of ϕ = 24 mm/h, determine (a) Total precipitation (b) Runoff (ha-m)

Problem 6: A catchment area having an average rate of infiltration of 20 mm/h experience the storm of the following intensities:

(a) 60 mm/h for 3

(b) 35 mm/h for 2

The resulting runoff was 11 × 10^6 m^3. Find the catchment area.

Problem 7: For a catchment area of 15 km^2, a 7 h storm was as follows:

Time (h)	1	2	3	4	5	6	7
Precipitation (mm)	24	45	2	35	55	46	9

The discharage observed at the gauge station was as follows:

Time (h)	0	1	2	3	4	5	6	7	8	9	10	11	12	13	14	15
Discharge (m^3/s)	0	9	19	35	70	58	48	40	31	24	19	15	10	5	3	0

Assume the evaporation losses to be 0.9 mm/h/m^2 and the seepage loss equal to be 60% of the evaporation loss. Find the ϕ-index and *W*-index.

Further Reading

Chow, V.T., Maidment, D.R., and Mays, L.W., *Applied Hydrology*, McGraw-Hill, Singapore, 1988.

Raghunath, H.M., *Hydrology—Principles, Analysis and Design*, New Age Publishers, New Delhi, 2014.

Subramanya, K., *Engineering Hydrology*, Tata McGraw-Hill, New Delhi, 2010.

Suresh, R., *Watershed Hydrology*, Standard Publishers Distributors, New Delhi, 2015.

Yadupathi Putty, Mysooru, R., *Principles of Hydrology*, I.K. International, New Delhi, 2013.

CHAPTER

4

Streamflow Measurement

4.1 Introduction

The water from atmosphere, overland flow, groundwater flow and subsurface flow which constitutes the flow in the stream is called *streamflow* or *discharge*. The two components of streamflow—velocity and volume—combine to determine the energy of the water. Precipitation falling on the ground may infiltrate into the ground or flow down on the surface to join the network of natural channels to finally drain out to the water bodies. A fraction of water infiltrating into the ground may also meet the stream (interflow) at a later stage as subsurface runoff and the rest may also meet the groundwater table or aquifers.

Streamflow is an important variable for estimating the rate of flow, and volume of the streamflow is used for designing of water resources projects. Therefore, it is necessary to analyse and measure streamflow to be utilised in the scientific investigations such as unit hydrograph studies, establishing the rainfall-runoff relations, flood control warning and protections, etc.

4.2 Measurement of Stage and Discharge

Stage can be defined as the height of the water surface above a datum plane, whereas *gauge height* can be defined as the water surface elevation to some pre-determined gauge datum. Further, gauge height and stage-discharge relationship are used for calculating the streamflow. It is necessary to have minimum error in gauge height measurement and also in establishing stage-discharge relationship to get the streamflow record with high accuracy. After stable establishment of stage-discharge relationship, only stage record is continued at the place of discharge record.

4.2.1 Non-recording Gauge

Non-recording gauge is that type of gauge where daily or sub-daily stages are recorded manually, with the reading shown on the installed gauges. Different types of non-recording gauges are discussed here.

Staff gauge

The standard vertical staff gauge consists of porcelain-enamelled iron sections of 4 inch wide and 3.4 ft long and graduated at every 0.02 ft. Staff gauge is generally attached to a bridge pier or any other structure near stream bank. The staff gauge can be vertical or inclined. In a high fluctuating river or stream, sometimes it becomes difficult to measure all the stage variation in one staff. In such a case, multiple staff concept is used, as shown in Figure 4.1.

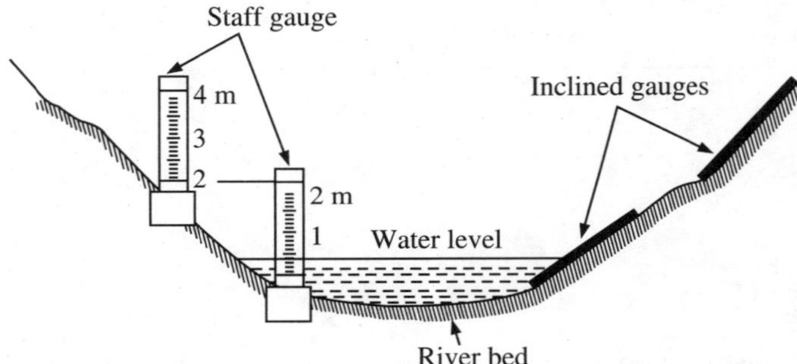

Figure 4.1 Staff gauge.

Electric tape gauge

An electric tape gauge consists of a tape graduated in feet, a frame and a voltmeter (Figure 4.2). In this tape, there are two terminals of which the negative terminal of the battery is attached to a ground connection, while the positive terminal is attached to the positive terminal of the voltmeter. The negative terminal of the voltmeter is connected to the weight, through the frame, reel and tape.

For measuring stage, the weight is dropped until it contacts the water surface. Once it comes in contact of water surface, it completes the electric circuit and produces a signal on the voltmeter. Then gauge height of the water level is noted by observing tape reading at the index marker.

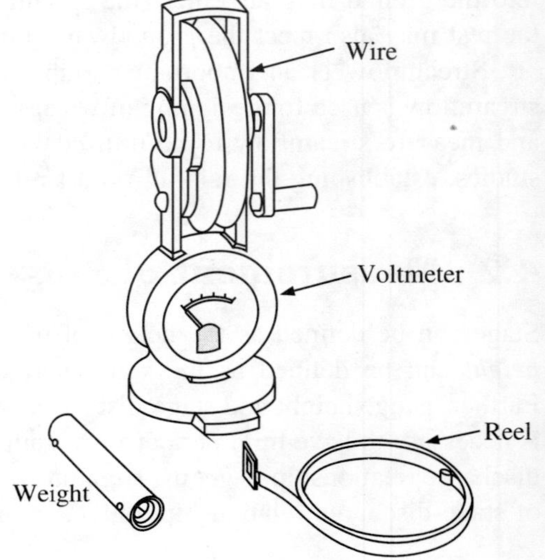

Figure 4.2 Electric tape gauge.

Wire-weight gauge

A wire-weight gauge consists of a drum wound with a layer of cable, which is guided to its position on the drum by a threading sheave. The wire-weight gauge is generally attached to a bridge. It can also be utilised as a reference gauge. The gauge is set by lowering the weight to a position a few feet above the water surface where levelling can be used to determine the elevation of the bottom of the weight. For calculating the gauge height of the water surface, weight is lowered to the water surface until it just touches it. In case of turbulence, it is suggested to take several readings at the crest and trough, and then, the average of these may be used for the water surface elevation.

Crest-stage gauge

It is the simplest and the most reliable gauge for measuring the level of streams. The most popular crest stage gauge is a vertical rod of a 2-inch galvanised pipe containing a wood or aluminium staff held in a fixed position with relation to a datum reference (Figure 4.3).

A perforated tin cup attached to the bottom of the staff contains regranulated cork. As the water rises inside the pipe, the cork floats on its surface. The cork sticks to the staff inside the pipe when the water reaches its peak and starts to deplete, thus achieving the highest stage of the flood. The gauge height of a peak is then calculated by measuring the interval between the reference point on the staff and the flood mark. Scaling can be simplified by graduating the staff.

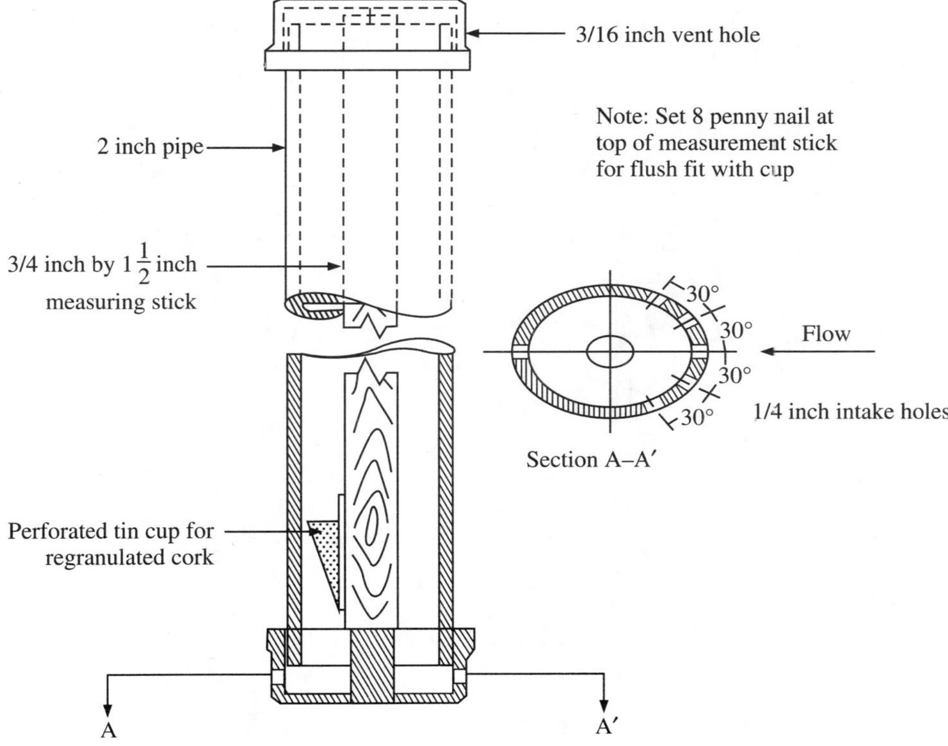

Figure 4.3 Crest stage gauge.

4.2.2 Recording Type Gauge

The recording rain gauge is generally used to measure the stage continuously with time and can be referred to as the *automatic water level recorder*. The automatic water level recorder works on the principle of a float. The most common type of automatic water level recorder utilises a float line, with a metal float at one end and small counter weight at the other end (Figure 4.4). The float line permits over an elevator and transfers the changes of water level to it. A recording stylus is attached to the pulley or elevator. It moves laterally and traces the water level fluctuations on a recorder chart. The recorder chart is a tracing quality strip paper wound over rollers or a drum. The recorder chart is connected to a clockwork mechanism which moves it at pre-determined speed continuously. The recorder chart, thus, gets wound slowly at a uniform rate and the recording stylus traces the water level fluctuations laterally on it to provide a continuous record of water levels. The water level recorder is a compact unit which is installed on a platform constructed over a stilling well. The stilling well may be constructed with corrugated iron sheet, concrete or masonry. On the top, a roofed shelter with locking arrangement is constructed to house and secure the recorder. The stilling well is connected to the river by means of intakes. A ladder is provided in the stilling well for inspection and cleaning purposes. An enamelled gauge plate is also fixed on the stilling well from outside to record the water level by visual observations during inspection visit. During each inspection visit, the recorder is inspected, the datum of the recorded water level is compared and suitably adjusted with the date and signature of the inspecting officer on the chart. It helps in computation of water levels in the office subsequently.

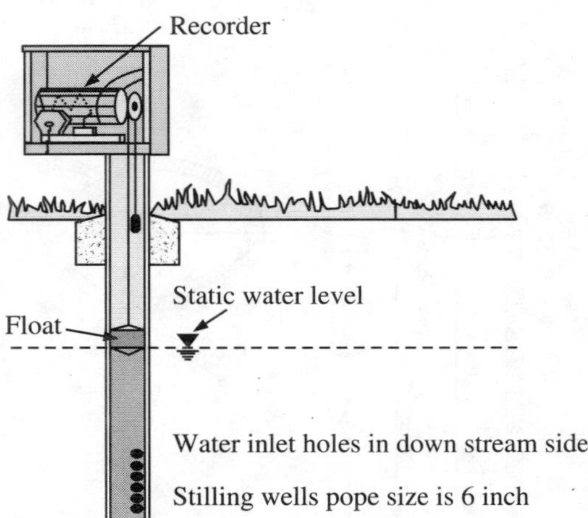

Figure 4.4 Automatic water level recorder.

4.3 Streamflow Measurement Using Area-Velocity Method

There are various methods available for the measurement of streamflow; perhaps, the most reliable among all of them is the area-velocity method. In this method, the stream discharge Q is computed by calculating the cross-sectional area A of flow within the river and the average velocity V of flow using continuity equation. The equation is given as:

$$Q = AV \tag{4.1}$$

To find mean velocity and area of flow, it is required to collect information at several point over the defined cross section. The number of points should be defined such that the computations are finished within a given time. This is necessarily required when the stage in the river is changing rapidly at the time of streamflow measurement. The procedure consists of dividing the flow area into a number of strips by means of verticals selected along the width of the stream and measuring the depths at these verticals. The velocities are measured at several points on the verticals to compute the mean velocity for any vertical. The streamflow velocities are generally measured by two types of instruments, namely, floats and current meters.

4.3.1 Floats

The use of floats for the measurement of flow velocities is very popular due to their low cost and convenience in handling the instrument. Any floating object can be used as float, however, normally the leak-proof and easily identifiable floats are utilised. The floats are generally of four types, namely, the surface floats, double canister floats, rod floats and twin floats (Figure 4.5). Surface floats are easy to handle. Any simple float moving on the surface of the stream is a surface float. The submerged floats are built of hollow cylinders. Generally, for higher accuracy, more than one float is used for the measurement of flow velocity. The floats are placed on the surface of flowing water at pre-determined spacing. The time taken by the float to reach a known distance is measured, and the velocity is calculated using Eq. (4.2).

$$Vs = \frac{S_d}{t} \tag{4.2}$$

where, V_s is surface velocity (m/s), S_d is distance travelled (m) and t is time(s).

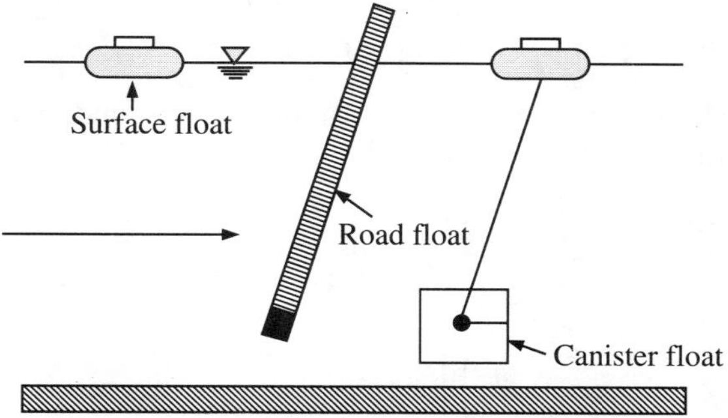

Figure 4.5 Floats.

4.3.2 Current Meters

Two types of current meters, namely cup type current meter and propeller type current meter, are generally used. The cup type current meter is also known as the *price current meter*. This current meter works on the principle such that the water flowing through the element of the meter makes it revolve due to the drag force acting on it and the speed of the rotating element is directly proportional to the velocity of water. The main features of cup type current meter are six conical cups rotating about a vertical axis and tail wings always align with the direction of flow (Figure 4.6). This instrument is recommended for a velocity range of 0.15 m/s to 4.0 m/s. The instrument is not recommended for use in situations where a vertical component of velocities exists.

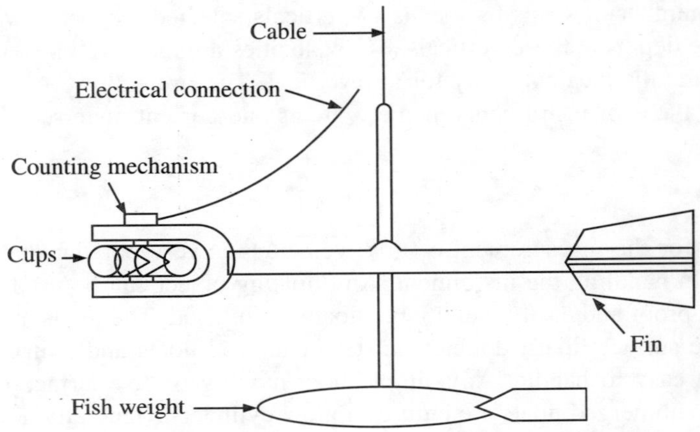

Figure 4.6 Cup type current meter.

Horizontal axis current meter (propeller type current meter) is fitted with a propeller at the end of its horizontal shafts (Figure 4.7). This kind of current meter is rugged and generally not affected by oblique flows of even 15°. The range of the propeller diameter is from 6 cm to 12 cm. The instrument can measure velocities from 0.15 m/s to 4.0 m/s, with accuracy in the range of 1% to 0.25%. The velocity can be calculated using current meter by the following formula:

$$V = aN_s + b \tag{4.3}$$

where, V is stream velocity, N_s is number of revolution per second of the current meter, and a, b are the constants for the meter.

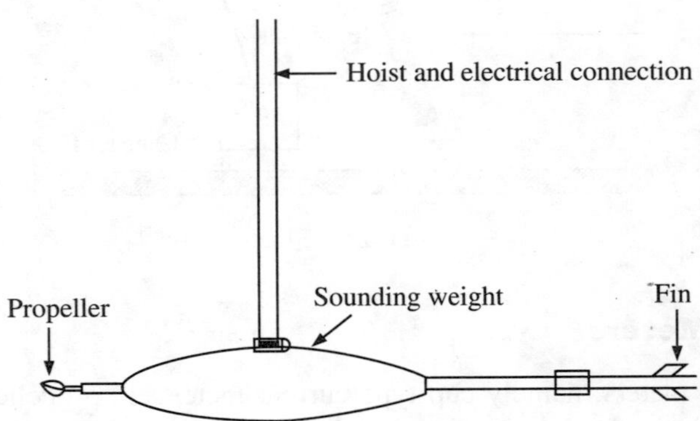

Figure 4.7 Propeller type current meter.

Table 4.1 shows the values of constants for current meters for use in the velocity formula.

Table 4.1 Values of Constants for Current Meters

Type of instrument	*Constant*	
	a	*b*
12.5 cm diameter (cup type), price meter	0.65	0.03
5 cm diameter (cup type), pigmy meter	0.3	0.003

Note: Each instrument has a threshold velocity below which the velocity does not apply.

4.3.3 Average Velocity Across a Vertical

The velocity in an open channel at any vertical varies from zero at the bottom to a maximum value at or slightly below the free surface (Figure 4.8). Under turbulent flow conditions, this variation can be described by the logarithmic law [Eq. (4.4)]:

$$v \propto \ln y \quad \text{or} \quad v \propto y^{(1/7)} \tag{4.4}$$

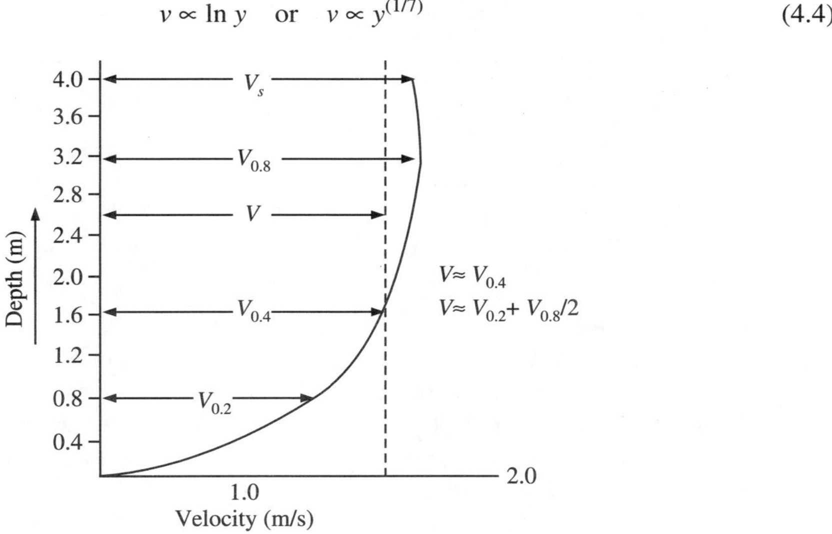

Figure 4.8 Velocity distribution at a vertical.

Integrating this variation over the depth of flow *d*, it can be shown that the velocity at a distance of 0.368*d* from the bottom is equal to the average velocity for the vertical in the case of logarithmic law, and the velocity at a distance of 0.393*d* from the bottom is equal to the average velocity for the vertical in the case of one-seventh power law. Thus, the general practice is to take the velocity measured at a distance of 0.4*d* from the bottom *a*, equal to the average velocity of the vertical. The arithmetic mean of the flow velocities at 0.2*d* and 0.8*d* is almost equal to the mean velocity of the vertical. The velocity distribution formula for a rough and turbulent flow is given below:

$$V = 5.75V^{*} \times \log_{10}\left(\frac{30y}{K_g}\right) \tag{4.5}$$

where, V is the velocity at point y above the bed, V^* is shear velocity and K_g is equivalent sand grain roughness.

For accuracy purposes, the mean flow velocity of a stream at a point is calculated by measuring the velocity of flow at several depths in a vertical cross section of the stream at that point and then taking the average of the measured velocities, as shown in Table 4.2.

Table 4.2 Velocity of Flow at Several Depths in a Vertical Cross Section of the Stream

Stream condition	*Depth*	*Velocity equation*
Shallow stream	At 0.6 × Depth of water from the stream surface	$\overline{V} = V_{0.6}$
Moderate depth stream	0.2 × Depth of the water from the stream surface, and also 0.8 × Depth of water from the stream surface	$\overline{V} = \dfrac{(V_{0.2} + V_{0.8})}{2}$
River and high velocity flows	Surface velocity V_s at about 0.5 m depth from the stream surface	$\overline{V} = KV_s$, where $K = 0.85$ to 0.95

4.3.4 Streamflow Computation

The distance between the two edges of a stream is divided into approximately 25 subsections (as per Figure 4.9). In the deeper parts of the stream, these subdivisions are again subdivided. It should be noted that not more than 10% of the total discharge occurs in any one subdivision. The streamflow velocity is measured in each subsection, as described below:

1. Select a number of cross section of the stream. Observe the depth of flow at these sections. Measure the vertical depth of water in each subsection of the stream to select the observation depths. If the stream depth is 0.6 m or less, then the current meter is set at 0.6 of the stream depth from the surface. When the depths are greater than 0.6 m, then the average of the measurements at 0.2 times the depth, and 0.8 times the depth from the water surface are taken.

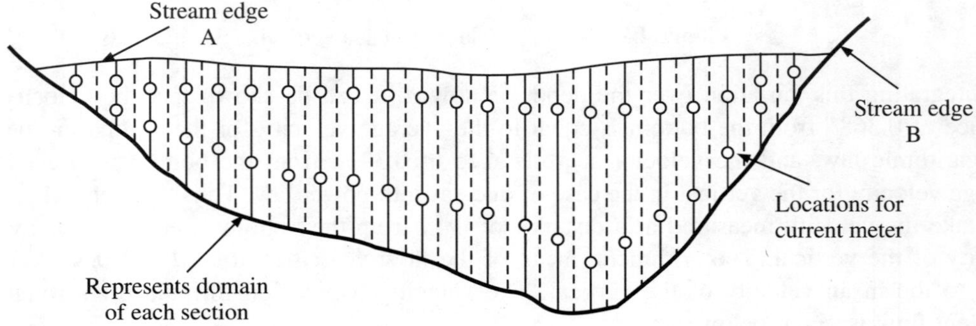

Figure 4.9 Discharge measurement using current meter.

2. Count the number of revolutions of the current meter for a period of 100 s or 50 s, and then, calculate the number of revolutions per second.
3. The stage of the stream should be observed and recorded periodically. If the moderate changes in the stage of the stream have been recorded because of floods, the value of the reduction factor K should be chosen carefully to apply Eq. (4.6).

$$V = KV_S \tag{4.6}$$

The area of each subsection is calculated by taking an average width for each subsection and multiplying it by the depth at that point.

This slope area method is based on the assumption that steady uniform flow exists between two cross sections. This method is not correct, especially during flood.

EXAMPLE 4.1 Determine the discharge in the stream for the given data at a gauging station below. The rating equation of current meter may be taken as $V = 0.45\ N_s + 0.025$ m/s, where N_s is the revolution per second.

Distance from left	*Depth* (m)	*Revolution of a current meter kept at a depth of* 0.6 m	*Duration of observation* (s)
0	0	0	0
1.1	1.2	41	110
3.2	2.7	62	110
5.3	2.7	115	160
7.1	2.3	95	153
9.2	1.9	47	105
11.6	1.7	35	107
12.5	0	0	0

Solution Given current meter equation, $v = 0.45\ N_s + 0.025$

Distance along with depth are given in the table. We need to calculate average width for the calculation.

The first distance average width $= \dfrac{(1.1 + 2.1/2)^2}{(2 \times 1.1)} = 2.101$

The second to fifth average width $= \dfrac{(W_{n-1} + W_n)}{2}$

$$\frac{(5.3 - 3.2) + (7.1 - 5.3)}{2} = 1.95$$

The last, i.e., 11.6 m average width $= \dfrac{(2.4/2 + 0.9)^2}{(2 \times 0.9)} = 2.45$

Average width along with N_s is given below in the table.

Here, revolution per second is calculated N_s = given N_s/Given time

Distance from left water edge (m)	*Average width* (m)	*Depth, y* (m)	*Revolution per second, N_s* (s)	*Velocity* (m/s)	*Segments discharge* (m^3/s)
0	0	0	0	0	0
1.1	2.10	1.2	0.372	0.1924	0.4848
3.2	2.1	2.7	0.563	0.2783	1.5779
5.3	1.95	2.7	0.718	0.3481	1.8327
7.1	1.95	2.3	0.620	0.304	1.3634
9.2	2.25	1.9	0.447	0.2261	0.9665
11.6	2.45	1.7	0.327	0.1721	0.7167
12.5	0	0	0	0	
					Sum = 6.942

So, discharge in the stream is 6.942 m^3/s.

4.4 Streamflow Measurement Using Moving-Boat Method

The area-velocity method of streamflow measurement described in Section 4.3 is costly and tedious method, especially during floods when site may be inundated. Therefore, the moving-boat method of streamflow measurement can be applied in case of rapid measurement in large streams. This method has been developed by the United States Geological Survey in 1969. This method is similar to the conventional method, as both use the area-velocity approach.

In the moving-boat method, the current meter moves with the same velocity as the boat moves. The current meter records the resultant of boat and stream velocities, which is the vectorial sum of these two velocities. During a traverse of the boat across the stream, a sonic sounder records the depth of flow at different points which define the geometry of the cross section. A continuously operating current meter records the resultant of boat and stream velocities. The traverse is made without stopping along a preselected path that is normal to the streamflow.

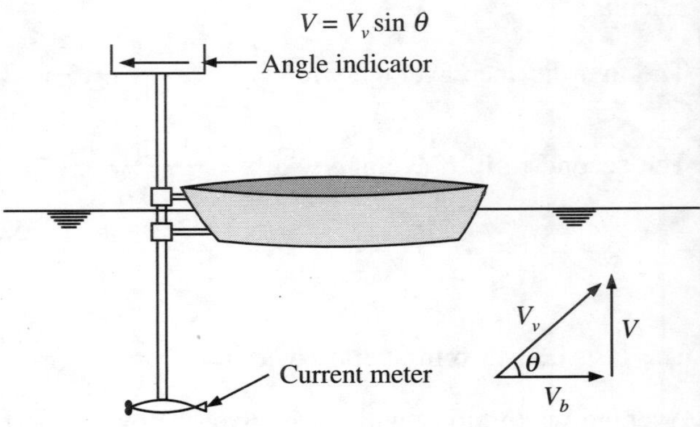

Figure 4.10 Moving-boat method.

In the velocity triangle at any point of observation, V, V_b and V_v denote the component of stream velocity normal to the cross section at the sampling point, the velocity of boat and the resultant velocity, respectively (Figure 4.10). Stream velocity V, which is perpendicular to the boat path, is computed as

$$V = V_v \sin \theta \tag{4.7}$$

At the same velocity triangle, the distance L_b travelled by the boat along the true course between two consecutive observation points may be obtained as

$$L_b = \int V_v \cos \theta \, dt \tag{4.8}$$

If θ is almost uniform over the relatively short distance in any one increment, then it should be treated as a constant. The mean of the readings at two consecutive points may be taken as the constant 6 for the interval. Thus, the equation can be written as

$$L_b = \cos \theta \int V_v \, dt \tag{4.9}$$

Here, d is the depth (stream) at each observation point and same can be obtained from the sonic sounder chart. After computations of V, L_b and d for each vertical, the mid-section method of computing discharge is applied. Several limitations are also involved in this method. This method cannot be used in shallow streams. In this method, the velocity coefficient is to be worked out for every stage of the river, as the ratio between stream depth and the depth of lowering current meter gets changed.

4.5 Streamflow Measurement Using Slope-Area Method

If the direct measurement is not possible to compute streamflow, especially during floods, slope-area method could be utilized as indirect method of streamflow measurement. In this method, streamflow is calculated on the basis of a uniform flow equation involving channel characteristics, water surface profile and a roughness coefficient. The streamflow can be computed by the following equation:

$$Q = AC\sqrt{RS} \tag{4.10}$$

$$Q = A\frac{1}{n}R^{(2/3)}S^{(1/2)} \tag{4.11}$$

where A is the cross-sectional area, S is the energy slope, R is the hydraulic radius, C is the Chezy's coefficient and n is the Manning's roughness coefficient. However, Manning's equation was developed for uniform flow and it is assumed to be valid for non-uniform reaches that are encountered in case of natural channel. The friction slope S to be used in Manning's equation is given by the following relation:

$$S = \frac{h_f}{L} = \Delta h + \Delta h_v - \frac{K(\Delta h_v)}{L} \tag{3.12}$$

where, Δh is the difference in water surface elevation at the two sections, L is the length of the reach, Δh_v is upstream velocity head minus downstream velocity head and $K(\Delta h_v)$ is energy loss due to acceleration or deceleration in a contracting or expanding reach (K being coefficient). Here, it can be difficult to calculate the value of the coefficient K. Therefore, value of S can be calculated as

$$S = \frac{\Delta h + [(\Delta h_v)/2]}{L} \quad \text{when } \Delta h_v \text{ is positive} \tag{4.13}$$

$$S = \frac{\Delta h + \Delta h_v}{L} \quad \text{when } \Delta h_v \text{ is negative} \tag{4.14}$$

In Manning's equation the quantity ($1/nAR^{2/3}$) is called *conveyance*, and is denoted by K. The conveyance for each section of the river is different. The mean value of the conveyance has, therefore, been taken as a geometric mean of the conveyance values of the two sections at both ends of the selected reach L. The discharge is, thus, given by Eq. (4.15).

$$Q = \sqrt{K_1 K_2 S} \tag{4.15}$$

Following are the steps to compute discharge from slope-area method:

1. There should be good high-water marks available, if staff gauges are not available to give gauge record.
2. The reaches should be as uniform as possible because it is difficult to correctly calculate coefficient K tó assess velocity head loss or gain.
3. It is desirable to have a reach in which flow is confined to a single trapezoidal channel so that single value of n can be used for the whole section.
4. Over the whole reach, the cross section should be as uniform as is practically possible. Sudden deepening of bed, constrictions, expansions, etc. should not be present.
5. The length of reach should not be less than 75 times the mean depth in the channel.
6. The fall of water profile in the reach should be equal to or greater than the velocity head,
7. The fall in the reach should not be less than 15 cm.

EXAMPLE 4.2 Compute the flood discharge in a stream by the slope-area method for the following data:

	Area of section (m^2)	*Wetted perimeter* (m)	*Roughness coefficient*, n
Section 1-1	206	65	0.045
Section 2-2	200	53.8	0.045

The drop in head and length between the two sections are 0.98 m and 125 m, respectively.

Solution *Step 1:* The discharge is given by the Manning's formula:

$$Q = \frac{1}{n} AR^{2/3} S^{1/2}$$

This formula is further modified to take average conveyance into account as follows:

$$Q = \sqrt{K_1 K_2 S}$$

where $K = \frac{1}{n} AR^{2/3}$ for each section and $S = \frac{h_f}{L}$

Step 2: Conveyance for Section 1-1,

$$K_1 = \frac{1}{0.045}\left(\frac{A}{P}\right)^{2/3} \times A$$

$$A = \frac{1}{0.045}\left(\frac{206}{65}\right)^{2/3} \times 206 = 9876.97 \cong 9877$$

Step 3: Conveyance for Section 2-2,

$$K_2 = \frac{1}{0.045}\left(\frac{200}{53.8}\right)^{2/3} \times 200 = 10665.53 \cong 10666$$

Step 4: As a first approximation,

$$S = \frac{\text{Fall in reach}}{\text{Length of fall}} = \frac{h_f}{L} = \frac{0.98}{125} = 7.84 \times 10^{-3}$$

Step 5: Peak discharge $Q = \sqrt{9877 \times 10666 \times 7.84 \times 10^{-3}}$ = 908.80 cumec

Step 6: With this discharge, average velocity of flow at both the sections can be calculated. The energy loss due to constriction/expansion.

$$V_1 = \frac{090}{206} = 4.41 \text{ m/s}$$

$$V_2 = \frac{909}{200} = 4.54 \text{ m/s}$$

$$\Delta h_v = \frac{V_1^2}{2g} - \frac{V_2^2}{2g} = \frac{4.41^2}{2 \times 9.81} - \frac{4.54^2}{2 \times 9.81} = -0.059$$

Step 7: Since Δh_v is negative,

$$S = \frac{\Delta h + \Delta h_v}{L} = \frac{0.98 + 0.59}{125} = 0.0083$$

Step 8: With the value of S, the discharge can be recalculated.

Therefore, $Q = \sqrt{9877 \times 10666 \times 0.0083} = 935.08$ cumec

4.6 Streamflow Measurement Using Venturi

Venturi meter is a flowmeter to measure mass/volumetric flow rate or velocity of flow. Venturi meter is also known as *variable head meter*. In venturi meter, there are converging conical inlet, a cylindrical throat and a diverging recovery cone. In a venturi meter (Figure 4.11), the area of the fluid stream reduces at converging section, causing increase in velocity and decrease in pressure. Pressure can be measured in the centre of the cylindrical throat, as the pressure is lower. As the fluid flow through the diverging section, the pressure is largely recovered lowering the velocity of the fluid.

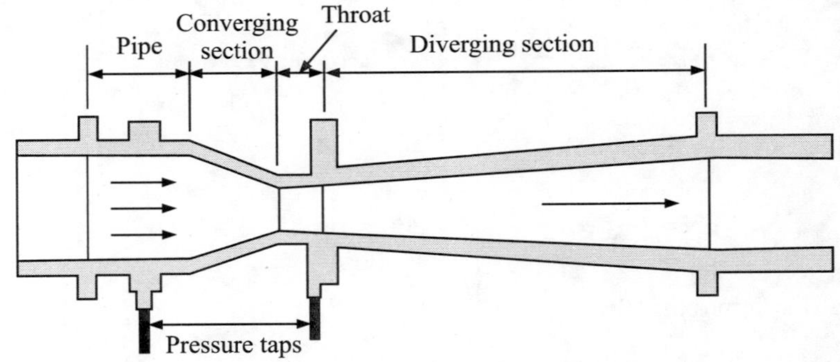

Figure 4.11 Venturi meter.

In order to satisfy continuity equation, the velocity of flowing fluid should increase once it passes through the converging portion of venturi, and using conservation of energy law, the pressure will decrease. If V_1 and V_2 are the mean upstream and downstream velocities and ρ is the density of the fluids, then applying Bernoulli's equation,

$$\alpha_2 v_2^2 - \alpha_1 V_1^2 = \frac{2g(P_1 - P_2)}{\rho} \tag{4.16}$$

where α_1 and α_2 are kinetic energy correction factors at two pressure tap positions P_1 and P_2. The continuity equation is given as

$$V_1 = \left(\frac{D_1}{D_2}\right)^2 V_2 \tag{4.17}$$

where D_1 and D_2 are diameter of pipe and throat, respectively. Excluding V_1 from Eq. (4.17),

$$V_2 = \frac{1}{\sqrt{(\alpha_2 - \alpha_1 \beta_4)}} \sqrt{\frac{2(P_1 - P_2)}{\rho}} \tag{4.18}$$

where α_1 and α_2 are approximated as unity and β is the ratio of the diameter of throat to that of diameter of pipe.

Here, a small friction loss between two pressure taps can be assumed and Eq. (4.18) can be corrected by introducing empirical factor C_V and the new equation can be written as

$$V_2 = \frac{C_V}{\sqrt{1 - \beta^4}} \sqrt{\frac{2(P_1 - P_2)}{\rho}} \tag{4.19}$$

The small effect of the kinetic energy factors α_1 and α_2 are also taken into account in the definition of C_v.

Volumetric flow rate Q_a can be computed as

$$Q_a = V_2 S_2 = \frac{C_V S_2}{\sqrt{1-\beta^4}} \sqrt{\frac{2(P_1 - P_2)}{\rho}} \tag{4.20}$$

where S_2 is the cross-sectional area of throat (m^2).

Substituting $(P_1 - P_2) = \rho g H$ in Eq. (4.20), we get,

$$Q_a = V_2 S_2 = \frac{C_V S_2}{\sqrt{1-\beta^4}} \sqrt{2h\Delta H} \tag{4.21}$$

where ΔH is the manometric height difference.

4.7 Stage-Discharge Relationship (Rating Curve)

The relationship between water-surface stage and the simultaneous flow discharge in an open channel or stream can be referred to as *rating curve* or *stage-discharge relation*. The stage-discharge relation plays an important role in hydrology because the accuracy of discharge or streamflow is mostly dependent on the rating curve at the gauging station. A calibration curve can be plotted between periodic measurements of flow and simultaneous stage observations to predict the flow discharge when the stage is known. For the computation of higher streamflow, these rating curves are required to be extended.

4.7.1 Extension of Rating Curves

The extension of rating curve involves extrapolation of the curve beyond the highest measured discharges. Three methods, namely, Logarithmic method, Stevens method and conveyance factor method are used for extension of rating curves.

Logarithmic method

The equation of the rating curve may be given as

$$Q = k(h - b)^n \tag{4.22}$$

where Q is the runoff discharge, k is constant, h is the stage of water in the channel, b is the elevation between the zero elevation of the gauge and the elevation of zero flow (gauge height).

The extension of stage-discharge curve may be done as follows:

1. A rating curve of runoff data is plotted and approximate value of b is estimated.
2. The depth of flow $(h - b)$ is determined for a number of values of b and a graph is plotted between log $(h - b)$ versus log Q.
3. The value of b is recalculated by selecting lines which are nearly straight line relationships.
4. A graph is plotted between log Q versus $(h - b)$.
5. A relationship nearest to straight line is chosen for the final analysis.
6. The constants b, n and k are determined such that, b is the constant value of the selected straight line, n is the slope of the selected straight line, and $k = Q$ when $h - b = 1$.

7. The values of these constants may be put in Eq. (4.22), and the rating curve may be extended.

Equation for a log normal straight line: The equation for a long normal straight line is defined as

$$Q = kD^b \tag{4.23}$$

where Q is the discharge, k is a constant, D is the depth (= $h - a$), and b is an exponent signifying the slope of the line. On expressing Eq. (4.24) in its logarithmic form,

$$\log Q = \log k + b \log D \tag{4.24}$$

using $\log Q = X$, $\log k = C$ and $\log D = Y$

So, $X = C + bY$

When $Y = 0$, $X = C$, i.e., $\log Q = \log k$

Hence, when $D = (h - b) = 1$, $\log D = Y = 0$

Stevens cross-sectional area of the stream-mean depth method

As per Chezy's formula,

$$Q = AC\sqrt{RS} \tag{4.25}$$

where A is the cross-sectional area of the stream/channel, R is the hydraulic radius, S is slope of the energy gradient, and C is a constant.

Here, it can be assumed $C\sqrt{S}$ is a constant k for the station. Then,

$$Q = kA\sqrt{D_m} \tag{4.26}$$

where D_m is mean depth.

To extrapolate the rating curve by this method, known values of Q against the calculated value of $A\sqrt{D_m}$ is plotted. This plot is a straight line curve. This line can be extrapolated for the desired value.

Conveyance factor method

In this procedure, the conveyance of water in a channel, with a non-uniform flow, can be defined as

$$Q = k\sqrt{S_f} \tag{4.27}$$

Here we get,

$$S_f = \frac{Q^2}{k^2} \tag{4.28}$$

where Q is discharge in the channel, S_f is slope of the energy line, k is the conveyance factor = $1/n\ AR^{2/3}$, where n is Manning's roughness coefficient, A is channel cross-sectional area and R is the hydraulic radius of the channel.

The stepwise procedure is given below:

1. The known values of stage h is plotted against the calculated values of $k(k = 1/n\ AR^{2/3})$ for the same stage.
2. Then, extend the curve to the desired point.
3. Now, the known values of stage h is plotted against the calculated value of $\sqrt{S_f}$ by using Eq. (4.28) for the same stage.
4. This curve is extended for the known value of stage h to compute the corresponding values of k and $\sqrt{S_f}$ and the value of Q is calculated.

EXAMPLE 4.3 Determine the discharge for the stage of 7.8 m using the given data in the table below at a station. Use $A\sqrt{D_m}$.

	Storage, g (m)	*Area, A* (m^2)	*Depth, D_m* (m)	*Discharge, Q* (m^3/s)
1	1.4	155	2.5	330
2	1.56	189	3.8	370
3	2.10	225	3.95	415
4	2.44	269	4.05	460
5	3.35	305	4.10	545
6	4.08	401	4.23	611
7	4.92	555	4.40	722
8	5.67	598	4.65	766
9	6.89	640	4.72	820
10	7.80	710	4.80	–

Solution The parameters $\sqrt{D_m}$ and $A\sqrt{D_m}$ are calculated as shown in the following table.

	Storage, g (m)	*Area, A* (m^2)	*Depth, D_m* (m)	Discharge, Q (m^3/s)	$\sqrt{D_m}$	$A\sqrt{D_m}$
1	1.4	155	2.5	330	1.5811	245.07
2	1.56	189	3.8	370	1.9493	368.42
3	2.10	225	3.95	415	1.874	447.17
4	2.44	269	4.05	460	2.0124	541.35
5	3.35	305	4.10	545	2.0248	617.577
6	4.08	401	4.23	611	2.0566	824.73
7	4.92	555	4.40	722	2.0976	1164.17
8	5.67	598	4.65	766	2.1563	1289.51
9	6.89	640	4.72	820	2.1725	1390.43
10	7.80	710	4.80	900.96 (estimated)	2.1908	1555.53

Then, on an ordinary graph paper, the relationships between $A\sqrt{D_m}$ and Q, and $A\sqrt{D_m}$ and g, are plotted, as shown in the figure below:

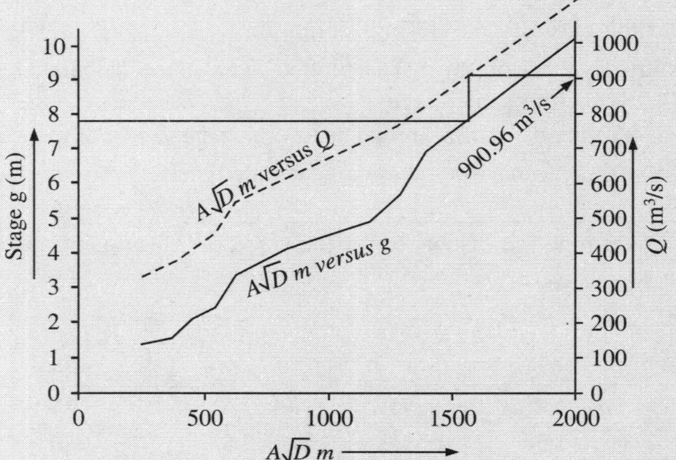

From the curve $A\sqrt{D_m}$ versus g, at g = 7.80, the value of $A\sqrt{D_m}$ = 1555.53 is found. For this value of $A\sqrt{D_m}$, the predicted value of Q is 900.96 m^3/s as found from $A\sqrt{D_m}$ versus Q curve.

EXAMPLE 4.4 At a stream gauging station, the daily discharge for each day of a week is 10, 14, 16, 21, 16, 11 and 9 m^3/s, respectively.

The catchment area of the stream is 200 km^2.

Determine the flow in cumec-day (m^3/s days) and ha-m, mean flow for the week and runoff volume in terms of depth of water.

Solution

Total volume of flow in the week = 10 + 14 + 16 + 21 + 16 + 11 + 9 = 97 cumec-days

$$= 97 \times 86400 = 8380800 \text{ m}^3 = 838.08 \text{ ha-m}$$

$$\text{Week-mean flow} = \frac{97 \text{ cumen-days}}{7 \text{ days}} = 13.85 \text{ m}^3/\text{s}$$

$$\text{Runoff volume in terms of depth} = \frac{\text{Runoff volume}}{\text{Catchment area}} = \frac{8380800}{200 \times 10^6} = 4.19 \text{ cm}$$

4.8 Selection of a Stream-gauging Site and Stream-gauging Network

The selected site on stream or channel to measure streamflow/discharge in a continuous manner is referred to as *hydro-observation station* or *stream-gauging station*. There are several criteria for selecting a site for a new stream-gauging station. The following considerations should be made:

1. Site should be accessible under all conditions, particularly during floods.
2. The channel at the stream-gauging site should not be subjected to significant scour.
3. The site should be able to cover the full range of discharge which may occur.
4. The site should be located far enough of a junction with another stream near enough to be affected by backwater from any project. It should not be near upstream from the confluence of another river to avoid being influenced by high water in that river.
5. Stream-gauging site should not have large amount of aquatic growth.
6. There should be little or no change in the stage-discharge relation at the site with time.
7. The reach of the stream should be straight.

Regarding stream gauging network, the following norms are followed as per the WMO guideline.

1. **Region I: Flat regions of temperate, mediterranean and tropical zones:** Minimum one station for 1000–2500 km^2 is recommended. However, one station for 3000–10000 km^2 is also acceptable,
2. **Region II: Mountainous regions of temperate, mediterranean and tropical zones:** Minimum one station for 300–1000 km^2 is recommended. However, one station for 1000–5000 km^2 is also acceptable.
3. **Region III: Arid:** Minimum one station for 5000–20000 km^2 is acceptable.

A stream-gauging station should be established at the following locations:

1. At the inlet or outlet of natural lake.
2. Downstream of a hydraulic dam.
3. At the head of alluvial plains and upstream of a city subject to flood damage.

4.9 Units of Streamflow

The basic unit of discharge/streamflow rate in the metric system is cubic metre per second (m^3/s), which can also be referred to as *cumec*. The flow rate per unit area contributing to the catchment can be expressed in cubic metre per second per square kilometre (m^3/s/km^2). When the streamflow hydrograph is to be compared with the rainfall hyetograph, the flow rate can be expressed in centimetre per hour (cm/h) or millimetre per hour (mm/h), as the ordinate of the rainfall hyetograph has such units. The total volume of streamflow in the period of 24 h at a rate of 1 m^3/s is called *one cumec-day*. The conversions among the various volume units just now mentioned may be listed as follows:

1 cumec-day = 86400 m^3 = 8.64 ha-m
1 million m^3 = 100 ha-m = 11.574 cumec-days
1 ha-m = 1 sq. km-cm = 10000 m^3 = 0.01 million m^3

Summary

- Stage is defined as the height of the water surface above a datum plane, whereas gauge height can be defined as the water surface elevation to some pre-determined gauge datum.

- Current meter is used to measure the velocity at a point in the flow cross section.
- The moving boat method for stream flow measurement could be used for monitoring and flood forecasting in downstream by observing discharges at the regular interval.
- Slope-area method is an indirect method of discharge estimation. In this method, discharge is estimated by observing the water surface slope and cross section area.
- A rating curve is a relation between stage and discharge. It is drawn for particular section.

Objective Type Questions

1. The top width and the depth of flow in a triangular channel were measured as 4 m and 1 m, respectively. The measured velocities on the centre line at the water surface 0.2 m and 0.8 m below the surface are 0.7 m/s, 0.6 m/s and 0.4 m/s, respectively. Using two-point method of velocity measurement, the discharge (m^3/s) in the channel is (GATE, 2012)

(a) 1.4 (b) 1.2 (c) 1.0 (d) 0.8

2. Group I contains parameters and Group II lists methods/instruments. (GATE, 2012)

Group I	Group II
A. Streamflow velocity	1. Anemometer
B. Evapotranspiration rate	2. Penman's method
C. Infiltration rate	3. Horton's method
D. Wind velocity	4. Current meter

The correct match of Group I with Group II is

Codes

	A	B	C	D
(a)	1	2	3	4
(b)	4	3	2	1
(c)	4	2	3	1
(d)	1	3	2	4

3. The slope area method is extensively used in (IES, 2002)

(a) development of rating curve
(b) estimation of flood discharge based on high-water marks
(c) cases where shifting control exists
(d) cases where back-water effect is present

4. In the case of large rivers, a number of equidistant vertical sections of the total width of flow are identified. For the purpose of finding, by numerical integration, the total discharge on any day on each section, the mean velocity is taken as the arithmetic average of two typical depths on that section. Then, the mean velocity is worked out for that section. Usually, the mean velocity of any section corresponds to which one of the following? (V represents the point velocity at the given section and the depth such as (0.1d, 0.2d, etc.) (IES, 2006)

(a) $(V_{0.1d} + V_{0.9d})/2$ (b) $(V_{0.2d} + V_{0.8d})/2$
(c) $(V_{0.3d} + V_{0.7d})/2$ (d) $(V_{0.4d} + V_{0.6d})/2$

5. How is the average velocity along the vertical in a wide stream obtained? (IES, 2007)
   (a) By averaging the velocities at 0.2 and 0.8 depth from surface.
   (b) By measuring velocity at 0.6 depth below the surface
   (c) By measuring velocity at half the depth
   (d) By measuring velocity at 0.1 times the depth below the surface
6. Consider the following with respect to measurement of streamflow during flood. (IES, 2009)
   1. Timing of the travel of floats released in the stream
   2. Use of weir formula for spillways provided on a dam
   3. Calculation of flow through a contracted opening at a bridge
   4. Using a current meter.

   Which of the above is/are reliable and accurate?
   (a) 1 only (b) 4 only (c) 3 and 4 (d) 2 and 3
7. Calibration of a current meter for use in channel flow measurement is done in a (IES, 2011)
   (a) Wind tunnel (b) Water tunnel (c) Towing tank (d) Flume
8. A stilling well is required when the stage measurement is made by employing (IES, 2010)
   (a) Bubble gauge (b) Float gauge recorder
   (c) Vertical staff gauge (d) Inclined staff gauge
9. Wading techniques is used: (IES, 2013)
   (a) To determine velocity of sea waves during Tsunami.
   (b) To determine thickness of canal lining in alluvial soils.
   (c) To measure the volume of dredging material in harbours.
   (d) To determine velocity of flow in shallow streams.

Answers

1. (c) **2.** (c) **3.** (b) **4.** (b) **5.** (c) **6.** (b) **7.** (c) **8.** (b) **9.** (d)

Explanations

1. The following figure shows

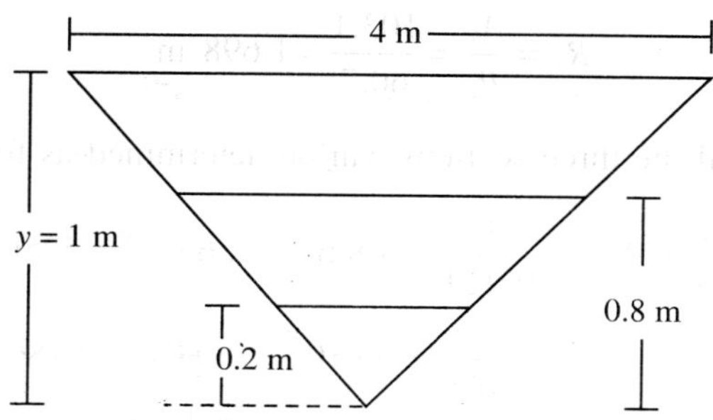

$$\text{Average velocity, } V = \frac{(V_{0.2} + V_{0.8})}{2}$$

$$V = \frac{(0.6 + 0.4)}{2} = 5 \text{ m/s}$$

$$\text{Discharge, } Q = \text{Area} \times \text{Velocity}$$

$$= \frac{1}{2} \times 1 \times 4 \times 0.5 = 1.0 \text{ m}^3/\text{s}$$

4. The average velocity for moderately deep streams may be taken as the average of the velocities at depths equal to 0.2 times the depths and 0.8 times the depths, i.e.,

$$V = \frac{V_{0.2d} + V_{0.8d}}{2}$$

IES CONVENTIONAL QUESTIONS

PROBLEM 4.1 In order to compute the flood discharge in a stream by the slope area method, the following data have been obtained:

	Upstream section	*Middle section*	*Downstream section*
Area (m^2)	108.6	103.1	99.8
Wetted perimeter (m)	65.3	60.7	59.4
Gauge reading (m)	316.8		316.55

Determine the flood discharge assuming Manning's n = 0.029 and length between downstream and upstream section as 250 m. (IES, 2000)

Solution Hydraulic mean depths R for all the three sections, namely, the upstream section, middle section, and downstream section, respectively, are calculated as

$$R_1 = \frac{A_1}{P_1} = \frac{108.6}{65.3} = 1.663 \text{ m}$$

$$R_2 = \frac{A_2}{P_2} = \frac{99.8}{59.4} = 1.680 \text{ m}$$

$$R_3 = \frac{A_3}{P_3} = \frac{103.1}{60.7} = 1.698 \text{ m}$$

Again, conveyance K of all the three sections can be determined as follows:

$$K_1 = \frac{1}{n} A_1 R_1^{2/3} = \frac{1}{0.029} \times 108.6 \times (1.663)^{2/3} = 5256.46$$

$$K_2 = \frac{1}{n} A_2 R_2^{2/3} = \frac{1}{0.029} \times 99.80 \times (1.680)^{2/3} = 4863.38$$

$$K_3 = \frac{1}{n} A_3 R_3^{2/3} = \frac{1}{0.029} \times 103.10 \times (1.698)^{2/3} = 5060.02$$

So, average conveyance can be determined as

$$K_{avg} = (K_1 \times K_2 \times K_3)^{1/3} = (5256.46 \times 4863.38 \times 5060.02)^{1/3} = 5057.40 \text{ m}$$

For flood discharge,

$$Q = K_{avg} \sqrt{\frac{h_f}{L}} \qquad \text{(i)}$$

where K_{avg} is to be determined by iteration process.

First iteration

Assuming $V_1 = V_2$

Friction loss, $H_f = h_1 - h_2 = (316.80 - 316.55) = 0.25$ m

Using Eq. (i), we get

$$Q = K_{avg} \sqrt{\frac{h_f}{L}} = 5057.40 \times \sqrt{\frac{0.25}{250}} = 159.93 \text{ m}^3/\text{s}$$

Therefore,

$$V_1 = \frac{Q}{A_1} = \frac{159.93}{108.6} = 1.47 \text{ m/s}$$

$$V_2 = \frac{Q}{A_2} = \frac{159.93}{99.8} = 1.602 \text{ m/s}$$

Second iteration

Take $K = 0.1$ for gradual contraction. $V_1 = 1.47$ m/s, $V_2 = 1.602$ m/s

$$h_f = (316.8 - 316.55) + \left(\frac{1.47^2}{2 \times 9.81}\right) - \left(\frac{1.602^2}{2 \times 9.81}\right) - 0.1 \left| \left(\frac{1.47^2}{2 \times 9.81}\right) - \left(\frac{1.602^2}{2 \times 9.81}\right) \right|$$

$$h_f = 0.25 - 0.021 - 0.0021 = 0.228 \text{ m}$$

$$Q = K_{avg} \sqrt{\frac{h_f}{L}} = 5057.40 \times \sqrt{\frac{0.228}{250}} = 152.73 \text{ m}^3/\text{s}$$

Therefore,

$$V_1 = \frac{Q}{A_1} = \frac{152.73}{108.6} = 1.406 \text{ m/s}$$

$$V_2 = \frac{Q}{A_2} = \frac{152.73}{99.8} = 1.53 \text{ m/s}$$

Third iteration

$$h_f = (0.25) + \left(\frac{1.406^2}{2 \times 9.81}\right) - \left(\frac{1.53^2}{2 \times 9.81}\right) - 0.1 \left| \left(\frac{1.406^2}{2 \times 9.81}\right) - \left(\frac{1.53^2}{2 \times 9.81}\right) \right|$$

$$h_f = 0.25 - 0.0186 - 0.00186 = 0.229 \text{ m}$$

Using Eq. (i), we get

$$Q = K_{avg}\sqrt{\frac{h_f}{L}} = 5057.40 \times \sqrt{\frac{0.229}{250}} = 153.24 \text{ m}^3/\text{s}$$

$$V_1 = \frac{Q}{A_1} = \frac{153.24}{108.6} = 1.41 \text{ m/s}$$

$$V_1 = \frac{Q}{A_2} = \frac{153.24}{99.8} = 1.53 \text{ m/s}$$

Fourth iteration

$$h_f = (0.25) + \left(\frac{1.41^2}{2\times 9.81}\right) - \left(\frac{1.53^2}{2\times 9.81}\right) - 0.1\left|\left(\frac{1.41^2}{2\times 9.81}\right) - \left(\frac{1.53^2}{2\times 9.81}\right)\right|$$

$$h_f = 0.25 - 0.018 - 0.0018 = 0.23 \text{ m}$$

Using Eq. (i), we get

$$Q = K_{avg}\sqrt{\frac{h_f}{L}} = 5057.40 \times \sqrt{\frac{0.23}{250}} = 153.39 \text{ m}^3/\text{s}$$

PROBLEM 4.2 Using equation V(m/s) = 0.65N + 0.03, obtain the velocity at 0.6 times the depth from free surface. Here, N stands for revolution per second. Data on current meter observations are given below in tabular form.

Distance from one bank (m)	*Depth* (m) (y)	*Current meter observation at* 0.6y *number of revolutions*	*Time* (s)
3	0.4	30	150
6	0.8	50	130
9	1.2	70	100
12	2	100	80
15	3	150	60
18	2.5	200	50
21	2.2	130	40
24	1	90	130

Also, compute the discharge through the section. (IES, 2008)

Solution The equation of velocity in terms of N is given below:

$$V \text{ (m/s)} = 0.65N + 0.03$$

where N is the number of revolution per second.

The method of mid sections has been used to determine the total discharge at that section. The average width of the first section can be determined by

$$\overline{W} = \frac{\left(W_1 + \frac{W_2}{2}\right)^2}{2W_1}$$

$$\overline{W} = \frac{\left(3 + \frac{3}{2}\right)^2}{2 \times 3}$$

$$\overline{W} = 3.375 \text{ m}$$

For the other segments,

$$\overline{W} = \frac{3}{2} + \frac{3}{2} = 3 \text{ m}$$

The discharge calculation of each segment has been shown in the table below:

Distance from one bank (m)	*Depth* (m) (*y*)	*Average width,* $\overline{W}$ (m)	*V*	*Segmental discharge* $\Delta Q = y \times V \times W$
3	0.4	3.375	0.16	0.216
6	0.8	3	0.28	0.672
9	1.2	3	0.485	1.746
12	2	3	0.843	5.058
15	3	3	1.655	14.895
18	2.5	3	2.63	19.725
21	2.2	3	2.14	14.124
24	1	3.375	0.48	1.62
		–	–	$\Sigma\Delta Q_i = 58.056$

So, the total discharge $Q = 58.056$ m^{3}s

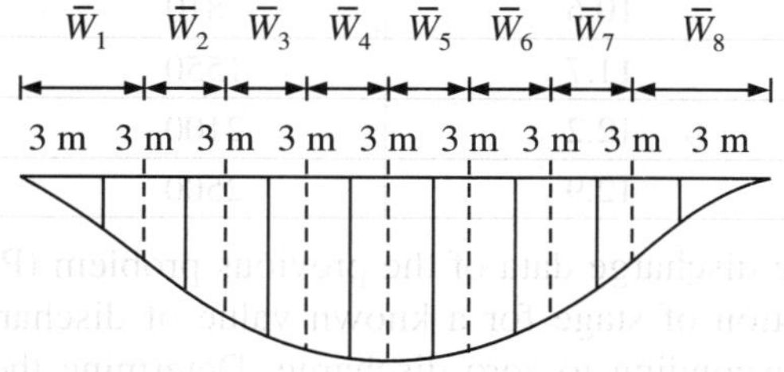

Theoretical Questions

1. What is a fluid? What are the different properties of fluid?
2. Discuss the various types of gauges.
3. Explain briefly the area-velocity method of streamflow measurement.
4. Explain the procedure of using current meter.
5. Discuss briefly the moving-boat method of streamflow measurement.
6. Describe briefly the stage-discharge relationship.

Unsolved Problems

Problem 1: Determine the flood discharge through the rectangular channel of 12 m depth. The depth of water during flood flow in this channel was found to be 3.5 m and 3.2 m at the sections 210 m apart and the drop in the water surface elevation was found to be 0.15 m. Assume Manning's coefficient as 0.027.

Problem 2: Develop the stage-discharge relationship for the following data of gauge and discharge collected at a particular section of a river by stream gauging operation (see the table below). Use the value of a = 8.5 m for the gauge reading of 11.5 m at the gauging station.

Gauge discharge (m)	*Discharge* (m^3/s)
8.67	18
8.7	35
8.77	60
8.85	42
8.92	65
8.96	110
9.2	160
9.6	178
10	420
10.4	610
10.6	810
11.7	1550
12.2	2100
12.9	2500

Problem 3: For the stage-discharge data of the previous problem (Problem 2), fit a regression equation for use in estimation of stage for a known value of discharge. Use a value of 8.5 m as the gauge reading corresponding to zero discharge. Determine the stage for a discharge of 3600 m^3/s.

Further Reading

Chow, V.T., Maidment, D.R., and Mays, L.W., *Applied Hydrology*, McGraw-Hill, Singapore, 1988.

Das, Ghanshyam, *Hydrology and Soil Conservation Engineering Including Watershed Management*, PHI Learning, Delhi, 2014.

Davie, Tim, *Fundamentals of Hydrology*, 2nd ed., Routledge, Taylor and Francis Group, London and New York, 2008.

Deodhar, M.J., *Elementary Engineering Hydrology*, Pearson, New Delhi, 2013.

Raghunath, H.M., *Hydrology—Principles, Analysis and Design*, New Age Publishers, New Delhi, 2014.

Reddy, P. Jaya Rami, *A Textbook of Hydrology*, 3rd ed., University Science Press, Delhi, 2014.

Subramanya, K., *Engineering Hydrology*, Tata McGraw-Hill, New Delhi, 2010.

Suresh, R., *Watershed Hydrology*, Standard Publishers Distributors, New Delhi, 2015.

Yadupathi Putty, Mysooru R., *Principles of Hydrology*, I.K. International, New Delhi, 2013.

CHAPTER

5

Runoff

5.1 Introduction

After the occurrence of rainfall, a part of it is intercepted by vegetation and buildings and other obstacles. Some portion of it may evaporate and infiltrate; the rest, which flows as the sheet of water over the land surface before it reaches the channel, is called *overland flow*. The draining off of precipitation from a catchment area through a surface channel is called *runoff*. It, thus, represents the output from the catchment in a given unit of time. The direct runoff is a part of runoff which enters the stream promptly.

5.2 Types of Runoff

There are three types of runoff:

1. Surface runoff: It is the amount of rainfall which enters the stream channels immediately after occurring the rainfall.

2. Sub-surface runoff: It is the part of rainfall which infiltrates into the ground surface, and then, laterally flows towards streams without joining water table.

3. Base flow: It is the portion of streamflow that results from percolation of water from the ground into a channel slowly over time. This is also called *fair weathered runoff*.

5.3 Factors Affecting Runoff

The types of runoff and its magnitude depend on a number of factors, which can be classified as follows:

1. Storm characteristics: Storm characteristics such as duration, intensity, areal extent have a great effect on runoff. A high-intensity runoff over small areal extent increases the runoff because the abstractions are less. If precipitation falls in the form of rain, its effect on the runoff is almost immediately seen in comparison to precipitation falls in the form of snow.

2. Meteorological/climatological characteristics: These include temperature, humidity, direction of prevailing wind, etc. Greater humidity decreases evaporation.

3. Basin characteristics: These include shape, size, slope and land use among others. A fan-shaped catchment generates higher flood intensity than a fern-shaped catchment, as shown in Figure 5.1. If the slope is high, then the runoff velocity is more and infiltration is less, thereby resulting is peak runoff immediately. In urban areas, infiltration, interception and evapotranspiration losses are less, thereby producing high discharge.

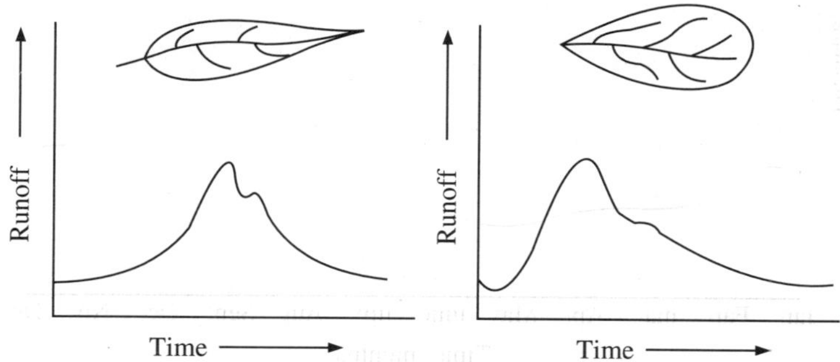

Figure 5.1 Basin characteristics affecting runoff.

5.4 Water Year

The time duration from the start of rainfall to the end of rainfall during a year is known as *water year.* In India, June 1st is the beginning of a water year, which ends on May 31st of the following calendar year. In a water year, a complete cycle of climatic changes is expected, and hence, the water budget has the least amount of carryover.

5.5 Classification of Streams

Based on the flow duration, the streams are classified into three classes—perennial, intermittent and ephemeral.

5.5.1 Perennial Streams

These streams (Figure 5.2) have continuous flow in parts of their stream beds all year round during years of normal rainfall. In such streams, there is considerable amount of groundwater flow throughout the year.

5.5.2 Intermittent Streams

These streams flow only briefly during and following a period of rainfall in the immediate locality, as shown in Figure 5.3. An intermittent stream has limited contribution from the groundwater. In rainy season, the water table is above the stream bed and there is a contribution of the baseflow to the streamflow, while in dry season, the water table drops to a level lower than that of the stream bed and the stream gets dry.

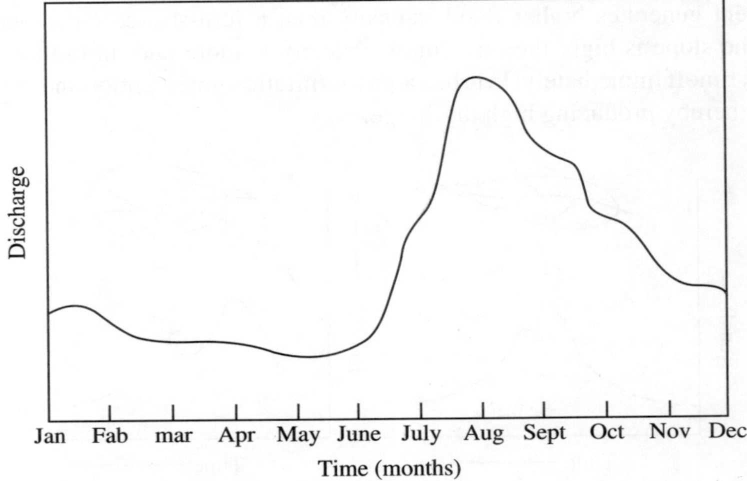

Figure 5.2 Perennial stream.

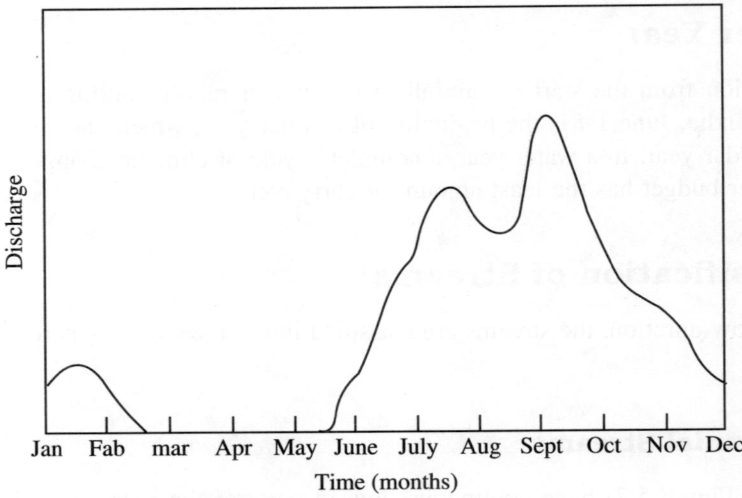

Figure 5.3 Intermittent stream.

5.5.3 Ephemeral Streams

Ephemeral streams are typically shallow and found in arid zones. These streams carry water in rainy season only and are normally dry for rest of the year, as shown in Figure 5.4. These streams do not have any baseflow contribution. The annual hydrograph of such a stream shows series of short-duration spikes, marking flash flows in response to storms.

5.6 Catchment Yield

The *yield from a catchment* is defined as the total quantity of water available as streamflow from it in a given period such as a year. It is usual for yield to be referred to the period of a year, and then, it represents the annual runoff volume. The *monthly yield* refers to the amount of water during a given month of the year. The yield of a catchment Y may be written using water balance equation as

$$Y = R_N + V_r = R_0 + \text{Losses} + \Delta S$$

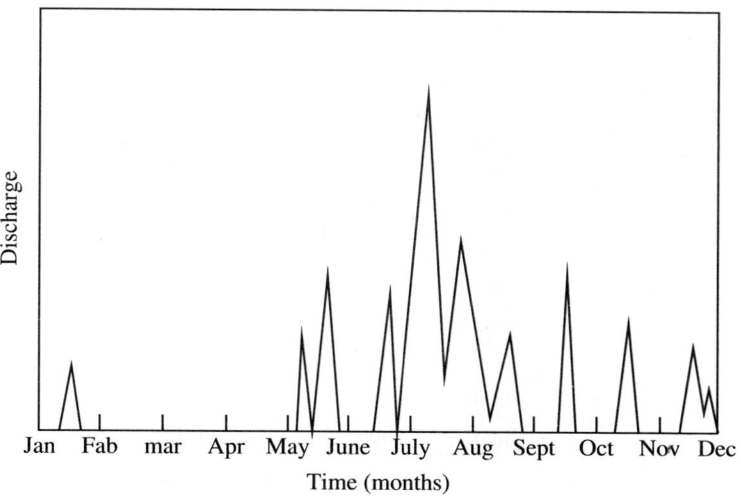

Figure 5.4 Ephemeral stream.

where ΔS is change in the storage volumes of water storage bodies on the stream, R_N is natural flow volume in time Δt, R_0 is observed flow volume in time Δt at outlet, and V_r is volume of return flow from irrigation, domestic water supply and industrial use. In the expression, losses denote losses in time Δt for irrigation, water supply and industrial use and inclusive of evaporation losses in surface water bodies.

Dependable yield

Annual yield varies year to year because it depends on several processes such as precipitation, evaporation, storage, etc. Therefore, for any year in future, annual yield can be estimated only with some degree of dependability. Usually, 50% to 75% dependable yields are relied upon. 50% dependable yield can be expected every alternate year on an average.

EXAMPLE 5.1 For a catchment, annual rainfall and runoff values (cm) spanning 20-year period are given in the table below. Determine 80% and 60% dependable annual yield of the catchment.

Year	*Rainfall* P	*Runoff* R
1985	120	60
1986	100	48
1987	115	55
1988	101	45
1989	89	26
1980	95	38
1991	145	72
1992	93	27
1993	110	45
1994	87	15
1995	109	35
1996	77	19
1997	114	42
1998	83	20
1999	96	35
2000	145	52
2001	166	82
2002	99	31
2003	89	22
2004	131	125
2005	100	30

Solution The years along with the annual rainfall and runoff are given.
To determine 80% and 60% dependable annual yield of the catchment, the annual rainfall values are arranged in descending order of magnitude and a rank *m* is assigned for each value starting from the highest value.

The excedence probability *P* is calculated for each runoff value as

$$P = \frac{m}{N} + 1$$

where *m* is rank *N* and is the observation.

To find 80% dependable annual yield, we can find that $P = 0.8$ in between runoff 26 and 22. So, by interpolation

$$P = 0.7727 \quad \text{when} \quad R = 26$$
$$P = 0.8181 \quad \text{when} \quad R = 22$$

So, for $P = 0.8$,

$$R = \frac{26 - 4 \times (0.8 - 0.7727)}{0.8181 - 0.7727}$$

$$= 23.59 = 23.6$$

To find 60% dependable annual yield,

$$P = 0.5909 \quad \text{when} \quad R = 35$$
$$P = 0.6363 \quad \text{when} \quad R = 31$$

So, for $P = 0.6$,

$$R = \frac{35 - 4 \times (0.6 - 0.5909)}{0.6363 - 0.5909} = 34.2$$

The necessary calculations are shown in the table below:

Year	*Rainfall* *P*	*Runoff* *R*	*Sorted annual runoff* (cm)	*Rank* (m)	*Excedence probability* *p*
1985	120	60	125	1	0.045
1986	100	48	82	2	0.0909
1987	115	55	72	3	0.1363
1988	101	45	60	4	0.1818
1989	89	26	55	5	0.2272
1990	95	38	52	6	0.2727
1991	145	72	48	7	0.3181
1992	93	27	45	8	0.3636
1993	110	45	45	9	0.409
1994	87	15	42	10	0.4545
1995	109	35	38	11	0.5
1996	77	19	35	12	0.5454
1997	114	42	35	13	0.5909
1998	83	20	31	14	0.6363
1999	96	35	30	15	0.6818
2000	145	52	27	16	0.7212
2001	166	82	26	17	0.7727
2002	99	31	22	18	0.8181
2003	89	22	20	19	0.863
2004	131	125	19	20	0.909
2005	100	30	15	21	0.95

5.7 Catchment Characteristics

1. Stream density: The stream density D_s of a catchment is defined as the ratio of number of the streams and area of basin.

$$D_s = \frac{N_s}{A} \tag{5.1}$$

where, N_s is the number of streams and A is the area of basin.

2. Drainage density: The drainage density D_d is expressed as the total length of all the channels divided by catchment area.

$$D_d = \frac{L_s}{A} \tag{5.2}$$

where L_s is the total length of all channels in the catchment and A is the area of basin.

3. Slope of the basin: Stream gradient or slope of the basin S is the grade measured by the ratio of drop in elevation of a stream per unit horizontal distance.

$$\text{Hence, average stream slope} = \frac{\text{Total fall of the longest water course}}{\text{Length of the longest water course}}$$

$$S = \frac{1.5(CI)N_c}{\sum L} \tag{5.3}$$

where, S is slope of the basin, CI is contour interval, N_c is the number of the contours crossed by all the subdividing lines, and ΣL is the total length of the subdividing lines.

5.8 Isochrone

An *isochrone* is a line joining all the points in a basin of equal travel time such as beginning of precipitation, as shown in Figure 5.5. Any point on a given isochrone takes the same time to reach the outlet of the basin. The isochrone are useful in deriving hydrographs.

5.9 Estimation of Runoff

The calculation of yield is of fundamental importance in all water resources development studies. The various methods used for the estimation of yield can be listed below:

1. Empirical formulae
2. Soil Conservative Services (SCS) curve number method
3. Rational method
4. Infiltration indices method
5. Linear and non-linear rainfall-runoff relationships
6. Overland flow hydrograph

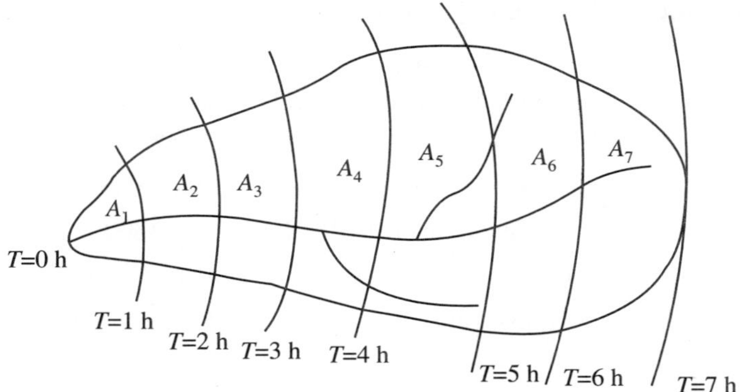

Figure 5.5 Isochrone.

5.9.1 Empirical Formulae

A number of empirical methods have been developed to relate runoff rainfall. These methods have a limitation that they cannot be used at all the places. Some typical examples of empirical formulas are shown below:

$$R = aP + b \tag{5.4}$$

$$R = aP^b \tag{5.5}$$

where, R is the runoff, P is rainfall, and a, b are constants.

Some widely used empirical formals are as follows:

Khosla's method

A.N. Khosla (1960) analysed the rainfall, runoff and temperature data for various catchments in India and the USA and developed an empirical relationship between runoff and rainfall. The relationship for monthly runoff is

$$R_m = P_m - L_m \tag{5.6}$$

$$L_m = 0.48T_m \quad \text{for} \quad T_m > 4.5 \tag{5.7}$$

where, R_m is monthly runoff (cm) and $R_m \geq 0$, P_m is monthly rainfall (cm), L_m shows monthly losses (cm) and T_m is the mean monthly temperature of the catchment (°C).

For $T_m \leq 4.5$ °C, the loss L_m may provisionally be assumed as

T °(C)	4.5	–1	–6.5
L_m (cm)	2.17	1.78	1.52

Further, the total annual runoff may be compouted as

$$\text{Annual runoff} = \sum R_m$$

Formulas for some of the drainage basins in India are given below:

$$\text{Ganga basin, } R = 2.14P^{0.64}$$

$$\text{Yamuna basin (Delhi), } R = 0.14P^{1.1}$$

$$\text{Rihand basin (UP), } R = P - 117P^{0.86}$$

$$\text{Chambal basin (Rajasthan), } R = 120P - 4945$$

$$\text{Tawa basin (MP), } R = 90.5 - 4800$$

$$\text{Tapti basin (Gujarat), } R = 435P - 17200$$

where R is the average annual runoff (cm), P is the average annual rainfall (cm) and T is the mean annual temperature (°C) for the entire drainage basin.

EXAMPLE 5.2 Determine the annual runoff and annual runoff coefficient for a given catchment by Khosla method. The mean monthly temperature and rainfall are given below in the table:

Month	January	February	March	April	May	June	July	August	September	October	November	December
Temperature (°C)	14	18	23	29	33	37	34	33	31	33	18	13
Rainfall (cm)	6	7	3	1	4	14	35	31	17	3	2	3

Solution In Khosla's formula applicable to this case, $R_m = \phi_m - L_m$ with $L_m = 0.48 \times T_m$ (°C) having a maximum value equal to corresponding P_m. The calculations are shown below in the table:

Month	January	February	March	April	May	June	July	August	September	October	November	December
Rainfall P_m (cm)	6	7	3	1	4	14	35	31	17	3	2	3
T (°C)	14	18	23	29	33	37	34	33	31	33	18	13
L_m (cm)	6	7	3	1	4	14	16.32	15.84	14.88	3	2	3
Runoff R_m (cm)	0	0	0	0	0	0	18.68	15.16	2.12	0	0	0

$$\text{Total annual runoff} = 18.68 + 15.16 + 2.12 = 35.96$$

$$\text{Annual runoff coefficient} = \frac{\text{Annual runoff}}{\text{Annual rainfall}} = \frac{35.96}{126} = 0.285$$

Inglis Formula

C.C. Inglis (1940) developed the following empirical formulae as follows:

1. For Ghat areas,

$$R = 0.85P - 30.5 \tag{5.8}$$

2. For plain areas,

$$R = \frac{(P - 17.8)}{254} \tag{5.9}$$

where P is rainfall (cm).

Dicken's formula

It is a regression model developed for determining runoff.

$$Q = CA^{3/4} \tag{5.10}$$

where Q is discharge in m^3/s and A is area of catchment in km^2.

The value of constant C varies from 2.80 to 5.6 for plain areas and 14 to 28 for hilly regions.

5.9.2 Soil Conservation Services (SCS) Curve Number Method

The Soil Conversion Services (SCS) curve number method relates the direct runoff Q with the rainfall P. The basic assumption of this methods is the ratio of actual runoff to the potential runoff during an event is equal to the ratio of actual infiltration to potential infiltration.

$$Q = \frac{(P - I_a)^2}{(P - I_a + S)} \tag{5.11}$$

$$F = P - I_a - Q \tag{5.12}$$

where, I_a shows initial abstractions, S is storage potential of soil and F is infiltration.

In Eq. (5.12), the value of I_a is taken as $0.2S$ as best approximation. Therefore,

$$Q = \frac{(P - 0.2S)^2}{(P + 0.8S)} \tag{5.13}$$

This expression has S as unknown variable, which may be determined by using curve numbers developed by the US SCS. The relationship between curve number and storage potential is given below:

$$S = \frac{25400}{\text{CN}} - 254$$

where CN is curve number.

EXAMPLE 5.3 The curve number of a watershed is 72. Complete the initial loss and retention capacity of the watershed.

Solution Given, curve number (CN) = 72

We know that,

$$S = \frac{25400}{\text{CN}} - 254$$

$$S = \frac{25400}{72} - 254 = 98.778 \text{ mm} = 9.88 \text{ cm}$$

$$I_a = 0.2 \times 9.88 = 1.976 \text{ cm}$$

5.9.3 Rational Method

This approach is used to compute the yield of a catchment by assuming a suitable runoff coefficient. The *runoff coefficient* is defined as the fraction of rainfall which becomes runoff from the catchment.

$$\text{Yield} = CAP \tag{5.14}$$

where, A is the area of catchment, P is precipitation and C is runoff coefficient.

The value of the runoff coefficient C varies depending on the soil type, vegetation geology, etc. and is given in Table 5.1.

The method assumes that rainfall intensity is constant during the rainfall.

Table 5.1 Runoff Coefficients for Various Types of Catchments

S.No.	*Types of catchment*	*Value of C*
1	Rockey and impermeable concrete streets	0.8–1.0
2	Slightly permeable, bare	0.6–0.8
3	Cultivated or absorbent with vegetation	0.4–0.6
4	Cultivated land	0.3–0.4
5	Heavy forest	0.05–0.25

5.9.4 Infiltration Indices Method

This method is suitable for computing the runoff from large catchment having uniform infiltration characteristics. From the total precipitation, losses due to infiltration are deducted. For this purpose, ϕ-index and W-index may be used. This method is largely empirical and the derived values are applicable only when the rainfall characteristics and initial soil moisture conditions are identical to those for which these are derived.

5.9.5 Linear and Non-linear Rainfall-Runoff Relationship

In case of linear regression, a statistical correlation between the observed monthly rainfall and the monthly runoff is established. The relationship is plotted for log-log graph for each month. Hence, straight line is fitted using linear relationship as follows:

$$Q = aP + b \tag{5.15}$$

where, Q is runoff (mm or ft), P is rainfall (mm or cm or ft), b is a coefficient to be determined that accounts for losses, and a is reduction factor which accounts for losses such as interception, depression storage, etc.

EXAMPLE 5.4 Annual rainfall and runoff values (cm) for a catchment spanning 20-year period are given in the table below. Develop a linear relation equation to estimate the annual runoff volume for a given annual rainfall value.

Year	*Rainfall*	*Runoff*	*Year*	*Rainfall*	*Runoff*
1985	120	60	1996	77	19
1986	100	48	1997	114	42
1987	115	55	1998	83	20
1988	101	45	1999	96	35
1989	89	26	2000	145	52
1990	95	38	2001	166	82
1991	145	72	2002	99	31
1992	93	27	2003	89	22
1993	110	45	2004	131	125
1994	87	15	2005	100	30
1995	109	35			

Solution The years along with the annual rainfall and runoff are given.

We have to find the linear correlation equation to estimate the runoff volume for a given annual rainfall value.

To develop a linear correlation equation, we have to use linear regression relationship. The following table shows the calculations of P, R, P^2, R^2 and PR:

Year	*Rainfall P*	*Runoff R*	P^2	R^2	*PR*
1975	120	60	14400	3600	7200
1976	100	48	10000	2304	4800
1977	115	55	13225	3025	6325
1978	101	45	10201	2025	4545
1979	89	26	7921	676	2314
1980	95	38	9025	1444	3610
1981	145	72	21025	5184	10440
1982	93	27	8649	729	2511
1983	110	45	12100	2025	4950
1984	87	15	7569	225	1305
1985	109	35	11881	1225	3815
1986	77	19	5929	361	1463
1987	114	42	12996	1764	4788
1988	83	20	6889	400	1660
1989	96	35	9216	1225	3360
1990	145	52	21025	2704	7540
1991	166	82	27556	6724	13612
1992	99	31	9801	961	3069
1993	89	22	7921	484	1958
1994	131	125	17161	15625	16375
1995	100	30	10000	900	3000
	Sum = 2264	Sum = 924	Sum = 254490	Sum=53610	Sum = 108640

So, $\sum P = 2264$; $\sum R = 924$; $\sum P^2 = 254490$; $\sum R^2 = 53610$; $\sum PR = 108640$.

The correlation equation can be written as

$$R = aP + b$$

where
$$a = \frac{N\sum(PR) - \left(\sum P\right)\left(\sum R\right)}{N\left(\sum P^2\right) - \left(\sum R\right)^2}$$

$$a = \frac{21 \times 108640 - 2264 \times 924}{21 \times 254490 - (924)^2}$$

$$a = 0.0422$$

$$b = \frac{\sum R - a\sum P}{N}$$

$$b = \frac{924 - 0.0422 \times 2264}{21}$$

$$b = 39.45$$

So, the required annual rainfall runoff relationship of the catchment is given by

$$R = 0.0422P + 39.45$$

Coefficient of correlation,
$$r = \frac{N \times \sum(PR) - \left(\sum P\right)\left(\sum R\right)}{\sqrt{N\left(\sum P^2\right) - \left(\sum P\right)^2 \times N \times \left(\sum R^2\right) - \left(\sum R\right)^2}}$$

$$= \frac{21 \times 108640 - 2264 \times 924}{\sqrt{(21 \times 254490 - 2264^2)(21 \times 53610 - 924^2)}}$$

$$= 0.771$$

Also, it can be obtained from the regression analysis curve as shown in the following graph:

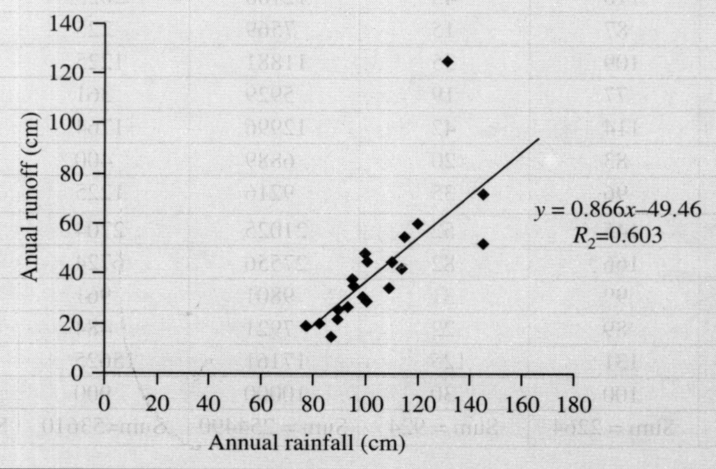

5.9.6 Overland Flow Hydrograph

Overland flow occurs as a thin sheet of water over the ground surface (soon after storm starts), joins a stream channel, and then, flows in a channel to the concentration point. *Time of concentration* T_c is the longest time required for a particle to travel from the watershed divide to the watershed outlet. For channel flow component of runoff, the Kirpich equation is

Time of concentration, $$T_c = K \times (L)^{0.77} \times (S)^{-385} \tag{5.16}$$

where, T_c is the time of concentration (min), K is conversion coefficient in which $K = 0.0078$ for traditional units and $K = 0.0195$ for SI units, L is the flow length of channel and S is the dimensionless main channel slope.

Time to peak T_p is given as

$$T_p = 0.6T_c + \sqrt{T_c} \tag{5.17}$$

Overland flow is an essential—uniform flow over the surface as developed by C.F. Izzard (1948). The Reynolds number is

$$Re = \frac{Vd}{\nu} = \frac{q}{\nu} \tag{5.18}$$

where, V is the velocity of flow, d is uniform depth flow, and q is discharge per unit width.

Note: Experiments indicate that the overland flow can be assumed

1. Laminar if $Re < 1000$.
2. For turbulent flow if $Re > 1000$
3. For a transition region of uncertainty in the vicinity of $Re = 1000$

EXAMPLE 5.5 Estimate the time of concentration of 350 ha area of watershed. The average slope of the area is 0.035 and maximum drainage course is 380 m.

Solution Given,

$$L = 380 \text{ m}$$
$$S = 0.035$$

$$\text{Time of concentration } T_c = 0.01947 \times (L)^{0.77} \times (S)^{-335}$$
$$T_c = 0.01947 \times (380)^{0.77} \times (0.035)^{-0.385}$$
$$T_c = 0.01947 \times 96.92 \times 3.63$$
$$T_c = 6.86 \text{ min}$$

EXAMPLE 5.6 A watershed of 180 ha area is divided into three parts.

(a) Calculation land (105 ha) with 1.2% slope and runoff coefficient of 0.60
(b) Pasture land (40 ha) with 6.9% slope and runoff coefficient of 0.42
(c) Forest land (35 ha) with 13.5% slope and runoff coefficient of 0.50

The maximum length of channel reach is 2400 m, average channel slope is 4.5% and rainfall depth is 4.2 cm in watershed. Calculate the peak runoff rate for 12-year return period.

Solution For cultivated land, runoff coefficient = 0.60

For pasture land, runoff coefficient = 0.42

For forest land, runoff coefficient = 0.50

$$\text{Weightage runoff coefficient} = \frac{105 \times 0.60 + 40 \times 0.42 + 35 \times 0.5}{105 + 40 + 35} = 0.54$$

$$\text{Time of concentration, } T_c = 0.01947 \times (L)^{0.77} \times (S)^{-0.385}$$

$$= 0.01947 \times (2400)^{0.77} \times 0.045^{-0.385}$$

$$= 25.74 \text{ min}$$

$$\text{Rainfall intensity, } i = \frac{\text{Rainfall depth}}{\text{Time of concentration}}$$

$$i = \frac{4.2}{0.429} = 9.79 \text{ cm/h}$$

$$\text{Pack runoff, } Q_{\text{peak}} = CiA = 0.54 \times 9.79 \frac{10^{-2}}{3600} \times 180 \times 10^4$$

$$= 26.433 \text{ m}^3\text{/s}$$

5.10 Flow-Duration Curve

The *flow-duration curve* is a plot that shows the percentage of time and certain value of flow is likely to equal or exceed some specified value of interest in the available number of years of record. Flow-duration curves are useful for several applications such as flood control, water resources planning, hydropower generation, computing the sediment load and dissolved solid load of a stream, design of the drainage system, etc. The derivation of flow-duration curve is as follows:

1. Arrange the flow data in descending order. Use class interval if the number of records is large.
2. Assign rank (m) to each value, with the largest discharge assigned the rank of 1.
3. Compute the probability of exceedance to each data item using the following formula.

$$P(\%) = \frac{m}{(n+1)} \times 100 \tag{5.19}$$

where m is rank and N is the number of observation.

4. Plot the values of probability of excedance P against their corresponding flow value. Join the points to get the flow-duration curve.

The flow-duration curve (Figure 5.6) represents the cumulative frequency distribution. A flat curve indicates a river with few floods with large groundwater contribution, while a steep curve indicates high variable flow, i.e., frequent floods and dry periods with little water contributions. Additionally, a flow-duration curve derived from monthly flow data has flat slope because of smooth flow peaks, while flow-duration curve derived from daily data has steep slope curve.

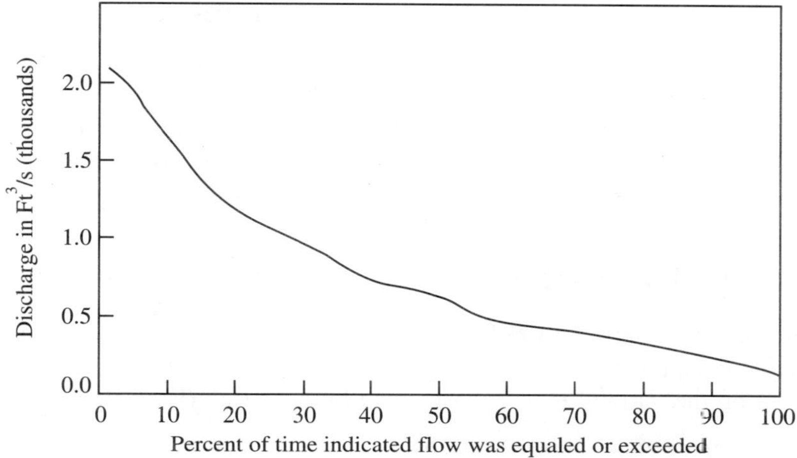

Figure 5.6 Flow-duration curve.

5.11 Flow-Mass Curve

The *flow-mass curve* is a plot of cumulative discharge volume against time plotted in sequential order, as shown in Figure 5.7. The flow-mass curve is an integral curve (summation curve) of the hydrograph and is given by

$$V = \int_{t_0}^{t_1} Q\, dt \tag{5.20}$$

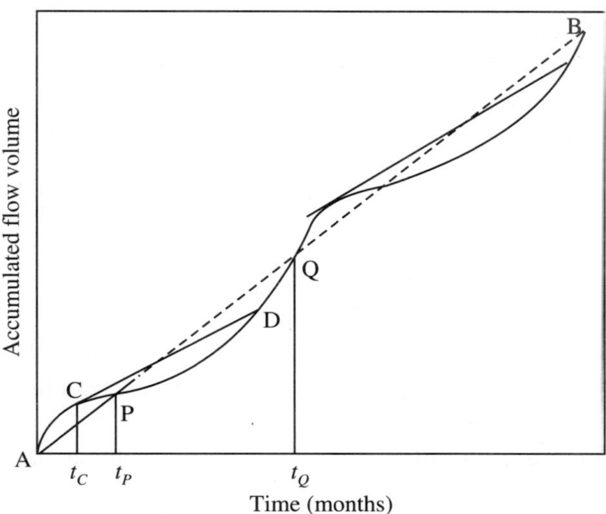

Figure 5.7 Flow-mass curve.

Here, V is the discharge volume, Q is discharge rate t_0, t_1 represent the time of beginning, end of the mass-curve, respectively. The slope of the flow-mass curve $\frac{dV}{dt}$ represent the flow rate.

In Figure 5.7, straight line joining two points P and Q, represents the average rate of flow that can be maintained between the time t_P and t_Q if a reservoir of adequate storage is available.

5.12 Surface Water Resources of India

Although India occupies only 3.29 million km^2 geographical area, which forms 2.4% of the world's land area, it supports over 15% of the world's population. The surface water potentials of country based on the various zones as estimated by K.L. Rao (Rao, 1979) are as follows:

1. The Ganga river zone: 500000 million m^3
2. The Brahmaputra-Barak river zone: 520000 million m^3
3. The east flowing river zone: 360000 million m^3
4. The west flowing river zone: 320000 million m^3
5. The Sindhu river zone: 80000 million m^3

Hence, the total annual water potential is estimated to be about 1650×10^9 m^3.

Summary

- A perennial stream always carries flow, while an ephemeral stream does not carry baseflow.
- The basin yield is the quantity of water available from a stream at a given point over specified duration of time.
- The Khosla's method of runoff computation is based on the runoff water balance approach and mean monthly temperature of the watershed.
- The SCS curve number method depends on land use pattern, hydrological condition of soil and hydrological soil group.
- The rational approach is used to compute the yield of a catchment by assuming a suitable runoff coefficient.
- The time of concentration is the function of watershed features such as length of the longest channel and its weighted slope.
- A flow-duration curve is the plot of discharge versus probability of exceedance of the data.
- A flow-duration curve is plotted to understand the percentage of time for which the various flow magnitudes in a stream are exceeded.
- A flow-mass curve is a plot of cumulative discharge volume and time in chronological order.

Objective Type Questions

1. What applying the rational formula for computing the design discharge, the rainfall duration is stipulated as the time of concentration because. (GATE, 2003)
   (a) This leads to the largest possible rainfall intensity.
   (b) This leads to the smallest possible rainfall intensity
   (c) The time of concentration is the smallest rainfall duration for which the rational formula is applicable.
   (d) The time of concentration is the largest rainfall duration for which the rational formula is applicable.
2. An isolated 4 h storm occurred over a catchment as follows:

Time	1st hour	2nd hour	3rd hour	4th hour
Rainfall (mm)	9	28	12	7

   The ϕ-index for the catchment is 10 mm/h. The estimated runoff depth from the catchment due to the above storm is (GATE, 2007)
   (a) 10 mm (b) 16 mm (c) 20 mm (d) 23 mm
3. Match List-I with List-II and select the correct answer using the codes given below the lists: (IES, 1996)

   List-I
   A. Conservation reservoirs B. Retarding basins
   C. Flood plains D. Flood walls

   List-II
   1. Uncontrolled outlets
   2. Flood fighting
   3. Temporary storage of flood water
   4. Controlled outlets

   Codes

	A	B	C	D
(a)	1	4	3	2
(b)	1	4	2	3
(c)	4	1	3	2
(d)	4	1	2	3

4. In a flow-mass curve study, the demand line drawn from a ridge does not intersect the mass curve again. This implies that (IES, 1999)
   (a) The reservoir is not full at the beginning
   (b) The storage is not adequate
   (c) The demand cannot be met by the inflow as the reservoir will nct refill
   (d) The reservoir is wasting by spill
5. If a tangent drawn parallel to the demand line from a ridge point of a mass curve does not intersect the mass curve again, it can be inferred that the (IES, 2000)
   (a) Frequency of flood entering into the reservoir is less
   (b) Inflow into the reservoir cannot meet the demand

(c) Reservoir is overflowing, resulting in wastage
(d) Reservoir can meet higher demand

6. Consider the following statements: (IES, 2002)
1. An ephemeral stream is one which has a baseflow contribution.
2. Flow characteristics of a stream depend on rainfall the catchment characteristics, and also, on the climatic factors which influence evapotranspiration.
3. Sequent peak algorithm is used for estimating runoff from rainfall.

Which of these statement is/are correct?
(a) 1, 2 and 3 (b) 1 and 3 (c) 2 and 3 (d) 2 alone

7. A 6 h rainstorm with hourly intensities of 7, 18, 25, 17, 11 and 3 mm/h produced a runoff of 39 mm. Then, the ϕ-index is (IES, 2002)
(a) 3 mm/h (b) 7 mm/h (c) 8 mm/h (d) 10 mm/h

8. Which one of the following characteristics describes a watershed system in system's parlance? (IES, 2003)
(a) Linear (b) Non-linear
(c) Linear and time-interval (d) Non-linear and time-variant

9. A catchment area of 90 ha has a runoff coefficient of 0.4. A storm of duration larger than the time of concentration of the catchment and of intensity 4.5 cm/h creates a peak discharge rate of (IES, 2003)
(a) 11.3 m^3/s (b) 0.45 m^3/s (c) 450 m^3/s (d) 4.5 m^3/s

10. A 6 h storm has 6 cm of rainfall and the resulting runoff is 3 cm. If ϕ-index remains at the same value, which one of the following is the runoff due to 12 cm of rainfall in 9 h in the catchment? (IES, 2005)
(a) 4.5 cm (b) 6.0 cm (c) 7.5 cm (d) 9.0 cm

11. A catchment has an area of 150 ha and a runoff/rainfall ratio of 0.40. If due to 10 cm rainfall over the catchment, a streamflow at the catchment outlet lasts for 10 h, what is the average stream flow in the period? (IES, 2005)
(a) 60,000 m^3/h (b) 100 m^3/min (c) 3.5 m^3/s (d) 1.33 m^3/s

12. For a given storm, other factors remaining same, (IES, 2005)
(a) Basins with large drainage densities give smaller flood peaks
(b) Low drainage density basins give shorter time bases of hydrographs
(c) The flood peak is independent of the drainage density.
(d) Basins having low drainage density give smaller peaks in flood hydrographs

13. The time of concentration at the outlet in an urban area catchment of 1.5 km^2 area with a runoff coefficient of 0.42 is 28 min. The maximum depth of rainfall with a 50-year return period for this time of concentration is 48 min. What is the peak flow rate at the outlet for this return period? (IES, 2006)
(a) 12 m^3/s (b) 14 m^3/s (c) 16 m^3/s (d) 18 m^3/s

14. Maximum possible discharge from a small catchment corresponding to a particular rainfall intensity is independent of which one of the following? (IES, 2008)
(a) Soil moisture conditions (b) Drainage characteristics of catchment
(c) Area of the catchment (d) Duration of the rainstorm

15. A 4 h storm had 4 cm of rainfall and the resulting direct runoff was 2 cm. If the Φ-index remains at the same value, the runoff due to 10 cm of rainfall in 8 h in the catchment is (IES, 2008)

(a) 6.0 cm (b) 7.5 cm (c) 2.3 cm (d) 2.8 cm

16. A catchment area of 60 ha has a runoff coefficient of 0.40. If a storm of intensity 3 cm/h and duration longer than the time of concentration occurs in the catchment, then what is the peak discharge? (IES, 2009)

(a) 2.0 m^3/s (b) 3.5 m^3/s (c) 4.5 m^3/s (d) 2.5 m^3

17. The land use of an area and the corresponding runoff coefficients are as follows: (IES, 2002)

S.No.	*Land use*	*Area (ha)*	*Runoff coefficient*
1	Roads	10	0.70
2	Lawn	20	0.10
3	Residential area	50	0.30
4	Industrial area	20	0.80

What is the equivalent runoff coefficient?

(a) 0.15 (b) 0.36 (c) 0.40 (d) 0.51

18. A catchment of area 200 ha has a runoff coefficient 0.5. A storm of duration larger than the time of concentration of the catchment and of intensity 3.6 cm/h causes a peak discharge of (IES, 2010)

(a) 5 m^3/s (b) 10 m^3/s (c) 100 m^3/s (d) 360 m^3/s

19. A catchment consists of 30% area with runoff coefficient 0.40 and the remaining 70% area with runoff coefficient 0.60. The equivalent runoff coefficient will be (IES, 2010)

(a) 0.48 (b) 0.54 (c) 0.63 (d) 0.76

20. The excess of runoff hydrograph from a catchment area of 10 km^2 due to a storm of 6 h duration has been observed to be 10 m^3/s and the base length is 20 h. The rainfall excess in the catchment is (IES, 2012)

(a) 5.1 cm (b) 3.6 cm (c) 4.5 cm (d) 2.5 cm

21. An urban area is located in plains having average climatic conditions. The impervious area thereof for which drainage must be provided is 3.6 ha and the design rainfall intensity is 2.0 cm/h. The drains will be designed for a runoff of (IES, 2012)

(a) 0.05 m^3/s (b) 0.10 m^3/s (c) 0.20 m^3/s (c) 0.40 m^3/s

22. The rainfalls on 5 successive days in a catchment were 2 cm, 7 cm, 8 cm, 4 cm and 3 cm. If the ϕ-index for the storm is 3 cm/day, the total direct runoff volume generated from a 195 km^2 catchment is (IES, 2012)

(a) 19.5 million m^3 (b) 23.4 million m^3
(c) 15.6 million m^3 (d) 32.5 million m^3

23. The best estimate of runoff represented by 57 mm of runoff depth from a basin area 3300 km^2 is nearly (IES, 2014)

(a) 2300 cumec-days (b) 2225 cumec-days
(c) 2175 cumec-days (d) 2020 cumec-days

24. A conventional flow-duration curve is a plot between (GATE, 2014)

(a) Flow and percentage time flow is exceeded.
(b) Duration of flooding and ground level elevation.
(c) Duration of water supply in a city and proportion of area receiving supply exceeding this duration
(d) Flow rate and duration of time taken to empty a reservoir at that flow rate

Answers

1. (a) **2.** (c) **3.** (c) **4.** (c) **5.** (b) **6.** (c) **7.** (b) **8.** (c) **9.** (d) **10.** (c)
11. (b) **12.** (d) **13.** (d) **14.** (d) **15.** (a) **16.** (a) **17.** (c) **18.** (b) **19.** (b) **20.** (b)
21. (c) **22.** (a) **23.** (c) **24.** (a)

Explanations

2.

$$\phi\text{-index}=\frac{\text{Total infiltration}}{\text{Time of rainfall excess}}$$

$$10 = \frac{(28+12)-\text{Runoff}}{\text{Time of rainfall excess}}$$

For 1st and 4th hours, the precipitation is less than Φ-index, which rainfall of these two hours is ineffective. Hence, the time of rainfall excess is 2 h only.

$$10 = \frac{40-\text{Runoff}}{2}$$

$$\text{Runoff} = 20 \text{ mm}$$

9. Peak discharge is given by

$$Q_p = \frac{K \times i \times A}{36}$$

$$Q_p = \frac{0.4 \times 4.5 \times 90}{36} = 4.5 \text{ m}^3/\text{s}$$

If area is in square kilometres and intensity is in millimetre per hour, then

$$Q_p = \frac{K \times i \times A}{3.6}$$

10. Amount of rainfall in 9 h = 12 cm

$$\Phi\text{-index} = \frac{(6-3)}{6} = 0.5 \text{ cm/h}$$

Infiltration for 9 h rainfall = 9 × 0.5 = 4.5 cm

Runoff due to 12 cm rainfall = Rainfall – Infiltration

= 12 – 4.5 = 7.5 cm

11. $$\text{Runoff} = 150 \times 10^4 \times 0.4 \times 0.1 = 60000 \text{ m}^3$$

$$\text{Average stream flow} = \frac{600000}{(10 \times 10)} = 100 \text{ m}^3/\text{min}$$

13. Time of connection = 28 min

The mean intensity of precipitation for duration equal to time of concentration and an exceedance probability p,

$$I_{t,\,cp} = \frac{48}{(28/60)} = \frac{720}{7} \text{ mm/h}$$

Calculating peak discharge from rational formula,

$$Q = Ci_{t,\,cp}\frac{A}{3.6} = 0.42 \times \left(\frac{720}{7}\right) \times \left(\frac{1.5}{3.6}\right) = 18 \text{ m}^3/\text{s}$$

Peak discharge = 18 m^3/s

19. The equivalent runoff coefficient is given by

$$K = \frac{K_1A_1 + K_2A_2}{A_1 + A_2}$$

$$K = \frac{0.4 \times 0.3 + 0.6 \times 0.7}{0.3 + 0.7}$$

$$K = 0.54$$

20. $$\text{Total runoff} = \frac{1}{2} \times 20 \times 60 \times 60 \times 10 = 360000 \text{ m}^3$$

Therefore, $$\text{rainfall excess} = \left(\frac{360000}{10 \times 10^6}\right)$$

$$= 0.036 \text{ m} = 3.6 \text{ cm}$$

21. Drains will be designed for a peak runoff,

$$Q_{pk} = CiA$$

For impervious area, $C = 1$.

Therefore, $$Q_{pk} = i.A = \frac{2 \times 10^{-2}}{3600} \times 3.6 \times 10^4 = 0.2 \text{ m}^2/\text{s}$$

22. $$\text{Total direct runoff} = 195 \times (0 + 4 + 5 + 1) \times 10^{-2}$$

$$= 19.5 \text{ million m}^3$$

23. $$\text{Best estimated runoff} = \frac{3300 \times 10^6 \times 57 \times 10^{-3}}{24 \times 60 \times 60}$$

$$= 2177 \text{ cumec-days}$$

Theoretical Questions

1. Define runoff.
2. What is catchment yield?
3. Describe the various factors affect the runoff from a catchment.
4. Discuss the various methods for computing runoff.
5. Explain how runoff is determined using Khosla's method.
6. What is the rational formula?
7. Write short notes on (a) isochrones (b) ephemeral streams.
8. Explain the derivation of flow-duration curves.

Unsolved Problems

Problem 1: In a 390 ha watershed, the CN value was assessed as 75. Determine the value of direct runoff volume for the following 4 days of rainfall using standard SCS-CN equations.

Date	June 21	June 22	June 23	June 24
Rainfall (mm)	59	25	35	16

Problem 2: A small watershed is 280 ha in size. The land cover can be classified as 35% open forest, with CN value of 60, and 65% poor quality pasture, with CN value of 86. Determine the direct runoff volume due to rainfall of 80 mm in one day.

Problem 3: A 150 ha catchment area has the following characteristics:

(a) The longest length of the stream of the catchment = 3000 m

(b) Difference in elevation between the most remote point on the catchment and the outlet = 60 m

(c) Landuse/cover details:

Landuse Landcover (LULC) classes	*Area* (ha)	*Runoff coefficient*
Forest	60	0.26
Grazing-barren land	40	0.20
Agricultural	50	0.45

The maximum intensity-duration relationship for the watershed is given by

$$i = \frac{(3.97T^{0.165})}{(D+0.15)^{0.733}}$$

where, i (intensity) is in centimetre per hour, T is return period in years, and D is duration of rainfall in hours. Calculate 30-year peak runoff from the catchment that can be expected at the outlet of the catchment.

Problem 4: Calculate the peak runoff rate (m^3/s) of a catchment if the area of catchment is 600 ha, runoff depth is 5.3 cm and time of concentration is 40 min.

Problem 5: Estimate the peak runoff of 280 ha size watershed. The curve number of watershed is 70, rainfall depth is 14 cm, maximum drainage course is 1400 m and the average slope is 0.45.

Further Reading

Chin, David A., *Water-Resources Engineering*, Pearson Prentice Hall, Upper Saddle River, NJ, 2000.

Chow, V.T., Maidment, David R., and Mays, Larry W., *Applied Hydrology*, Tata McGraw-Hill, New Delhi, 1988.

Corbitt, Robert A., *Standard Handbook of Environmental Engineering*. 2nd ed., McGraw-Hill, New York, 1999.

Das, Ghanshyam, *Hydrology and Soil Conservation Engineering Including Watershed Management*, PHI Learning, Delhi, 2014.

Deodhar, M.J., *Elementary Engineering Hydrology*, Pearson, New Delhi, 2013.

Lindsley, Ray K., Franzini, Joseph B., Freyberg, David L. and Tchobanoglous, George, *Water-Resources Engineering*, 4th ed., McGraw-Hill, New York, 1992.

McCuen, Richard H., *Hydrologic Analysis and Design*, 2nd ed., Pearson Prentice-Hall, Upper Saddle River, NJ, 1998.

Raghunath, H.M., *Hydrology—Principles, Analysis and Design*, New Age Publishers, New Delhi, 2014.

R., Mysooru, Yadupathi Putty, *Principles of Hydrology*, I.K. International, New Delhi, 2013.

Singh, Vijay P., *Elementary Hydrology*, Prentice-Hall, 1992.

Suresh, R., *Watershed Hydrology*, Standard Publishers Distributors, New Delhi, 2015.

CHAPTER

6

Hydrograph

6.1 Introduction

A *hydrograph* is a plot of the relationship of discharge (streamflow) with respect to time after a storm. The interval at which ordinates are drawn depends mainly on the data available. This hydrograph results from a combination of physiographic, geomorphological and meteorological conditions in a catchment and represents the integrated effects of abstraction losses, surface runoff, interflow, and groundwater flow. Since hydrograph is a plot of discharge versus time, therefore the area beneath the hydrograph between any two points in time gives the total volume of water passing through the point of interest during the time interval. Hydrograph analysis is usually important in flood damage mitigation, flood forecasting, or establishing design flows for structures that convey floodwaters. In case of annual hydrograph, the hydrograph is plotted between discharge and time over a year.

6.2 Components of Hydrograph

The shape of hydrograph depends on meteorological factors, catchment properties and human factors. Meteorological factors include rainfall intensity, duration of storm and spatial distribution or rainfall over the basin. The catchment properties (physiographic factors) include size and shape of the drainage area, channel morphology and drainage type, soil type, channel slope, storage detention structure in the catchment, while human factors include the effects of land use and land cover.

To understand the different parts of a hydrograph, a typical single-peaked hydrograph, as shown in Figure 6.1, is divided into three limbs as follows:

1. Rising limb: This is an ascending portion of hydrograph. Its shape depends on storm and basin characteristics. This is also know as *concentration curve*. This rising limb BC has a well-defined point of rise B followed by increasing discharge.

2. Crest segment: This is the most important segment of the hydrograph because the crest segment CE contains the peak D of the hydrograph. Generally, the peak occurs after the rain has stopped. Sometimes, it is possible that the hydrograph has multiple peaks due to occurrence of two or more storms of different intensities in close interval.

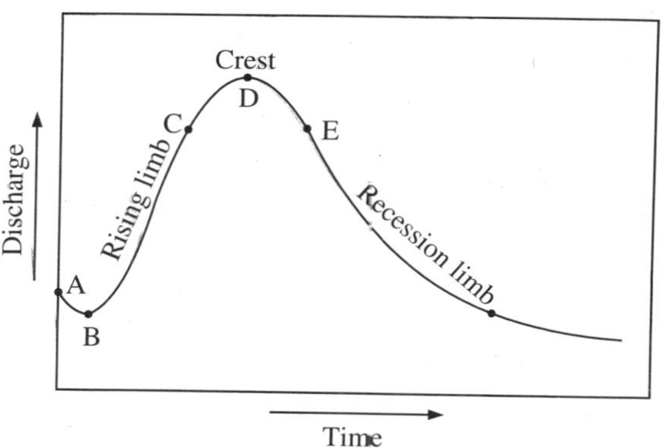

Figure 6.1 A typical hydrograph.

3. Falling limb: The descending portion of the hydrograph is called *falling limb*. This is also known as *recession limb*, which is extended from point of inflection E to the start of ground-water flow.

The total runoff hydrograph is a sum of baseflow and direct runoff (surface runoff). The direct runoff is the summation of all other contributions to runoff, except groundwater.

6.3 Recession Constants

Recession curves are used to separate a total runoff hydrograph's direct surface runoff and baseflow components. The recession curve shown in Figure 6.1 has been found to be an exponential function. Barnes (1940) proposed the following formula for determining the recession storage constant:

$$Q_t = Q_0 K_r^t \tag{6.1}$$

where, Q_t is discharge at any time interval t, Q_0 is initial discharge, and K_r is recession constant ($K_r < 1$).

Equation (6.1) can be expressed in different form as

$$Q_t = Q_0 e^{-at} \tag{6.2}$$

where $a = -\ln K_r$.

A plot of Eq. (6.2) results in straight line on semi-logarithmic paper with t-plotted on linear scale and Q plotted on logarithmic scale. The quantity of water storage S_t at time t can be obtained as

$$S_t = \int_t^{\infty} Q_t \, dt = \frac{Q_t}{a}$$

$$S = \frac{Q_t}{-\ln K_r}$$

The recession constant K_r is a product of three types of basin storage constant and is expressed as

$$K_r = K_{rs} K_{ri} K_{rb} \tag{6.3}$$

where, K_{rs} is recession constant for surface storage (0.05 to 0.20), K_{ri} is recession constant for interflow (0.5 to 0.85), and K_{rb} is recession constant for baseflow (0.85 to 0.99).

When contribution of interflow is not significant, then K_{ri} may be assumed as unity.

6.4 Baseflow Separation

The idea of hydrograph separation is to distinguish between surface runoff and baseflow so that the amount of water resulting from a storm can be calculated. For many engineering applications, it is necessary to separate total runoff hydrograph into surface runoff hydrograph [or direct runoff hydrograph (DRH)] and baseflow. The direct runoff hydrograph is derived by separating the baseflow from runoff hydrograph. Some of the baseflow separation methods are given below:

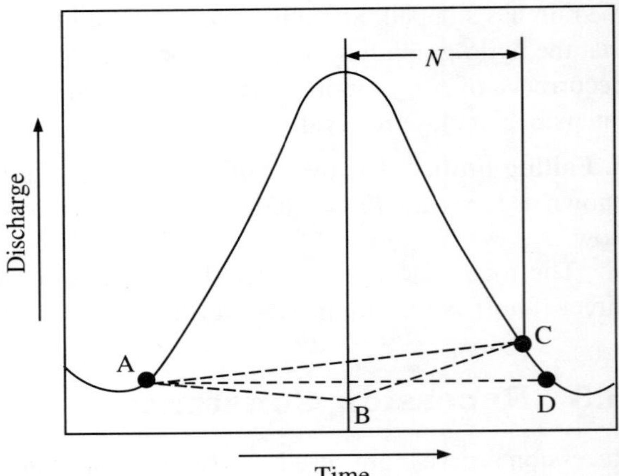

Figure 6.2 Methods of baseflow separation.

1. Straight line method: In this method, a straight line AD is drawn by joining the starting and end points of the surface runoff, as shown in Figure 6.2. The area of hydrograph above the line represents the surface runoff and below the line represents the baseflow. This method is simple and easy to use, but it is approximate.

2. Fixed base method: In this method,it is assumed that the baseflow decreases while the streamflow increases and the surface runoff ends at fixed time N after peak. An empirical equation for the time interval (in days) from peak to point C is given as

$$N = 0.827A^{0.2} \tag{6.4}$$

where A is drainage area (km^2).

A line AB, as shown in Figure 6.2, is drawn from baseflow recession to a point directly below the hydrograph peak, and then, another line BC is drawn, connecting a point N time periods after the peak. This is most widely used method for separating the baseflow.

3. Constant slope method: In this method, a line is drawn connecting the beginning of hydrograph to the inflection point on the receding limb of hydrograph. Line AC represents the baseflow separation line.

6.5 Hydrograph Time Relationships and Effective Rainfall

It is necessary to understand how the runoff is distributed over time in a catchment, i.e., timing of storm, the time required to peak flow, etc. The *travel time* for any point in catchment is defined as the time required for a particle of water to travel in a channel from that point to the outlet of the catchment.

Time of concentration T_c is defined as the time required for a particle of water to travel from the most hydraulically remote point in the watershed to the point of collection (outlet). The *hydraulically most distant point* is the point with the longest travel time to the outlet. Time of concentration depends on the flow path, slope and character of the catchment.

Effective rainfall or rainfall excess is a part of rainfall that becomes direct runoff at the outlet of the catchment. This is also known as *excess rainfall*. This part of rainfall is neither retained on the land surface nor infiltrated into the soil, as shown in Figure 6.3.

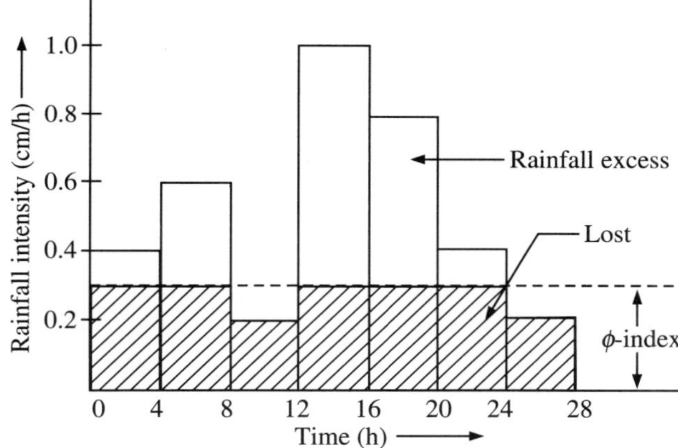

Figure 6.3 Rainfall excess.

6.6 Unit Hydrograph (UH)

A *unit hydrograph* is defined as the hydrograph of direct runoff resulting from one unit (generally 1 cm) of effective rainfall (runoff after abstractions) occurring uniformly over a basin at a uniform rate for a specified duration. This was first proposed by Sherman (1932). The term *unit* has nothing to do with the duration of rain that produces the hydrograph, whether an hour or a day. Unit hydrograph is a widely used method for predicting the flood hydrograph resulting from a known storm.

6.6.1 Assumptions Made in Unit Hydrograph

The following assumptions are made for the development of unit hydrograph:

1. Time invariance: It states that direct runoff due to the rainfall excess over the watershed is time invariant. This means that unit hydrograph coordinates do not change with respect to period and time of rainfall occurrence has no effect.

2. Linear response: It states that there is a linear relationship between direct runoff and excess rainfall. The coordinates of the resulting surface runoff hydrograph are directly proportional to the excess rainfall (runoff after abstractions). This assumption is also known as *principle of superposition* or *principle of proportionality*.

3. The excess rainfall is uniformly distributed over the entire watershed.

4. The excess rainfall is uniformly distributed within its duration.

5. The duration of the direct runoff hydrograph, i.e., its time base is independent of the effective rainfall intensity and depends only on the effective rainfall duration.

6.6.2 Derivation of Unit Hydrograph

The following rules should be followed in developing a unit hydrograph from gauged watersheds:

1. Isolated storms occurring individually should be preferred relatively having uniform spatial and temporal distributions.
2. Unit hydrographs are best suited to areas less than 5000 km^3.
3. Rainfall duration should be from 1/5th to 1/3rd of the basin lag of the watershed.
4. A high rainfall excess producing storm should be preferred.
5. The multi-peaked storms are selected only when the single-peaked storms are not there for analysis.

For a given watershed, streamflow and rainfall data are obtained. The procedure to develop unit hydrograph is simple. A unit hydrograph is derived using the following essential steps from a single storm:

1. Separate the baseflow and obtain the direct runoff hydrograph (DRH).
2. Determine the total volume of direct runoff under the hydrograph and convert this volume into equivalent depth of effective rainfall (in millimetres or in inches) over the entire watershed.
3. Divide the ordinates of the direct runoff hydrograph by the volume in inches (mm).
4. Plot the time interval on x-axis and ordinate of unit hydrograph on y-axis for the basin. The plotted points are joined by smooth line.
5. Check the volume of the unit hydrograph to make sure it is 1.0 and graphically adjust ordinates as required.

EXAMPLE 6.1 The 6 h unit hydrograph of a watershed is as follows:

Time (hr)	0	6	12	18	24	30	36	42
UH Ordinates (cumec)	0	1.8	30.9	85.6	41.8	14.6	5.5	1.8

Two rainstorms of each 6 h duration occurred successively. The first rainstorm had excess rainfall of 5 cm for the first 6 h and the second rainstorm had excess rainfall of 15 cm. Compute the storm hydrograph ordinates if the baseflow is 100 cumec.

Solution By linear response additivity concept, we will get 12 h storm hydrograph ordinates. First 6 h direct runoff ordinates can be found by multiplying 6 h UH ordinates by excess runoff of 5 cm. Similarly second, 6 h DRH ordinates can be found by multiplying with 15 cm. Then, by adding these two, we will get 12 h DRH ordinates. Then, by adding baseflow to these 12 h DRH ordinates, we get 12 h storm hydrograph ordinates. A table showing these calculations is given below:

Time (h) (Col. 1)	*UH* (cumec) (Col. 2)	*First* 6 h *DRH ordinates* (Col. 3) = (Col. 2) × 5	*Second* 6 h *DRH ordinates* (Col. 4) = (Col. 2) × 15	12 h *DRH ordinates* (Col. 4 + Col. 5) Col. 5 = (Col. 3 + Col. 4)	*Baseflow* (cumec) Col.6	12 h *storm hydrograph ordinates* (Col. 5 + Col. 6)
0	0	0		0	100	100
6	1.8	9	0	9	100	109
12	30.9	154.5	27	181.5	100	281.5
18	85.6	428	463.5	891.5	100	991.5
24	41.8	209	1284	1493	100	1593
30	14.6	73	627	700	100	800
36	5.5	27.5	219	246.5	100	346.5
42	1.8	9	82.5	91.5	100	191.5
48	0	0	27	27	100	127

EXAMPLE 6.2 Compute the size of catchment area for given discharge value for 30-min UH.

Time (min)	0	15	30	45	60	75	90
Discharge (m^3/s)	0	1.5	4	6.5	5	2.5	0

Solution

$$\text{Total discharge} = 19.5 \text{ m}^3/\text{s}$$

$$\Delta T = 15 \text{ min} = 15 \times 60 \text{ s} = 900 \text{ s}$$

$$Q = 19.5 \text{ m}^3/\text{s}$$

$$\text{Volume of direct runoff} = Q \times \Delta T = 19.5 \times 900 = 17550 \text{ m}^3$$

$$P_{\text{eff}} \times \text{Area} = \text{Volume of direct runoff}$$

or

$$A = \frac{17550}{1} = 17550 \text{ m}^2$$

Thus, area of the catchment is 17550 m^2.

6.6.3 Derivation of Unit Hydrograph for Different Durations

Superposition method

This method is based on the assumption that linear response of the catchment is not influenced by previous storms, i.e., one can superimpose hydrographs offset in time and flow will be directly additive. If a T-h unit hydrograph is available for the given watershed, and it is required to develop a unit hydrograph of *n*D-h, where *n* is an integer and not a fraction, then the method of superposition is used, as shown in Figure 6.4. The following steps are followed:

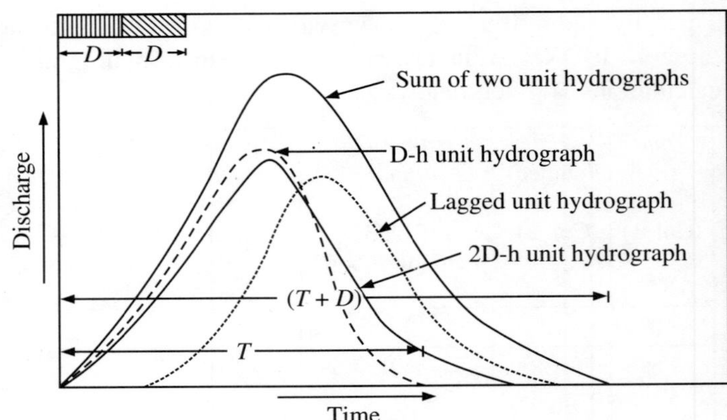

Figure 6.4 Conversion of D-h UH into 2D-UH.

1. Take the given T-h unit hydrograph and determine the *n*.
2. Superimpose *n* unit hydrographs, with each graph separated from the previous one by T-h, and add the ordinates of *n* unit hydrograph.
3. The obtained ordinates are the ordinates of DRH for ($n \times 1$) cm (*n* cm) rainfall excess of *n*T-h duration.
4. Calculate the ordinates of unit hydrograph of *n*D-h duration by dividing the ordinates of DRH of *n*T-h by *n*.

EXAMPLE 6.3 Given the ordinates of a 2 h UH as below. Derive the ordinate of a 6 h hydrograph for the same catchment.

Time (h)	*Ordinate of* 2 h UH
0	0
2	3
4	5
6	8
8	14
10	19
12	26
14	34
16	20
18	8
20	6
22	0

Solution The calculations are shown in the table below:

Time (h)	*Ordinate of* 2 h UH	*Ordinate of* 2 h UH *lagged by* 2 h	*Ordinate of* 2 h UH *lagged by* 4 h	*Ordinates of* 6 h *Hydrograph*
0	0			0
2	3	0		3
4	5	3	0	8
6	8	5	3	16
8	14	8	5	27
10	19	14	8	41
12	26	19	14	59
14	34	26	19	79
16	20	34	26	80
18	8	20	34	62
20	6	8	20	34
22	0	6	8	14
24		0	6	6
26			0	0

The following figure shows 2 h and 6 h UH.

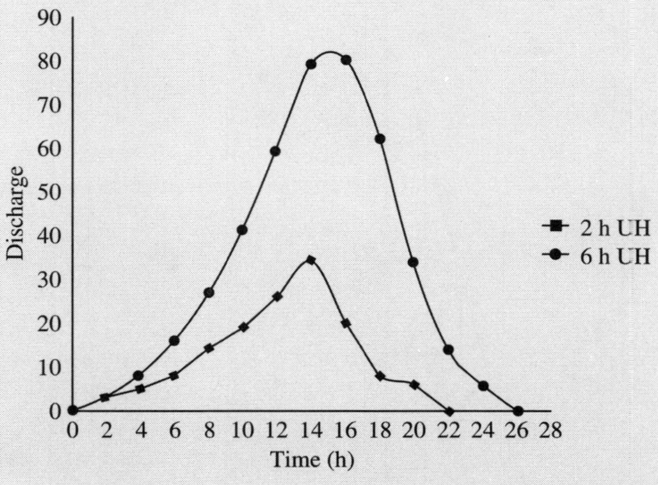

S-hydrograph

The limitation of superposition method is that if it is desired to develop a unit hydrograph of duration mT, where m is a fraction, the method of superposition cannot be used. In such cases, S-curve method is used.

The *S-curve* or *S-hydrograph* is a hydrograph produced by a continuous effective rainfall occurring at a constant rate for an infinite period, as shown in Figure 6.5. It is derived by summation of an infinite series of T-h unit hydrograph at a spacing of T-h duration. Thus, the ordinate of an S-curve at any time is the summation of the ordinates of unit hydrograph at lag T-h.

The S-curve has an initial steep portion, and reaches a maximum equilibrium discharge at a time equal to the base time of the first unit hydrograph. The average intensity of effective rainfall of S-curve in centimetre per hour is given by

$$I_{avg} = \frac{1}{D} \tag{6.5}$$

The equilibrium discharge of S-curve in cubic metre per hour can be determined as

$$Q_s = 2.778\frac{A}{D} \tag{6.6}$$

where, A is area of watershed (km^2) and D is time of unit rainfall excess (h).

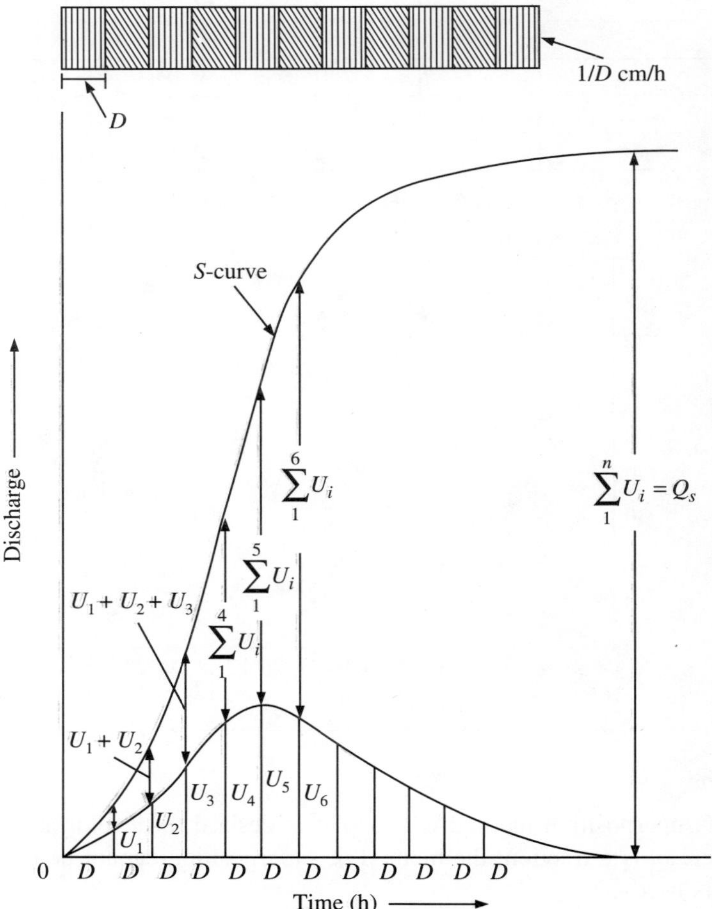

Figure 6.5 S-curve hydrograph.

EXAMPLE 6.4 Derive the S-curve for 6 h UH as follows:

Time (h)	0	6	12	18	24	30	36	42	48	54	60	66
Ordinate of 6 h UH	0	15	25	40	48	62	78	34	18	8	4	0

Solution At 6 h, ordinate of 6 h UH is given as 15.

S-curve addition = Ordinate of 6 h UH at t = (6 – 6) = 0 h = 15 + 0 = 15

At 12 h, Ordinate of 6 h UH is given as 25.

S-curve addition = Ordinate of 6 h UH at t = (12 – 6) = 0 h = 25 + 15 = 40

This calculation may be repeated for all other time intervals, as shown in the following table:

Time (h)	*Ordinate of* 6 h UH	*S-curve addition*	*S-curve ordinates*
0	0		0
6	15	0	15
12	25	15	40
18	40	40	80
24	48	80	128
30	62	128	190
36	78	190	268
42	34	268	302
48	18	302	320
54	8	320	328
60	4	328	332
66	0	332	332

The following figure shows S-curve.

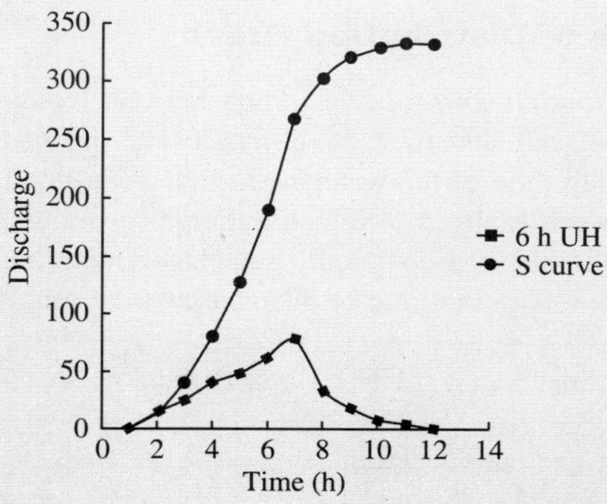

6.6.4 Limitations of Unit Hydrograph Theory

The limitations of unit hydrograph theory are as follows:

1. The basic assumptions of uniform rainfall intensity and uniform distribution of rainfall over the catchment are never strictly satisfied.
2. The unit hydrograph theory is applicable for catchments less than 5000 km^2.
3. Precipitation must be from rainfall only. Snowmelt runoff or other form of precipitation cannot be satisfactorily represented by unit hydrograph.
4. The watershed should not have large storage, which affects the linear relationship assumption between storage and discharge.

6.6.5 Uses of Unit Hydrograph Theory

The uses of unit hydrograph theory are as follows:

1. Development of direct runoff hydrograph (DRH) and unit hydrograph (UH) of different durations
2. Development of flood forecasting and warning systems based on rainfall in the watershed
3. Extension of flood flow records based on rainfall records

6.7 Distribution Graph

Bernard (1935) developed a special type of unit hydrograph where ordinates of a T-h UH are shown as the percentage of the surface runoff occurring in successive periods of equal time intervals of T h. It is a dimensionless representation of a unit hydrograph in the form of bar chart, as shown in Figure 6.6. The total area under the distribution graph adds up to 100%. The distribution graphs are free from the effects of size of catchment, and therefore, are useful for comparing the runoff characteristics of different catchments.

6.7.1 S-hydrograph of Distribution Graph

An *S-hydrograph of distribution graph* is the graph between cumulative percentage of total runoff volume, on ordinate scale and time, on abscissa. The use of this type of hydrograph is to determine basin lag or lag time of the watershed for a given flood. The *basin lag* is defined as the time difference between centroid of effective rainfall hyetograph (ERH) and hydrograph of surface runoff. From the summation graph, time lag is determined as the time from the beginning of the surface runoff to the time of 50% cumulative runoff.

EXAMPLE 6.5 The ordinates of 1 h UH for a catchment of area 940 km^2 are given below:

Time (h)	1	2	3	4	5	6	7	8	9	10	11	12
UH ordinates (m^3/s)	2.1	5.8	9.7	13.7	18.7	21.1	29.1	22.8	15.7	6.9	1.7	0.3

(a) Determine the S-curve hydrograph for the excess rainfall of unit intensity (1 mm/h).
(b) Compute the ordinates of 2 h UH (1 mm) using S-curve and superposition methods.

Solution (a) The S-curve hydrograph is shown below:

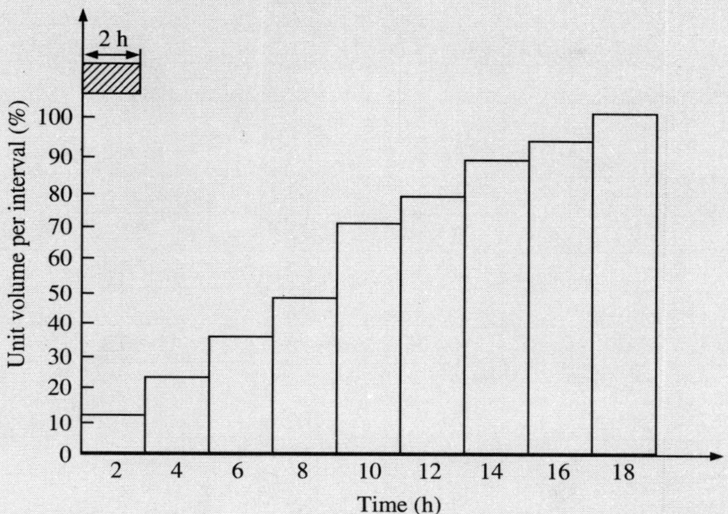

The following table shows computation of S-curve ordinate.

Time (h)	*UH ordinates* (m^3/s)	*S-curve addition*	*S-curve ordinate*
0	0		0
1	2.1	0	2.1
2	5.8	2.1	7.9
3	9.7	7.9	17.6
4	13.7	17.6	31.3
5	18.7	31.3	50
6	21.1	50	71.1
7	29.1	71.1	100.2
8	22.8	100.2	123
9	15.7	123	138.7
10	6.9	138.7	145.6
11	1.7	145.6	147.3
12	0.3	147.3	147.6
13	0	147.6	147.6

Computation of ordinates of 2 h UH (1 mm) using S-curve.

Time (h)	*UH ordinates* (m^3/s)	*S-curve addition*	*S-curve ordinate*	*S-curve lagged by* 2 h (m^3/s)	Col. 4 – Col. 5	Col. 6/(2) = 2 h *UH ordinates* (m^3/s)
Col. 1	Col. 2	Col. 3	Col. 4	Col. 5	Col. 6	
0	0		0		0	0
1	2.1	0	2.1		2.1	1.05
2	5.8	2.1	7.9	0	7.9	3.95
3	9.7	7.9	17.6	2.1	15.5	7.75
4	13.7	17.6	31.3	7.9	23.4	11.7
5	18.7	31.3	50	17.6	32.4	16.2
6	21.1	50	71.1	31.3	39.8	19.9
7	29.1	71.1	100.2	50	50.2	25.1
8	22.8	100.2	123	71.1	51.9	25.95
9	15.7	123	138.7	100.2	38.5	19.25
10	6.9	138.7	145.6	123	22.6	11.3
11	1.7	145.6	147.3	138.7	8.6	4.3
12	0.3	147.3	147.6	145,6	2	1
13	0	147.6	147.6	147.3	0.3	0.15
14			147.6	147.6	0	0

Computation of ordinates of 2 h UH (1 mm) using superposition method.

Time (h)	*UH ordinates* (m^3/s)	*Ordinates of* 1 h *UH* (m^3/s) *lagged by* 1 h	*lagged by* 2 h	*DRH of* 2 mm *in* 2 h (m^3/s) Col. 1 + Col. 2 + Col. 3	*Ordinate of* 2 h *UH* (m^3/s) Col. 5/2
Col. 1	Col. 2	Col. 3	Col. 4	Col. 5	Col. 6
0	0			0	0
1	2.1	0		2.1	1.05
2	5.8	2.1	0	7.9	3.95
3	9.7	5.8	2.1	17.6	8.8
4	13.7	9.7	5.8	29.2	14.6
5	18.7	13.7	9.7	42.1	21.05
6	21.1	18.7	13.7	53.5	26.75
7	29.1	21.1	18.7	68.9	34.45
8	22.8	29.1	21.1	73	36.5
9	15.7	22.8	29.1	67.6	33.8
10	6.9	15.7	22.8	45.4	22.7

11	1.7	6.9	15.7	24.3	12.15
12	0.3	1.7	6.9	8.9	4.45
13	0	0.3	1.7	2	1
14		0	0.3	0.3	0.15
15			0	0	0

EXAMPLE 6.6 The S-curve ordinates of intensity 1 cm/h derived from 4 h, 1 cm UH are given below. Derive a UH of 2 h and 1 cm depth.

Time (h)	0	2	4	6	8	10	12	14	16	18	20
S-hydrograph ordinates (m^3/s)	0	100	350	800	1230	1650	1830	2150	2650	2850	3000

Solution The step-wise procedure is as follows:

(a) The S-curve ordinate of 4 h UH is given in a time interval of 2 h. From this, S-hydrograph lagged by 2 h is prepared by the following formula and displayed in Col. 3.

$$g'(t) = g(t - \Delta t')$$

(b) Finally, the 2 h unit hydrograph is obtained by

$$h'(t) = \frac{1}{\Delta t'}[g(t) - g(t - \Delta t')]$$

i.e., $$\text{Col. 4} = \frac{1}{2} \times (\text{Col. 2} - \text{Col. 3})$$

Time (h)	*S-hydrograph* (m^3)	*S-hydrograph lagged by* 2 h (m^3)	2 h *unit hydrograph* (m^3/s)
t	$g(t)$	$g(t - \Delta t)$	$h'(t)$
Col. 1	Col. 2	Col. 3	Col. 4
0	0	0	0
2	100	0	50
4	350	100	125
6	800	350	225
8	1230	800	215
10	1650	1230	210
12	1830	1650	90
14	2150	1830	160
16	2650	2150	250
18	2850	2650	100
20	3000	2850	75

6.8 Synthetic Unit Hydrograph (SUH)

Synthetic unit hydrograph was first proposed by Snyder (1938) in easter part of the USA. Synthetic hydrographs typically represent basins without adequate stream gauge information and are developed from the information we have from numerous instrumented basins. The methodology consists of following sets of equations for computing different components of the SUH based on watershed characteristics. It allows the computation of parameters such as lag time, time base, duration of unit hydrograph, peak discharge, and hydrograph time widths of 50% and 75% of peak flow. These seven parameters are discussed below:

1. Basin lag (t_p): It is the time interval (in hours) from the mid-point of the unit rainfall excess to the peak of unit hydrograph, as shown in Figure 6.6. It can be computed as

$$t_p = C_t(LL_{ca})^{0.30} \tag{6.7}$$

where C_t is regional constant, L is length (km) of the main watercourse from the basin divide to the outlet and L_{ca} is length (km) of the watercourse from the outlet to the point on the stream opposite to the centroid of watershed. The regional constant is based on the watershed and storage.

The values of C_t given by various researchers are as follows:

(a) As per Snyder: 1.35 to 1.65
(b) As per Sokolov: 0.30 to 6.0
(c) 0.65 (Himalayan catchment of Ramganga river)

It is assumed that lag time is constant for a given watershed.

2. Standard duration of effective rainfall (t_r): It is used to determine the peak flows and function of lag time. Snyder proposed the following formula for computation:

$$t_r = \frac{t_p}{5.5} \tag{6.8}$$

3. Peak discharge (Q_{ps}): To computer peak discharge, Snyder proposed the following formula:

$$Q_{ps} = \frac{2.78C_pA}{t_p} \tag{6.9}$$

where, A is area of watershed km^2 and C_p is a regional that depends on retention and storage capacities of the watershed.

The value of C_p given by various researchers are as follows:

(a) As per Snyder: 0.56 to 0.69
(b) As per other researchers: 0.31 to 0.93
(c) 0.94 (Himalayan catchment of Ramganga river)

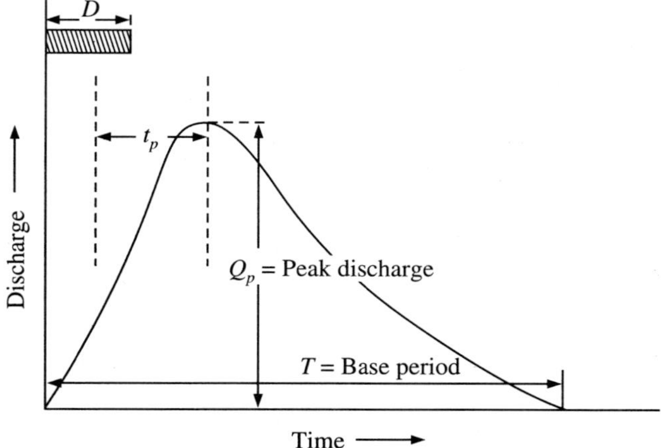

Figure 6.6 Synthetic unit hydrograph.

4. Other durations of effective rainfall: When instead of the standard duration of effective rainfall (t_r), non-standard duration of effective rainfall is considered, the it can be computed by using the following formula:

$$t_{pr} = t_p + \frac{(t_R - t_r)}{4} = \frac{21}{22}t_p + \frac{t_R}{4} \tag{6.10}$$

where, t_m is the modified basin lag for non-standard duration of effective rainfall and t_R is the non-standard duration of effective rainfall.

Accordingly, the modified formula for peak discharge is given below:

$$Q_{PR} = \frac{2.78C_pA}{t_{pr}} \tag{6.11}$$

where Q_{PR} is the modified peak discharge for time duration t_R.

5. Time base (T_b): Snyder suggested the following formula for time base T_b, which represents the duration of direct runoff.

$$T_b = 3 + \frac{t_{pr}}{8} \text{ days} \tag{6.12}$$

$$T_b = 72 + 3t_{pr} \text{ hours} \tag{6.13}$$

Taylor and Schwartz (1957) suggested the following formula:

$$T_b = 5\left(t_{pr} + \frac{t_R}{2}\right) \tag{6.14}$$

Mutreja (1986) suggested the following formula for small Indian watersheds:

$$T_b = t_{\text{peak}} \times (3 \text{ to } 5) \tag{6.15}$$

6. Unit hydrograph shape: The US Corps of Engineer developed the following set of formula for determining the widths of the unit hydrograph at 50% and 75% of the peak flow rate:

(a) Width of unit hydrograph at 50% peak flow rate (W_{50})

$$W_{50} = \frac{5.87}{q^{1.08}} \tag{6.16}$$

where q is peak discharge per unit watershed area m^3/3/km^2.

(b) Width of unit hydrograph at 75% peak flow rate (W_{75})

$$W_{75} = \frac{W_{50}}{1.75} \tag{6.17}$$

7. Time to peak: This is given as follows:

$$t_{\text{peak}} = t_{pr} + \frac{t_R}{2} \tag{6.18}$$

EXAMPLE 6.7 The Snyder's parameters for a catchment are given as $C_p = 0.62$, $C_t = 1.6$, distance from the outlet to a point on the stream, nearest to centroid of the catchment = 15 km and catchment area = 980 km^2. Using Snyder's method, develop a 2 h UH of the catchment.

Solution Given,

$C_p = 0.62$, $C_t = 1.6$, $L_{ca} = 15$ km, $A = 980$ km^2, $t_R = 2$ and $L = 30$ km/assumed

Now, basin lag, $t_p = C_t(L.L_{ca})^{0.3} = 10$ h

Duration of effective rainfall,

$$t_r = \frac{t_p}{5.5} = 1.82 \text{ h}$$

Using $t_R = 2$ h, i.e., for a 2 h unit hydrograph,

$$t_{pr} = t_p\left(\frac{21}{22}\right) + \frac{t_R}{4}$$

$$t_{pr} = 10\left(\frac{21}{22}\right) + \frac{2}{4} = 10.05 \text{ h}$$

Peak discharge, $$Q_p = \frac{(2.78 \times C_p \times A)}{t_p} = 168.07 \text{ m}^3/\text{s}$$

Width at 50% peak discharge, $$W_{50} = \frac{5.87}{(Q_P/A)^{1.08}} = 39.41 \text{ h}$$

Width at 75% peak discharge, $$W_{75} = \frac{W_{50}}{1.75} = 22.52 \text{ h}$$

Time base, $$T_b = 5\left(t_{pr} + \frac{t_R}{2}\right) = 55.25 \text{ h}$$

(after Taylor Schwartz)

The following figure shows 2 h synthetic unit hydrograph.

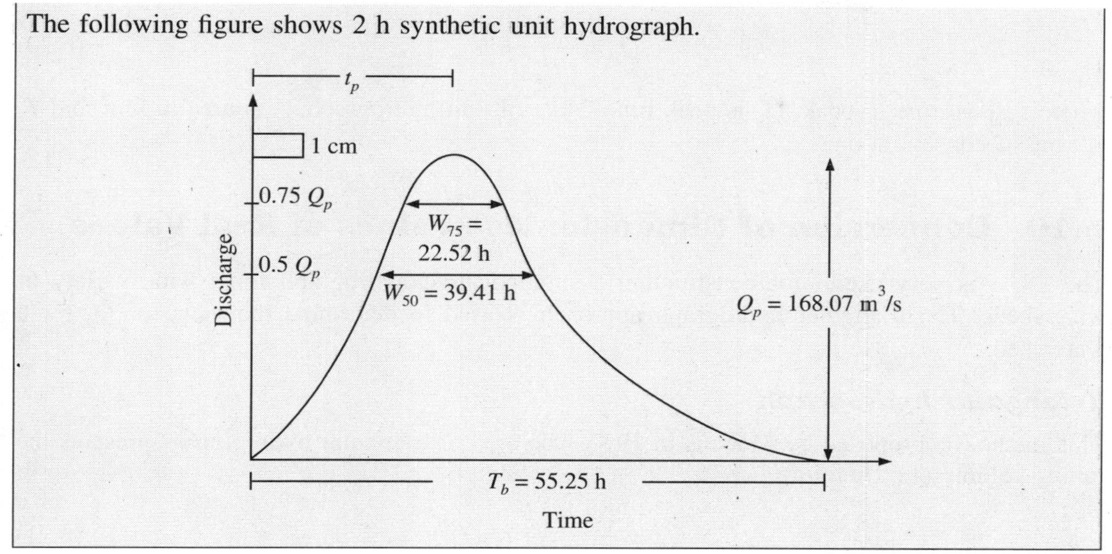

6.9 Dimensionless Unit Hydrograph

The US Soil Conservation Service (US SCS) developed a dimensionless unit hydrograph (DUH) based on the analysis of a large number of watersheds. The ordinate of this unit hydrograph is expressed as the ratio of discharge to peak discharge $\left(\frac{Q}{Q_p}\right)$ and abscissa in the ratio of time (t) to time to peak $\left(\frac{t}{t_{\text{peak}}}\right)$, as shown in Figure 6.7. The following equations are proposed by

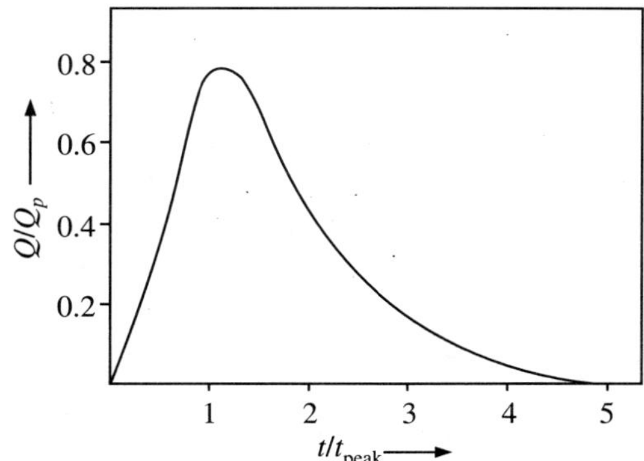

Figure 6.7 SCS dimensionless unit hydrograph.

US SCS for the development of unit hydrograph by using dimensionless unit hydrograph. This method requires only determination of the time to peak and the peak runoff as follows:

$$t_r = 1.33t_c \tag{6.19}$$

$$Q_p = \frac{5.36A}{t_{peak}} \tag{6.20}$$

where t_{peak} is time to peak, Q_p is peak runoff rate of unit hydrograph, A is area in km^2 and T_c is time of concentration

6.10 Conversion of Dimensionless Values of Real Values

The DUH is very useful for constructing a synthetic unit hydrograph for a wide variety of watersheds. The triangular hydrograph approach is used to determine the value of Q_p for a watershed.

Triangular hydrograph

This method is proposed by Mockus in 1957. The area of triangular hydrograph represents the runoff volume Q_v. Therefore,

$$Q_v = \frac{1}{2}(t_{peak} + t_b)Q_p \tag{6.21}$$

$$\therefore \quad Q_p = \frac{2Q_v}{(t_{peak} + t_b)} \tag{6.22}$$

The value of t_{peak} from dimensionless unit hydrograph is given by

$$t_{peak} = 0.7t_c = \frac{\text{Effective rainfall duration}}{2} + 0.6t_c \tag{6.23}$$

$$t_b = 1.67t_{peak} \tag{6.24}$$

$$t_b = 1.67 \times 0.7t_c = 1.169t_c \tag{6.25}$$

$$\therefore \quad Q_p = \frac{2Q_v}{(t_{peak} + t_b)} \tag{6.26}$$

$$Q_p = \frac{2Q_v}{(0.7t_c + 1.169t_c)} \tag{6.27}$$

$$Q_p = \frac{2Q_v}{1.869t_c} \tag{6.28}$$

$$\therefore \quad Q_p = \frac{0.75Q_v}{t_{peak}} \tag{6.29}$$

If area is in square kilometres and discharge is in cubic metres, then

$$Q_p = \frac{0.75 \times 0.2778AQ_v}{t_{peak}} \tag{6.30}$$

$$Q_p = 0.208\frac{AQ_v}{t_{peak}} \tag{6.31}$$

6.11 Instantaneous Unit Hydrograph (IUH)

Instantaneous unit hydrograph is a fictitious, conceptual unit hydrograph, resulting from an instantaneous unit precipitation of infinitesimally small duration, uniformly distributed over the entire watershed. In other words, a unit hydrograph of time of precipitation of $t - 0$ is an IUH. The advantage of the IUH over unit hydrograph (Sherman, 1932) is that the IUH is independent of duration of effective rainfall and the restriction on uniform distribution of rainfall. It is also useful for theoretical analysis of excess rainfall-runoff relationship of watershed.

6.11.1 Properties of IUH

The properties of IUH are as follows:

1. $0 \leq u(t) \leq$ some positive peak value
2. $u(t) = 0$ for $t \leq 0$
3. $u(t) \rightarrow 0$ for $t \leq \infty$
4. $\int_0^\infty u(t)\, du = 1$ (unit depth over catchment)
5. $\int_0^\infty u(t)\, dt$ = lag time of IUH

SUMMARY

- Hydrograph is a plot of discharge and time.
- The essential components of hydrograph are rising limb, crest segment and recession limb.
- Area of hydrograph represents the total rainfall volume due to a given storm.
- Time of concentration is the longest time taken by a drop of water to reach at the outlet from the farthest point of the catchment.
- The direct runoff hydrograph (DRH) is obtained by separating the baseflow from hydrograph.
- A unit hydrograph is the direct runoff hydrograph resulting from a unit depth of rainfall excess generated uniformly over the basin for a specified duration (D-hours).
- A unit hydrograph relates the direct runoff and the rainfall excess.
- The unit hydrograph theory is applicable for catchment area varying from 2 km^2 to 5000 km^2.
- The unit hydrograph theory can be used for computation of flood hydrograph for design of a structure, extension of flow records at site and flow forecasting models.
- The S-hydrograph is used to determine basin lag or lag time of the watershed for a given flood.
- The synthetic unit hydrograph is developed for ungauged basins.

- Instantaneous unit hydrograph (IUH) is a fictitious, conceptual unit hydrograph, resulting from an instantaneous unit precipitation of infinitesimally small duration, uniformly distributed over the entire watershed.
- The IUH is a single peak unit hydrograph with a finite base width.

Objective Type Questions

Common data for questions 1 and 2

An average rainfall of 16 cm occurs over a catchment during a period of 12 h with uniform intensity. The unit hydrograph (unit depth = 1 cm, duration = 6 h) of the catchment rises linearly from 0 to 30 cumecs in 6 h and then falls linearly from 30 to 0 cumecs in the next 12 h. ϕ-index of the catchment is known to be 0.5 cm/h. Baseflow in the river is known to be 5 cumecs.

1. Peak discharge of the resulting direct-runoff hydrograph shall be (GATE, 2003)
(a) 150 cumec (b) 225 cumec (c) 230 cumec (d) 360 cumec

2. Area of the catchment in hectares is (GATE, 2003)
(a) 97.20 (b) 270 (c) 9720 (d) 2700

3. The average rainfall for a 3 h duration storm is 2.7 cm and the loss rate is 0.3 cm/h. The flood hydrograph has a baseflow of 20 m^3/s and produces a peak flow of 210 m^3/s. The peak of a 3 h unit hydrograph is (GATE, 2004)
(a) 125.50 m^3/s (b) 105.50 m^3/s (c) 77.77 m^3/s (d) 70.37 m^3/s

4. When the outflow from a storage reservoir is uncontrolled as in a freely operating spillway, the peak of outflow hydrograph occurs at (GATE, 2005)
(a) A point of intersection of the inflow and outflow hydrographs
(b) A point after the intersection of the inflow and outflow hydrographs
(c) The tail of inflow hydrographs
(d) A point before the intersection of the inflow and outflow hydrographs

Statement for linked answer questions 5 and 6

A 4 h unit hydrograph of a catchment is triangular in shape with base of 80 h. The area of the catchment is 720 km^2. The baseflow and ϕ-index are 30 m^3/s and 1 mm/h, respectively. A storm of 4 cm occurs uniformly in 4 h over the catchment.

5. The peak discharge of 4 h unit hydrograph is (GATE, 2005)
(a) 40 m^3/s (b) 50 m^3/s (c) 60 m^3/s (d) 70 m^3/s

6. The peak flood discharge due to the storm is (GATE, 2005)
(a) 210 m^3/s (b) 230 m^3/s (c) 260 m^3/s (d) 720 m^3/s

Common data for questions 7 and 8

For a catchment, the S-curve (or S-hydrograph) due to a rainfall of intensity 1 cm/h is given by $Q = 1 - (1 + t) \exp(-t)$ (t in hour and Q in cubic metre per second)

7. What is the area of the catchment? (GATE, 2006)
(a) 0.01 km^2 (b) 0.36 km^2 (c) 1.00 km^2 (d) 1.28 km^2

8. What will be the ordinate of a 2 h unit hydrograph for this catchment at $t = 3$ h? (GATE, 2006)
(a) 0.13 m^3/s (b) 0.20 m^3/s (c) 0.27 m^3/s (d) 0.54 m^3/s

Common data for questions 9 and 10

Ordinates of a 1 h unit hydrograph at 1 h intervals, starting from time $t = 0$, are 0, 2, 6, 4, 2, 1 and 0 m^3/s.

9. Catchment area represented by this unit hydrograph is (GATE, 2007)
(a) 1.0 km^2 (b) 2.0 km^2 (c) 3.2 km^2 (d) 5.4 km^2

10. Ordinate of a 3 h unit hydrograph for the catchment at $t = 3$ h is (GATE, 2007)
(a) 2.0 m^3/s (b) 3.0 m^3/s (c) 4.0 m^3/s (d) 5.0 m^3/s

Common data for questions 11 and 12

1 h triangular unit hydrograph of a watershed has the peak discharge of 60 m^3/s cm at 10 h and time base of 30 h. The ϕ-index is 0.4 cm/h and base flow is 15 m^3/s.

11. The catchment area of the watershed is (GATE, 2009)
(a) 3.24 km^2 (b) 32.4 km^2 (c) 324 km^2 (d) 3240 km^2

12. If these are rainfall of 5.4 cm in 1 h, the ordinate of the flood hydrograph at 15th hour is (GATE, 2009)
(a) 225 m^3/s (b) 240 m^3/s (c) 249 m^3/s (d) 258 m^3/s

13. A watershed got transformed from rural to urban over a period of time. The effect of urbanisation on storm runoff hydrograph from the watershed is to (GATE, 2011)
(a) Decrease the volume of runoff
(b) Increase the time to peak discharge
(c) Decrease the time base
(d) Decrease the peak discharge

Common data for Questions 14 and 15

The ordinates of a 2 h unit hydrograph at 1 h interval, starting from time $t = 0$, are 0, 3, 8, 6, 3, 2 and 0 m^3/s. Use trapezoidal rule for numerical integration, if required.

14. What is the catchment area represented by the unit hydrograph? (GATE, 2011)
(a) 1.00 km^2 (b) 2.00 km^2 (c) 7.92 km^2 (d) 8.64 km^2

15. A storm of 6.6 cm occurs uniformly over the catchment in 3 h. If ϕ-index is equal to 2 mm/h and baseflow is 5 m^3/s, what is the peak flow due to the storm? (GATE, 2011)
(a) 41.0 m^3/s (b) 43.4 m^3/s (c) 53.0 m^3/s (d) 56.2 m^3/s

Statement for linked answer questions 16 and 17

The drainage area of a watershed is 50 km^2. The ϕ-index is 0.5 cm/h and the baseflow at the outlet is 10 m^3/s. 1 h unit hydrograph (unit depth = 1 cm) of the watershed is triangular in shape with a time base of 15 h. The peak ordinate occurs at 5 h.

16. The peak ordinate (m^3/s/cm) of the unit hydrograph is (GATE, 2012)
(a) 10.00 (b) 18.52 (c) 37.03 (d) 185.20

17. For a storm of depth of 5.5 cm and duration of 1 h, the peak ordinate (m^3/s) of the hydrograph is (GATE, 2012)
(a) 55.00 (b) 82.60 (c) 92.60 (d) 102.60

18. Which one of the following constitute the basic assumption of unit hydrograph theory? (IES, 1995)
(a) Non-linear response and time invariance
(b) Non-linear time variance and linear response
(c) Linear response and linear time variance
(d) Time invariance and linear response

19. The following four hydrological features have to be estimated or taken as inputs before one can compute the flood hydrograph at any catchment outlet: (IES, 1996)
1. Unit hydrograph
2. Rainfall hydrograph
3. Infiltration index
4. Baseflow

The correct order in which they have to be employed in the computations is
(a) 1, 2, 3, 4
(b) 2, 1, 4, 3
(c) 2, 3, 1, 4
(d) 4, 1, 3, 2

20. The following steps are involved in arriving at a unit hydrograph: (IES, 1997)
1. Estimating the surface runoff in depth
2. Estimating the surface runoff in volume
3. Separation of baseflow
4. Dividing surface runoff ordinates by depth of runoff. The correct sequence of these steps is

(a) 3, 2, 1, 4
(b) 2, 3, 4, 1
(c) 3, 1, 2, 4
(d) 4, 3, 2, 1

21. If a 4 h unit hydrograph of a certain basin has a peak ordinate of 80 m^3/s, the peak ordinate of a 2 h unit hydrograph for the same basin will be (IES, 1999)
(a) Equal to 80 m^3/s (b) Greater than 80 m^3/s
(c) Less than 80 m^3/s (d) Between 40 m^3/s to 80 m^3/s

22. If the base period of a 6 h hydrograph of a basin is 84 h, then a 12 h unit hydrograph derived from this 6 h unit hydrograph will have a base period of (IES, 2000)
(a) 72 h (b) 78 h (c) 84 h (d) 90 h

23. Which of the following principles relate to a unit hydrograph?
1. The hydrographs of direct runoff due to effective rainfall of equal duration have the same time base.

2. Effective rainfall is not uniformly distributed within its duration.
3. Effective rainfall is uniformly distributed throughout the whole area of drainage basin.
4. Hydrograph of direct runoff from a basin due to a given period of effective rainfall reflects the combination of all the physical characteristics of the basin.

Select the correct answer using the codes given below: (IES, 2001)

(a) 1, 2 and 3 (b) 1, 2 and 4 (c) 2, 3 and 4 (d) 1, 3 and 4

24. The unit hydrograph theory is based on the assumption of (IES, 2002)
(a) Non-linear response and time invariance
(b) Linear response and non-linear time variance
(c) Time invariance and linear response
(d) Non-linear response and non-linear time variance

25. A 2 h storm hydrograph has 5 units of direct runoff. The 2 h unit hydrograph for this storm can be obtained by dividing the ordinates of the storm hydrograph by (IES, 2002)
(a) 2 (b) 2/5 (c) 5 (d) 5/2

Directions for questions 26 and 27

The following items consist of two statements, one labelled as 'Assertion (A)' and the other as 'Reason (R)'. Select the answer to these items using the codes given below:
Codes:
(a) Both A and R are true and R is the correct explanation of A
(b) Both A and R are true but R is not a correct explanation of A
(c) A is true but R is false
(d) A is false but R is true

26. **Assertion (A):** Unit hydrograph theory is not applicable to catchment areas larger than 5000 km^2.

Reason (R): Rainfall is not uniformly distributed on large catchment areas. (IES, 2002)

27. **Assertion (A):** In routing a flood hydrograph through a reservoir, the peak of the outflow hydrograph will be smaller than the inflow hydrograph and it occurs after the peak of the inflow.

Reason (R): In linear reservoir routing, the storage is a function of both outflow and inflow discharges. (IES, 2002)

28. A 6 h storm with hourly intensities of 7, 18, 25, 12, 10 and 3 mm/h produces a runoff of 33 mm. Then, the Φ-index is (IES, 2003)
(a) 7 mm/h (b) 3 mm/h (c) 10 mm/h (d) 8 mm/h

29. In Snyder's method of synthetic unit hydrograph development, basin lag is taken as (IES, 2004)
(a) The time interval between centroid of the rainfall excess and surface runoff
(b) The time interval from mid-point of the unit rainfall excess to the peak of the unit hydrograph
(c) Independent of rainfall duration
(d) Independent of catchment characteristics

30. Match List-I with List-II and select the correct answer using the codes given below the lists. (IES, 2004)

List-1

A. Rising limb of a hydrograph
B. Falling limb of a hydrograph
C. Peak rate of flow
D. Drainage density

List-II

1. Depends on intensity of rainfall
2. Function of total channel length
3. Function of catchment slope
4. Function of storage characteristics

Codes

	A	B	C	D
(a)	3	4	1	2
(b)	1	4	3	2
(c)	3	2	1	4
(d)	1	2	3	4

31. Match List-I with List-II and select the correct answer using the codes given below the lists: (IES, 2004)

List-I

A. Unit Hydrograph
B. Synthetic unit Hydrograph
C. Darcy's law
D. Rational method

List-II

1. Design flood
2. Permeability
3. Ungauged basin
4. 1 cm runoff

Codes

	A	B	C	D
(a)	2	3	4	1
(b)	2	1	4	3
(c)	4	3	2	1
(d)	4	1	2	3

32. Viewing watershed as a system, which one of the following assumptions is made in the unit hydrograph theory? (IES, 2005)

(a) Non-linearity
(b) Both linearity and time variance
(c) Both time invariance and non-linearity
(d) Both linearity and time invariance

33. A DRH due to a storm over a basin has a time base of 90 h with straight line portions of the hydrograph having flow rates of 0, 10, 70, 90, 40 and 0 m^3/s at elapsed durations of 0, 10, 20, 30, 50 and 90 h as indicated is the following diagram, respectively. The catchment area is 300 km^2. What is the rainfall excess in the storm? (IES, 2006)

(a) 2.83 cm (b) 3.46 cm (c) 3.87 cm (d) 4.02 cm

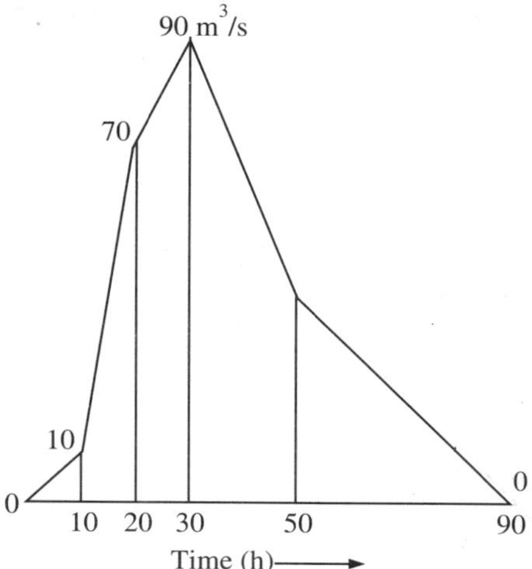

34. A 252 km^2 catchment area has a 6 h UH, which is a triangle with time base of 35 h. What is the peak discharge of the DRH due to 5 cm effective rainfall in 6 h from that catchment? (IES, 2006)

(a) 45 cumec (b) 115 cumec (c) 200 cumec (d) 256 cumec

35. A 4 h unit hydrograph of a basin can be approximated as a triangle with a base period of 48 h and peak ordinate of 300 m^3/s. What is the area of the catchment basin? (IES, 2006)

(a) 7776 km^2 (b) 5184 km^2 (c) 2592 km^2 (d) 1294 km^2

36. Consider the following statements:

Assertion (A): The unit hydrograph cannot be applied for areas less than 5000 km^2.

Reason (R): The runoff hydrograph reflects the physiographic factors of a catchment. Of these statements (IES, 2006)

(a) both A and R are true and R is the correct explanation of A
(b) both A and R are true but R is not a correct explanation of A
(c) A is true but R is false
(d) A is false but R is true

37. In constructing a 4 h synthetic unit hygrograph for a basin, the lag time is estimated to be 40 h. When will the peak discharge in the synthetic unit hydrograph occur from the start of the storm? (IES, 2007)

(a) 36 h (b) 40 h (c) 42 h (d) 44 h

38. Consider the following statements:

1. Only the surface flow constitutes the flood hydrograph due to an isolated storm.
2. For a given storm, the flood peak is dependent on the drainage density.
3. Fan-shaped catchments give narrow hydrograph with low peak.

Which of these statements is/are correct? (IES, 2007)

(a) 1, 2 and 3 (b) 1 and 3 (c) 2 only (d) 3 only

39. A unit hydrograph for a watershed is triangular in shape with base period of 20 h. The area of the watershed is 500 ha. What is the peak discharge in cubic metre per hour? (IES, 2007)
(a) 7000 (b) 6000 (c) 5000 (d) 4000

40. A triangular direct runoff hydrograph due to a storm has a time base of 60 h and a peak flow of 30 m^3/s occurring at 20 h from the start. If the catchment area is 300 km^2, what is the rainfall excess in the storm? (IES, 2009)
(a) 50 mm (b) 20 mm (c) 10.8 mm (d) 8.3 mm

41. A 3 h unit hydrograph U_1 of a catchment of area 235 km^2 is in the form of a triangle with peak discharge 30 m^3/s. Another 3 h unit hydrograph U_2 is also triangular in shape and has the same base width as U_1, but has a peak flow of 90 m^3/s. What is the catchment area of U_2? (IES, 2009)
(a) 117.5 km^2 (b) 235 km^2 (c) 470 km^2 (d) 705 km^2

42. A direct runoff hydrograph due to isolated storm is triangular in shape with a base of 80 h and peak of 200 m^3/s. If the catchment area is 1440 km^2, the effective rainfall of the storm is (IES, 2010)
(a) 20 cm (b) 10 cm (c) 5 cm (d) 2 cm

43. The shape of the recession limb of a hydrograph depends on (IES, 2010)
(a) Basin as well as storm characteristics (b) Storm characteristics only
(c) Basin characteristics only (d) Baseflow only

44. An S-curve hydrograph has been obtained for catchments of 270 km^2 from a 3 h unit hydrograph. The equilibrium discharge for the S-curve is (IES, 2010)
(a) 750 m^3/s (b) 277.8 m^3/s (c) 250 m^3/s (d) 187 m^3/s

45. The theory of synthetic hydrograph based on flood routing techniques is based on the principle that rainfall impulse(net storm rain) is modified by the factors: (IES, 2013)
1. The time of travel of the flow volume in channel and overland flow
2. Storage
3. Both translation and storage

(a) 1 and 2 only (b) 1,2 and 3 (c) 1 and 3 only (d) 2 and 3 only

46. A direct runoff hydrograph due to a storm idealised into a triangular shape has a peak flow rate of 60 m^3/s occurring at 25 h from its start. If the base width of this hydrograph is 72 h, and the catchment area is 777.6 km^2, the runoff from the storm is: (IES, 2014)
(a) 1 cm (b) 2 cm (c) 5 cm (d) 10 cm

47. The peak magnitude of a flood hydrograph during 4 h study duration over a catchment is 300 m^3/s. The total depth of rainfall is 6 cm and the infiltration loss during the said 4 h period is 2 cm. A constant uniform baseflow of 20 m^3/s is premised throughout. The peak value of the corresponding 4 h unit hydrograph is (IES, 2014)
(a) 75 m^3/s (b) 70 m^3/s (c) 50 m^3/s (d) 40 m^3/s

Directions for questions 48 and 49

The following items consist of two statements, one labelled as 'Assertion (A)' and the other as 'Reason (R)'. You are to examine these two statements carefully and decide if the Assertion A and the Reason R are individually true and if so, whether the Reason is a correct explanation of the Assertion. Select your answers to these items using the codes given below:

(a) Both A and R are true and R is correct explanation of A
(b) Both A and R are true but R is not a correct explanation of A
(c) A is true but R is false
(d) A is false but R is true

48. **Assertion (A):** Instantaneous unit hydrograph (IUH) is used in theoretical analysis of rainfall excess-runoff characteristics of a catchment.

Reason (R): For a given catchment, IUH, being independent of rainfall characteristic, is indicative of the catchment storage characteristics. (IES, 2014)

49. **Assertion (A):** Theoretically, an infinite number of unit graphs are possible for a given basin.
Reason (R): The rainfall duration and its areal distribution affect the hydrograph. (IES, 2014)

50. In reservoirs with an uncontrolled spillway, the peak of the plotted outflow hydrograph (GATE, 2014)

(a) Lies outside the plotted inflow hydrograph
(b) Lies on the recession limb of the plotted inflow hydrograph
(c) Lies on the peak of the inflow hydrograph
(d) Is higher than the peak of the plotted inflow hydrograph

51. An isolated 3 h rainfall event on a small catchment produces a hydrograph peak and point of inflection on the falling limb of the hydrograph at 7 h and 8.5 h respectively, after the start of the rainfall. Assuming, no losses and no baseflow contribution, the time of concentration (in hours) for this catchment is approximately (GATE, 2014)
(a) 8.5 (B) 7.0 (c) 6.5 (d) 5.5

52. The 4 h unit hydrograph for a catchment is given in the table below. What would be the maximum ordinate of the S-curve (in metre per second) derived from this hydrograph?

Time (h)	0	2	4	6	8	10	12	14	16	18	20	22	24
Unit hydrograph ordinate (m^3/s)	0	0.6	3.1	10	13	9	5	2	0.7	0.3	0.2	0.1	0

(GATE, 2015)

53. The direct runoff hydrograph in response to 5 cm rainfall excess in a catchment is shown in the figure. The area of the catchment is expressed in hectares is .... (GATE, 2016)

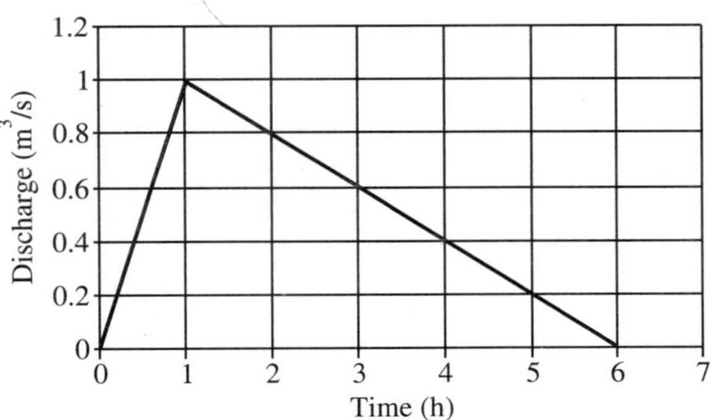

54. The ordinates of a one-hour unit hydrograph at sixty minute interval are 0, 3, 12, 8, 6, 3 and 0 m^3/s. A two-hour storm of 4 cm excess rainfall occured in the basin from 10 AM. Considering constant base flow of 20 m^3/s, the flow of the river (expressed in m^3/s) at 1 PM is .... (GATE, 2016)

Answers

1. (b)	**2.** (c)	**3.** (b)	**4.** (a)	**5.** (b)	**6.** (a)	**7.** (b)	**8.** (c)	**9.** (d)	**10.** (c)
11. (c)	**12.** (b)	**13.** (c)	**14.** (c)	**15.** (a)	**16.** (b)	**17.** (d)	**18.** (d)	**19.** (c)	**20.** (a)
21. (b)	**22.** (d)	**23.** (d)	**24.** (c)	**25.** (c)	**26.** (a)	**27.** (c)	**28.** (d)	**29.** (b)	**30.** (a)
31. (c)	**32.** (d)	**33.** (d)	**34.** (c)	**35.** (c)	**36.** (d)	**37.** (c)	**38.** (c)	**39.** (c)	**40.** (c)
41. (d)	**42.** (d)	**43.** (c)	**44.** (c)	**45.** (b)	**46.** (a)	**47.** (b)	**48.** (a)	**49.** (a)	**50.** (b)
51. (d)	**52.** 22	**53.** 21.5:21.7		**54.** 59:61					

Explanations

1. A 6 h UH can be superimposed on another 6 h UH lagged by 6 h to obtain the 12 h DRH of a storm.

$$\text{Rainfall excess} = 16 - 0.5 \times 12 = 10 \text{ cm}$$

The peak value of discharge of resulting direct runoff is given by

$$Q_p = 22.5 \times \text{rainfall excess} = 22.5 \times 10 = 225 \text{ cumec}$$

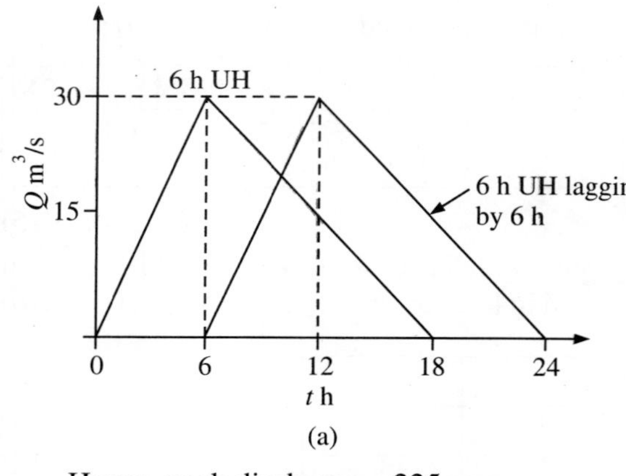

(a)

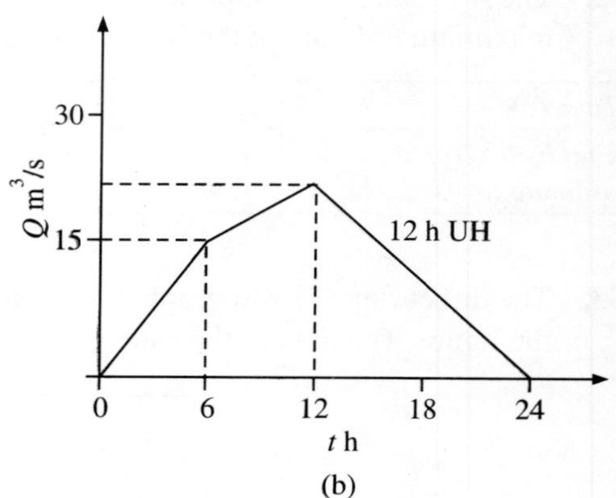

(b)

Hence, peak discharge = 225 cumec.

2.

$$\text{Area under UH} = \text{Catchment area (A)} \times \text{Rainfall excess}$$

$$\frac{(30 \times 18 \times 60 \times 60)}{2} = A \times \frac{1}{100}$$

$$A = 9720 \times 10^4 \text{ m}^2$$

$$A = 9720 \text{ ha}$$

Catchment area, $A = 9720$ ha

3. Rainfall excess = 2.7 – 0.3 × 3 = 1.8 cm

Peak discharge of DRH = Peak flow – Base flow = 210 – 20 = 190 m^3/s

$$\text{Peak discharge of 3 h unit hydrograph} = \frac{\text{Peak discharge}}{\text{Rainfall excess}} = \frac{190}{1.8} = 105.55 \text{ m}^3/\text{s}$$

5.

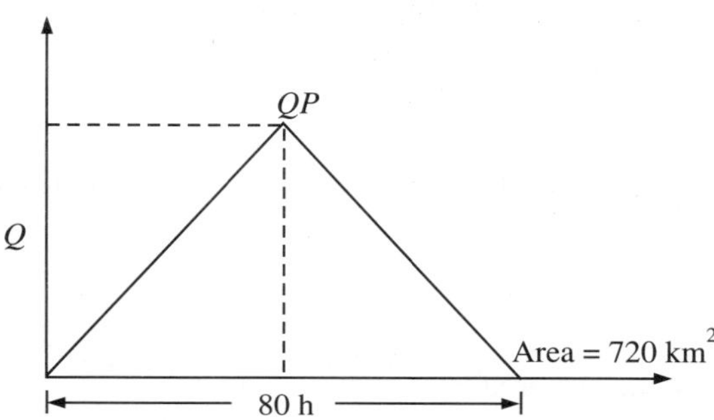

Equating the volume of water

$$\frac{1}{2} \times Q_P \times 80 \times 60 \times 60 = 720 \times 10^6 \times 1 \times 10^{-2}$$

The peak discharge, Q_P = 50 m^3/s

6. Infiltration occurred 4 h, $I = 4 \times 1 = 4$ mm = 0.4 cm

Round depth, $R = P - I = 4 - 0.4 = 3.6$ cm

Peak of DRH of 4 h = $Q_P \times R = 50 \times 3.6 = 180$ m^3/s

Peak flood discharge, $Q = 180 + 30 = 210$ m^3/s.

7. The saturation discharge for S-curve is given by

$$Q_s = \lim_{t\alpha} Q$$

$$Q = 1 \text{ m/s}$$

Given rainfall intensity,

$$I = 1 \text{ cm/h} = \frac{1}{360000} \text{ m/s}$$

$$\text{Catchment area} \times 1 = Q_s$$

$$\text{Catchment area} = \frac{1}{\left(\dfrac{1}{360000}\right)}$$

$$\text{Catchment area} = 360000 \text{ m}^2 = 0.36 \text{ km}^2$$

8. The duration of S-curve = 1 h

Steps for calculating ordinate of 2 h UH are given below:

Step 1: S-curve ordinate at t = 3 h,

$$S_1 = 1 - (1 + 3)e^{-3} = 0.8 \text{ m}^3/\text{s}$$

Step 2: S-curve ordinate at $t = 3 - 2 = 1$ h,

$$S_2 = 1 - (1 + 1)e^{-1} = 0.26 \text{ m}^3/\text{s}$$

Step 3: Substance S_1 and S_2,

$$S_1 - S_2 = 0.8 - 0.26 = 0.54 \text{ m}^3/\text{s}$$

Step 4: The ordinate of unit hydrograph at $t = 3$ h is

$$\frac{(S_1 - S_2)}{2} = 0.27 \text{ m}^3/\text{s}$$

9.

$$\text{Volume of UH} = 60 \times 60 \times 1 \times (0 + 2 + 6 + 4 + 2 + 1 + 0) = 54000 \text{ m}^3$$

$$\text{Catchment area} \times 1 \times 10^{-2} = 54000 \text{ m}^3$$

$$\text{Catchment area} = \frac{54000}{100} = 5.4 \text{ km}^2$$

10. The calculations are shown below in the table:

Time (h)	*Ordinates of* 1 h *UH* (m^3/s)			*DRH of* 3 cm *in* 3 h (m^3/s)	*Ordinate of* 3 h *UH* (m^3/s)
	0 h	1 h	2 h		
0	0	–	–	0	0/3 = 0
1	2	0	–	2	2/3 = 0.67
2	6	2	0	8	8/3 = 2.67
3	4	6	2	12	12/3 = 4
4	2	4	6	12	12/3 = 4
5	1	2	4	7	7/3 = 2.33
6	0	1	2	3	3/3 = 1
7	–	0	1	1	1/3 = 0.33
8	–	–	0	0	0/3 = 0

Thus, a time $t = 3$ h, the ordinate of 3 h UH is 4 m^3/s.

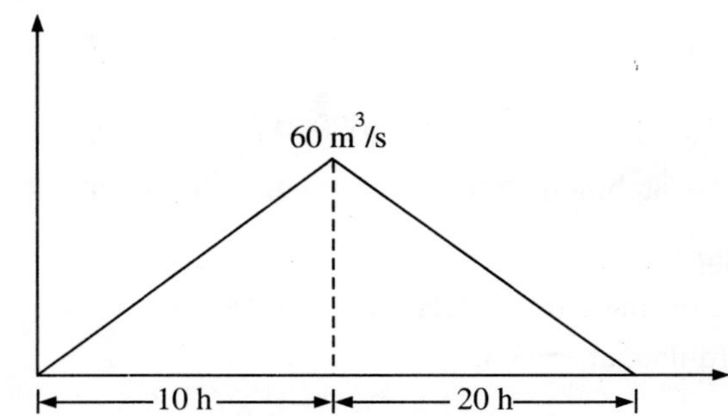

11. Depth of water = 1 cm

$$\text{Runoff volume} = \text{Area} \times \text{Depth} = A \times 1 \times 10^{-2}\ \text{m}^2$$

But, volume of runoff = Area of the unit hydrograph

$$= \frac{1 \times 30 \times 60 \times 3600}{2}$$

$$= 324 \times 10^4\ \text{m}^3$$

$$A \times 10^{-2} = 324 \times 10^4$$

$$A = 324 \times 10^4\ \text{m}^2$$

$$A = 324\ \text{km}^2$$

12. Total rainfall in 1 h = 5.4 cm

Total infiltration in 1 h = ϕ-index × 1

$$= 0.4 \times 1 = 0.4\ \text{cm}$$

Runoff depth = Total rainfall – Infiltration = 5.4 – 0.4 = 5 cm

Discharge has decreased linearly from 60 m^3/s to 0 in 20 h. Discharge at 15th hour can be calculated by linear interpolation.

$$\text{Discharge at the 15th hour} = 60 - \frac{(60 \times 5)}{2} = 45\ \text{m}^3/\text{s}$$

$$\text{Ordinate of first hour DRH} = 45 \times 5 = 225\ \text{m}^3/\text{s}$$

Ordinate of flood hydrograph, Q = Ordinate of DRH + Baseflow

$$Q = 225 + 15 = 240\ \text{m}^3/\text{s}$$

Ordinate of flood hydrograph at 15th hour = 240 m^3/s

15. The calculations are shown in the table below:

t	2 h *UH*	*Imaginary offsetted S-curve lagged by* 2 h	*S-curve ordinates*	*S-curve lagged by* 3 h	Col. 4 – Col. 5	3 h *UH* Col. 6 × (2/3)
Col. 1	Col. 2	Col. 3	Col. 4	Col. 5	Col 6	Col. 7
0	0	–	0	–	0	0
1	3	–	3	–	3	2
2	8	0	8	–	8	5.33
3	6	3	9	0	9	6
4	3	8	11	3	8	5.33
5	2	9	11	8	3	2
6	0	11	11	9	2	1.33
7	0	11	11	11	0	0
8	0	11	11	11	0	0

Peak of UH of 3 h = 6 m^3/s.

$$\text{Runoff due to storm} = 6.6 - \frac{(2 \times 3)}{10} = 6 \text{ cm}$$

$$\text{Peak discharge due to storm} = 6 \times 6 \times 5 = 41 \text{ m}^3/\text{s}$$

16.

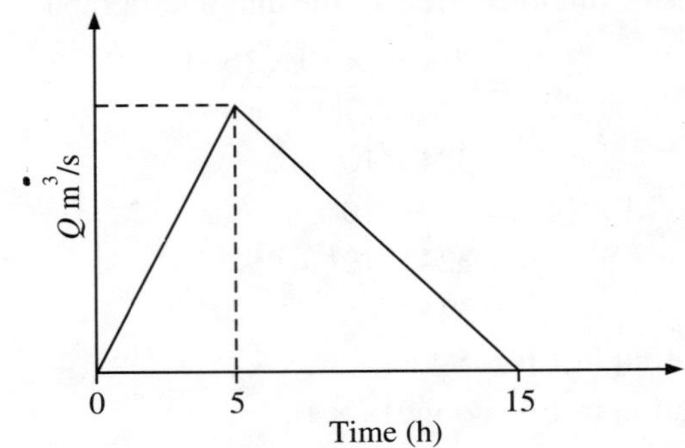

The area of triangle can be equated to volume of water from catchment in order to calculate the ordinate of peak discharge.

$$\frac{1}{2} \times Q_P \times (15 \times 3600) = 50 \times 10^6 \times 1 \times 10^{-2}$$

$$Q_p = 18.52 \text{ m}^2/\text{s}$$

$$\text{Peak discharge} = 18.53 \text{ m}^3/\text{s}$$

17.

$$\text{Depth of total infiltration} = 0.5 \times 1 = 0.5 \text{ cm}$$

$$\text{Total runoff depth} = 5.5 - 0.5 = 5.0 \text{ cm}$$

Peak ordinate of the direct runoff hydrograph (DRH) due to 5 cm of rainfall excess depth $= 18.25 \times 5 = 92.60$ cm

Peak of the flood hydrograph can be obtained by adding peak of DRH and baseflow = Peak of DRH + Baseflow = 92.60 + 10 = 102.60 m³/s

21. 1 cm of excess rainfall will occur with the reduction of duration of unit hydrograph. So, the time base of the UH will decrease, whereas the peak ordinate will increase. Hence, the peak ordinate of 2 h UH will be greater than 80 m³/s.

27. The routing storage is a function of outflow only in gear reservoir, i e., $S = kQ$.

28.

$$\text{Amount of total rainfall} = 7 + 18 + 25 + 12 + 10 + 3 = 75 \text{ mm}$$

$$\text{Effective rainfall} = 33 \text{ mm}$$

$$\text{Infiltration} = \text{Total rainfall} - \text{Effective rainfall}$$

$$= 75 - 33 = 42 \text{ mm}$$

$$\text{Assuming } t_e = 6 \text{ h}, \ \Phi\text{-index} = \frac{42}{6} = 7 \text{ mm/h}$$

This value is greater than 3 mm/h and 7 mm/h.

So, infiltration for 4 h would be 42 – 10 = 32 mm.

Hence, the Φ-index $= \frac{32}{4} = 8$ mm/h.

30. Concentration curve is the raising limb of a hydrograph. It represents the increase in discharge due to the gradual building up of storage in channels and over the catchment surface.

33. Area under DRH

$$= 0.5 \times 10 \times 10 + \frac{(10+70) \times 10}{2} + \frac{(70+90)}{2} \times 10 + \frac{(90+40)}{2} \times 20 + \frac{40}{2} \times 40$$

$$3350 \text{ m}^3/\text{h/s} = 22.06 \times 10^6$$

$$\text{Rainfall excess} = \frac{12.06 \times 10^6}{300 \times 10^6}$$

$$= 0.0402 \text{ m} = 4.02 \text{ cm}$$

34.

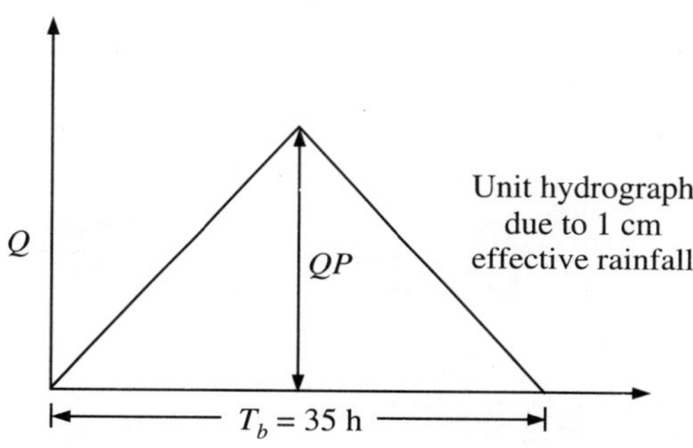

From the unit hydrograph, $$\frac{Q_p T_b}{2} = A \times \left(\frac{1}{100}\right)$$

$$\frac{Q_p \times 35 \times 3600}{2} = \frac{252 \times 10^6}{100}$$

Therefore,

$$Q_p = 40 \text{ m}^3/\text{s}$$

The peak discharge due to 5 cm effective rainfall is $5 \times 40 = 200$ m^3/s.

35. $$\frac{Q_p T_b}{2} = A \times \frac{1}{100}$$

$$A = \frac{300 \times 48 \times 3600 \times 100}{2 \times 10^6} = 2592 \text{ km}^2$$

39. From this unit hydrograph,

$$\frac{Q_p T_b}{2} = A \times \left(\frac{1}{100}\right)$$

$$\Rightarrow \quad \frac{Q_p \times 20}{2} = 500 \times 10^4 \times \frac{1}{100}$$

$$\Rightarrow \quad Q_p = 5000 \text{ m}^3/\text{s}$$

40.

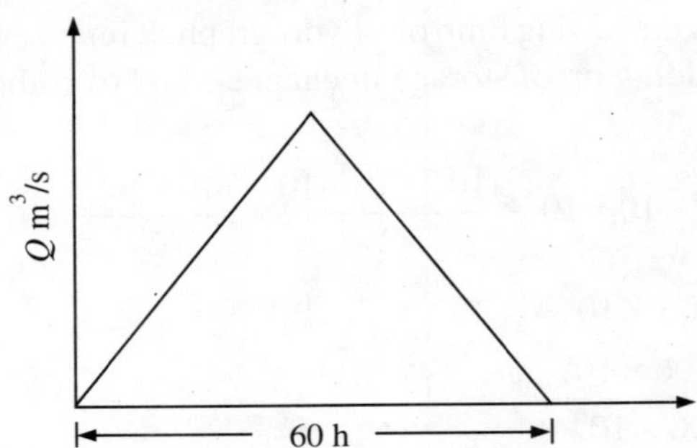

Area of direct runoff hydrograph = Total volume of excess rainfall

$$\Rightarrow \qquad \frac{1}{2} \times 30 \times 60 \times 60 \times 60 = V$$

$$\Rightarrow \qquad \text{Depth of excess rainfall} = \frac{V}{\text{Catchment area}}$$

$$= \frac{30 \times 60 \times 60 \times 60 \times 10^3}{2 \times 300 \times 10^6}$$

$$= 10.8 \text{ mm}$$

41. Time base of both the unit hydrographs is same. Let it be t.

$$\Rightarrow \qquad A_2 = 705 \text{ km}^2$$

42.

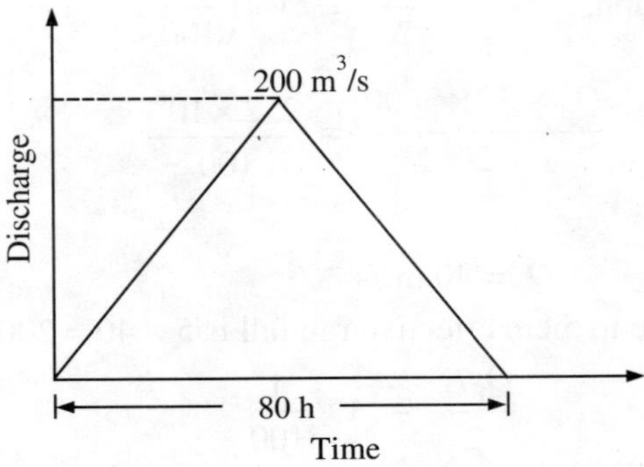

$$\text{The runoff} = \text{Area of triangle}$$

$$= \frac{1}{2} \times 80 \times 3600 \times 200$$

$$= 28.8 \times 10^6 \text{ m}^3$$

$$\text{Area of catchment} = 1440 \text{ km}^2$$

Therefore, $\quad \text{effective rainfall} = \dfrac{\text{Total runoff}}{\text{Area of catchment}} = \dfrac{28.8 \times 10^6}{1440 \times 10^6} = 2 \text{ cm}$

44. The equilibrium discharge is expressed as

$$Q_s = \frac{A}{D} \times 10^4 \text{ m}^3/\text{h}$$

where A is area of catchment in square kilometre and d is duration in hours.

Therefore,
$$Q_s = \frac{270}{3} \times 10^4$$

$$Q_s = 90 \times 10^4 \text{ m}^3/\text{h}$$

$$Q_s = \frac{90 \times 10^4}{3600} \text{ m}^3/\text{s}$$

$$Q_s = 250 \text{ m}^3/\text{s}$$

46. Applying mass conservation,

$$777.6 \times 10^6 \times h = 0.5 \times 60 \times 72 \times 3600$$

$$h = 0.01 \text{ m}$$

$$= 1 \text{ cm}$$

47.
$$\text{Peak discharge of DRH} = 300 - 20$$

$$= 280 \text{ m}^3/\text{s}$$

$$\text{Direct runoff} = 6 - 2$$

$$= 4 \text{ cm}$$

$$\text{The peak value of 4 h unit hydrograph} = \frac{280}{4}$$

$$= 70 \text{ m}^3/\text{s}$$

51. For small catchment, time of concentration is equal to lag time of peak flow.

$$T_c = 71.5$$

$$= 5.5 \text{ h}$$

52.

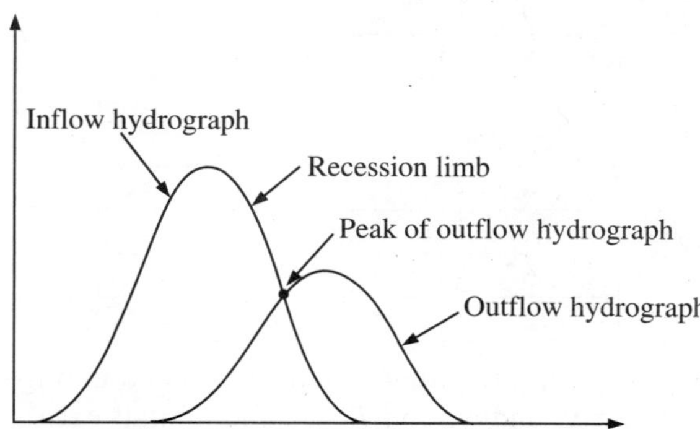

Time	*UHO*	*S-curve addition*	*SA*
0	0	0	0
2	0.6	0	0.6
4	3.1	0	3.1
6	10	0.6	10.6
8	13	3.1	16.1
10	9	10.6	19.6
12	5	16.1	21.1
14	2	19.6	21.6
16	0.7	21.1	21.8
18	0.3	21.6	21.9
20	0.2	21.8	22
22	0.1	21.9	22
24	0	22	22

The maximum S-curve ordinate is 22.

IES Conventional Questions

PROBLEM 6.1 The ordinates of flood hydrograph, resulting from two successive storms each of 1 cm rainfall excess and 6 h duration are tabulated below. Find a 6 h unit hydrograph.

Time (h)	*Ordinate of* 6 h *flood hydrograph* (m^3/s)
0	10
6	30
12	90
18	220
24	280
30	220
36	166
42	126
48	92
54	62
60	40
66	20
72	10

Solution Here, flood hydrograph, direct runoff hydrograph and unit hydrograph are expressed as FH, DRD and U, respectively, and P_1 and P_2 are the rainfall excess due to two successive 1 cm storms, respectively.

The calculation of 6 h unit hydrograph are tabulated below. The amount of baseflow is assumed as 10 m^3/s.

Time (h)	*FH*	*Baseflow*	*DRH* Col. 2 – Col. 3	P_1U	P_2U	Col. 4 $= P_1U + P_2U$	*U*
Col. 1	Col. 2	Col. 3	Col. 4	Col. 5	Col. 6	Col. 7	Col. 8
0	10	10	0	U_1		$U_1 = 0$	0
6	30	10	20	U_2	U_1	$U_1 + U_2 = 20$	20
12	90	10	80	U_3	U_2	$U_2 + U_3 = 80$	60
18	220	10	210	U_4	U_3	$U_3 + U_4 = 210$	150
24	280	10	270	U_5	U_4	$U_4 + U_5 = 270$	120
30	220	10	210	U_6	U_5	$U_5 + U_6 = 210$	90
36	166	10	156	U_7	U_6	$U_6 + U_7 = 156$	66
42	126	10	116	U_8	U_7	$U_7 + U_8 = 116$	50
48	92	10	82	U_9	U_8	$U_8 + U_9 = 82$	32
54	62	10	52	U_{10}	U_9	$U_9 + U_{10} = 52$	20
60	40	10	30	U_{11}	U_{10}	$U_{10} + U_{11} = 30$	10
66	20	10	10	U_{12}	U_{11}	$U_{11} + U_{12} = 10$	0
72	10	10	0	U_{13}	U_{12}	$U_{12} + U_{13} = 0$	0
					U_{13}	$U_1 = 0$	0

The ordinates of 6 h unit hydrograph are given in the following table:

Time (h)	0	6	12	18	24	30	36	42	48	54	60	66	72
UH	0	20	60	150	120	90	66	50	32	20	10	0	0

PROBLEM 6.2 What are the assumptions and limitations of Sherman's unit hydrograph theory? (IES, 1999)

Solution Please see the text.

PROBLEM 6.3 The following are the ordinates of a flood hydrograph resulting from an isolated storm of 6 h duration:

Time (h)	*Ordinate of flood hydrograph* (m^3/s)
0	5
12	15
24	40
36	80
48	60
60	50
72	25
84	15
96	5

Determine the ordinates of 1 cm, 6 h unit hydrograph if the catchment area is 450 km^2. (IES, 2001)

Solution Assuming a bsaeflow of 5 m^3/s, the calculations for determining the ordinates of 1 cm, 6 h unit hydrograph are given in the table below:

Time (h)	*FH* (m^3/s)	*Baseflow* (m^3/s)	*DRH* (m^3/s)	1 cm UH = $\frac{\text{Col. 4}}{DRD}$
Col. 1	Col. 2	Col. 3	Col. 4	Col. 5
0	5	5	0	0
12	15	5	10	4.17
24	40	5	35	14.58
36	80	5	75	31.25
48	60	5	55	22.92
60	50	5	45	18.75
72	25	5	20	8.33
84	15	5	10	4.17
96	5	5	0	0
			$\sum O = 250$	$\sum U = 104.17$

Now, direct runoff depth (DRD) can be determined by the following formula:

$$\text{DRD} = \frac{0.36 \times \sum O \times t}{A} \tag{i}$$

From the above table, we know that:

$$\sum O = 250 \text{ m}^3\text{/s}$$

Again, we know from the given data,
Time t = 12 h; area of the catchment A = 450 km^2
From Eq. (i), we can get

$$\text{DRD} = \frac{0.36 \times 250 \times 12}{450}$$
$$= 2.4 \text{ cm}$$

DRD of unit hydrograph,

$$\text{DRD} = \frac{0.36 \times 104.17 \times 12}{450}$$
$$= 100 \text{ cm}$$

PROBLEM 6.4 In a catchment, the average rainfalls for a storm at two successive 6 h intervals were 3.0 cm and 6.0 cm, respectively. The abstraction losses ϕ-index were estimated to be 0.20 cm/h. For the same catchment, the calculated data for a 6 h unit hydrograph is given below. Find the direct runoff hydrograph due to the storm.

Time (h)	*Unit hydrograph ordinate* (m^3/s)
0	0
6	10
12	25
18	40
24	100
30	150
36	100
42	75
48	25
54	15
60	0

(IES, 2004)

Solution From the given data ϕ-index = 0.20 cm/h

Let us consider that P_1, P_2 and R_1, R_2 are the average rainfalls and runoffs, respectively, at two successive 6 h intervals.

Again, we know

$$P_1 = 3 \text{ cm}, P_2 = 6 \text{ cm}, R_1 = 3 - (0.20 \times 6) = 1.8 \text{ cm}, R_2 = 6 - (0.20 \times 6) = 4.8 \text{ cm}$$

The calculations for determining direct runoff hydrograph (DRH) are given in the table below:

Time (h)	*Unit hydrograph ordinates,* *U* (m^3/s)	R_1U	R_2U	$R_1U + R_2U$ *(DRH)*
0	0	0	–	0
6	10	18	0	18
12	25	45	48	93
18	40	72	120	192
24	100	180	192	372
30	150	270	480	750
36	100	180	720	900
42	75	135	480	615
48	25	45	360	405
54	15	27	120	147
60	0	0	72	72

PROBLEM 6.5 What are the assumptions made in unit hydrograph theory? (IES, 2005)

Solution Please see the text.

PROBLEM 6.6 A 30 min unit hydrograph for a catchment is given in table (a). The ϕ-index is 4 mm/h. The storm details are given table (b). Obtain the runoff hydrograph for the storm.

Time (min)	*Runoff* (m^3/s)
0	0
30	1.2
60	2.8
90	1.7
120	1.4
150	1.2
180	1.1
210	0.91
240	0.74
270	0.61
300	0.5
330	0.28
360	0.17
390	0

(a)

Time (min)	*Rainfall* (cm)
0–30	3.3
30–60	2.7
60–90	1.9

(IES, 2005)

(b)

Solution Let the precipitation and runoff be denoted by P and R, respectively. From the two tables (a) and (b), we can get the following data:

$$P_1 = 3.3 \text{ cm, } \phi\text{-index} = 4 \text{ mm/h, } t_1 = 30 \text{ min} = 0.5 \text{ h}$$

$$P_2 = 2.7 \text{ cm, } t_2 = 30 \text{ min}$$

$$P_3 = 1.9 \text{ cm, } t_3 = 30 \text{ min}$$

$$R_1 = 3.3 - \left(\frac{4}{10} \times 0.5\right) = 3.1 \text{ cm}$$

$$R_2 = 2.7 - \left(\frac{4}{10} \times 0.5\right) = 2.5 \text{ cm}$$

$$R_3 = 1.9 - \left(\frac{4}{10} \times 0.5\right) = 1.7 \text{ cm}$$

The value of direct runoffs (R_1, R_2, R_3) of the three consecutive intervals which lagged by 30 min and 60 min, respectively, are multiple by the coordinates of unit hydrograph and added together to obtained the DRH, as given in the following table:

Time (min)	*Runoff* (m³/sec)	R_1U	R_2U	R_3U	$R_1U + R_2U + R_3U$ (*DRH*)
0	0	0	–	–	0
30	1.2	3.72	0	–	3.72
60	2.8	8.68	3	0	11.68
90	1.7	5.27	7	2.04	14.31
120	1.4	4.34	4.25	4.76	13.35
150	1.2	3.72	3.5	2.89	10.11
180	1.1	3.41	3	2.38	8.79
210	0.91	2.821	2.75	2.04	7.611
240	0.74	2.294	2.275	1.87	6.439
270	0.61	1.891	1.85	1.547	5.288
300	0.5	1.55	1.525	1.258	4.333
330	0.28	0.868	1.25	1.037	3.155
360	0.17	0.527	0.7	0.85	2.077
390	0	0	0.425	0.476	0.901
			0	0.289	0.289
				0	0

PROBLEM 6.7 The catchment area of a drainage basin is 2100 km^2. The length of mainstream L is 80 km. The distance along the main stream from the basin outlet to a point on the stream which is nearer to the centroid of the basin L_c is 50 km. Compute the width of the 3 h synthetic unit hydrograph at 50% and 75% of peak discharge using Snyder's method.

Take the coefficients $C_t = 1.85$; $C_p = 0.45$,
where, C_t is coefficient accounting catchment slope and C_p is dimensionless coefficient.

(IES, 2006)

Solution Basing lag,

$$t_p = C_t(LL_{ca})^{0.3}$$

$$t_p = 1.85 \times (80 \times 50)^{0.3} = 22.27 \text{ h}$$

Now, standard duration of effective rainfall can be found by following formula for a non-standard rainfall duration t_R:

$$t_r = \frac{t_p}{5.5} = \frac{22.27}{5.5} = 4.05 \text{ h}$$

Again, the modified basin lag can be found by the formula given below:

$$\overline{t_p} = \frac{21}{22} \times t_p + \frac{t_R}{4} = \frac{21}{22} \times 22.27 + \frac{3}{4} = 22.01 \text{ h}$$

To compute peak discharge for a duration t_R,

$$Q_p = 2.78 \times \frac{C_p A}{\overline{t_p}} = 2.78 \times \frac{0.45 \times 2100}{22.01} = 119.35 \text{ m}^3\text{/s}$$

Now, the peak discharge per unit catchment area ($m^3/s/km^2$) can be determined by the following formula:

$$q = \frac{Q_p}{A} = \frac{119.36}{2100} = 0.568 \text{ m}^3/\text{s/km}^2$$

Widths of 3 h synthetic unit hydrograph at 50% and 75% of peak discharge using Snyder's method are given below:

$$W_{50} = \frac{5.87}{q^{1.08}} = \frac{5.87}{0.0568^{1.08}} = 129.99 \text{ h}$$

$$W_{75} = \frac{W_{50}}{1.75} = \frac{129.99}{1.75} = 74.28 \text{ h}$$

$t_b = 72 + 3\overline{t}_p = 72 + (3 \times 22.01) = 138.03$ h.

PROBLEM 6.8 The runoff data at a stream gauging station for a flood in cubic metre per second three hourly intervals are as follows:

50, 50, 75, 125, 225, 290, 270, 145, 110, 90, 80, 70, 60, 55, 51, 50.

The drainage area is 40 km^2. The duration of the rainfall is 3 h. Derive the 3 h unit hydrograph for the basin. Assume a constant baseflow of 50 m^3/s throughout the duration. (IES, 2007)

Solution The value of baseflow = 50 m^3/s, drainage area = 40 km^2. Now, the calculations of 3 h unit hydrograph are tabulated below:

Time (h)	*Ordinate of storm hydrograph*	*Baseflow*	*DRH* (Col. 2 – Col. 3)	3 h *unit hydrograph* (Col. 4/26.9)
0	50	50	0	0
3	50	50	0	0
6	75	50	25	0.9
9	125	50	75	2.8
12	225	50	175	6.5
15	290	50	240	8.9
18	270	50	220	8.2
21	145	50	95	3.5
24	110	50	60	2.2
27	90	50	40	1.5
30	80	50	30	1.1
33	70	50	20	0.7
36	60	50	10	0.4
39	55	50	5	0.2
42	51	50	1	0.04
45	50	50	0	0
		Total	996	

$$\text{Direct runoff depth} = \frac{\text{Sum of ordinates} \times \text{Time of rainfall}}{\text{Drainage area}}$$

$$= \frac{996 \times 3 \times 60 \times 60}{40 \times (10)^6}$$

$$= 26.9 \text{ cm}$$

Thus, the ordinates of DRH (Col. 4) are divided by 26.9 cm to obtain the ordinates of 3 h unit hydrograph.

PROBLEM 6.9 The ordinates of 6 h unit hydrograph of a catchment are as follows:

Time (h)	0	6	12	18	24	30	36	42
Discharge (m^3 /sec)	0	10	40	55	45	30	7	0

The unit depth of the hydrograph is 1 cm. Arrive at the direct runoff hydrograph resulting from the following excess-rainfall hyetograph occurring over the catchment.

Time (h)	0–6	6–12
Rainfall intensity (cm/h)	1	0.5

(IES, 2009)

Solution Excess rainfall in first 6 h = 1 × 6 6 cm
Excess rainfall in next 6 h = 0.5 × 6 = 3 cm

It may be noted that 6 cm DRH occurs before 3 cm DRH. Thus, the ordinates of 3 cm DRH are lagged by 6 h.

The 6 h unit hydrograph of a catchment is given. Therefore, in order to compute the DRH due to 6 cm and 3 cm excess rainfall, ordinates of the unit hydograph need to be multiplied by 6 and 3, respectively.

Using the method of superposition, the ordinates of the resulting DRH (9 cm) are obtained by combining the ordinates of the 6 cm and 3 cm DRH, as shown below:

Time	*Ordinates of* 6 h *UH* (m^3/s)	*Ordinates of* 6 cm *DRH* Col. 2 × 6	*Ordinates of* 3 cm *DRH* (Col. 2 *lagged by* 6 h) × 3	*Ordinates of* 9 cm *DRH* (Col. 3 + Col. 4)(m^3/s)
Col. 1	Col. 2	Col. 3	Col. 4	Col. 5
0	0	0	0	0
6	10	60	0	60
12	40	240	30	270
18	55	330	120	450
24	45	270	165	435
30	30	180	135	315
36	7	42	90	132
42	0	0	21	21
45	0	0	0	0

PROBLEM 6.10 A 12 h rainfall with uniform intensity of 4 cm/h produces a storm hydrograph of peak discharge 1000 m^3/s. The abstractions of rainfall is at the rate of 1 cm/h and baseflow is 30 m^3/s. Compute the peak discharge of 12 h unit hydrograph. (IES, 2013)

Solution Given,

Intensity of rainfall = 4 cm/h

Peak discharge of storm hydrograph = 1000 m^3/s

Abstarction ϕ-index = 1 cm/h

Baseflow = 30 m^3/s

Effective rainfall intensity = (4 – 1) cm/h

= 3 cm/h

Effective rainfall depth = (3 × 12) cm = 36 cm

Peak discharge of DRH = Peak discharge of storm hydrograph – Baseflow

= (1000 – 30) m^3/s

= 970 m^3/s

$$\text{Peak discharge of 12 h UH} = \frac{\text{Peak discharge of 12 h DRH}}{\text{Effective rainfall depth}}$$

$$= \frac{970}{36}\ m^3/s$$

$$= 26.94\ m^3/s$$

Therefore, peak discharge of 12 h UH = 26.94 m^3/s

PROBLEM 6.11 Inflow and outflow hydrographs of a channel reach are triangular in shape and are plotted simultaneously as shown in the figures (a) and (b). The peak of inflow hydrograph is 10000 m^3/h and it occurs after 1 h from the starting. The base is 96 h. Similarly, the peak of the outflow hydrograph is 8000 m^3/h and falls on recession limb of inflow hydrograph. Determine the channel storage after 1 h (i.e., under the peak of inflow hydrograph) and the maximum storage and the time at which it occurs along with the principle on which these are computed.

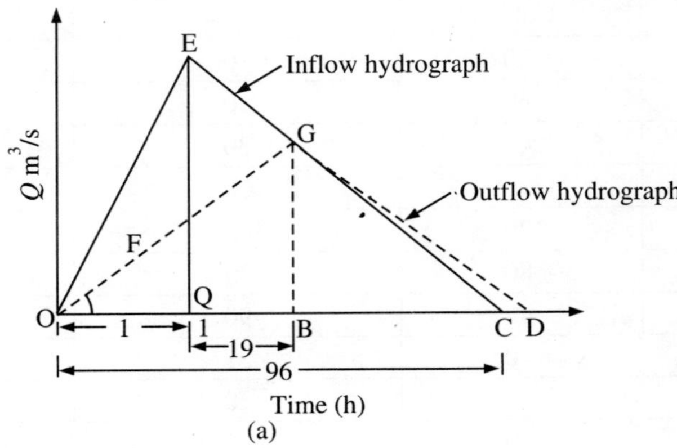

(a)

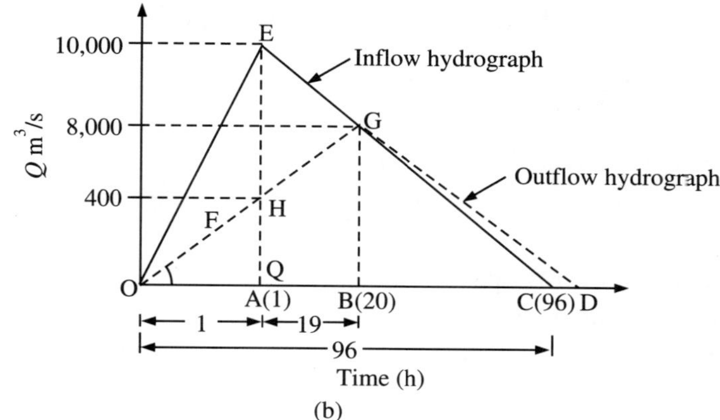

(b)

(IES, 2014)

Solution Given,

Time base of inflow hydrograph = 96 h

Peak of inflow hydrograph = 10000 m^3/h

Peak of outflow hydrograph = 8000 m^3/h

Channel storage after 1 h = Total inflow 1 h – Total outflow in 1 h

= (Area of OEA – Area of OAH) Ordinate of H

= 400 m^3/h

Therefore, storage after 1 h $= \left(\frac{1}{2} \times 10000 \times 1\right) - \left(\frac{1}{2} \times 400 \times 1\right)$

= 4800 m^3

Therefore, channel storage after 1 = 4800 m^3

Time at which peak of outflow hydrograph occurs

$= 1 + \frac{95}{10000}(10000 - 8000)$

= 1 + 19 = 20 h

Maximum storage will occur in the channel at the time of intersection cf inflow and outflow hydrograph.

Hence, maximum storage will occur at t = 20 h from the starting.

Maximum storage = (Area of OEGB) – (Area of OGB)

= (Area of OEA + Area of OGB) – (Area of OGB)

$= \left\{\frac{1}{2} \times 10000 \times 1 + \frac{1}{2}(10000 + 8000) \times 19\right\} - \frac{1}{2} \times 8000 \times 2$

= (5000 + 171000) – 80,000

= 96000 m^3

Therefore, maximum storage = 96000 m^3

Theoretical Questions

1. Define hydrograph and discuss its salient features.
2. What are the methods of baseflow separation?
3. Define unit hydrograph. State its assumptions and limitations.
4. Describe the procedure for the development of unit hydrograph.
5. Discuss the applications of unit hydrograph theory.
6. What is synthetic unit hydrograph?
7. Define dimensionless unit hydrograph.
8. What is IUH? Discuss the advantages of the IUH over the UH.
9. Discuss properties of the IUH.

Unsolved Problems

Problem 1: The ordinates of discharge for a typical storm of a catchment having an area of 380 km^2 are given below. Estimate the ordinates of direct surface runoff hydrograph using straight line method for base flow separation.

Time (h)	0	3	6	9	12	15	18	21
Q_{total} (m^3/s)	8	12	16	20	18	12	9	6

Problem 2: The ordinates of 4 h unit hydrograph are given below. Derive 8 h UH for the same unit volume as of the 4 h UH.

Time (h)	0	2	4	6	8	10	12	14	16	18	20	22
4 h *UH* (m^3/s)	0	50	80	140	220	360	500	650	350	120	30	0

Problem 3: The discharge observed in a river during and after uniform rainfall of 30 mm in a day is given as follows:

Time (days)	0	1	2	3	4	5	6
Q (m^3/s)	3	6	8	10	7	2	1

If the catchment area is 250 sq km and base is assumed constant =1 m^3/s

(a) Compute the direct runoff during the period (in m^3/s)
(b) Estimate the total losses for the rainfall
(c) What was the maximum discharge in the watercourse during the period?

Problem 4: The 4-hour UH of a catchment is triangular in shape with a base width of 36 hours and a peak ordinate of 20 cumec. Calculate the equilibrium discharge of the S-curve of the basin.

Problem 5: For a typical storm the ordinates of 8 h UH of 10 mm volume are given below in the table. If the equilibrium discharge is 2240 m^3/s, compute the S-curve ordinates and draw the diagram.

Day	*Time* (h)	*Time interval* (h)	8 h *UH of* 10 mm
21 July	0800	0	0
	1600	8	60
	2400	16	130
22 July	0800	24	200
	1600	32	280
	2400	40	360
23 July	0800	48	440
	1600	56	380
	2400	64	270
24 July	0800	72	100
	1600	80	20
	2400	88	0

Problem 6: The Snyder's parameters for a catchment are given as C_p = 0.59, C_r = 1.4, distance from the outlet to a point on the stream, nearest to centroid of the catchment= 23 km and catchment area = 1560 km^2. Using Snyder's method, develop a 2 h UH for the catchment.

Problem 7: The ordinates of 2 h. UH for a catchment area (= 630 km^2) are given below. Derive 1 h UH using the S-curve method.

Time (h)	1	2	3	4	5	6	7	8	9	10	11	12	13	14
UH ordinates (m^3/s)	0.9	1.2	2.8	5.4	9.8	13.7	18.9	22.8	16.5	12.6	8.5	3.6	1.8	0.3

Further Reading

Chow, V.T., Maidment, D.R., and Mays, L.W., *Applied Hydrology*, McGraw-Hill, Singapore, 1988.

Das, Ghanshyam, *Hydrology and Soil Conservation Engineering Including Watershed Management*, PHI Learning, Delhi, 2014.

Deodhar, M.J., *Elementary Engineering Hydrology*, Pearson, New Delhi, 2013.

Raghunath, H.M., *Hydrology—Principles, Analysis and Design*, New Age Publishers, New Delhi, 2014.

Suresh, R, *Watershed Hydrology*, Standard Publishers Distributors, New Delhi, 2015.

Yadupathi Putty, Mysooru R., *Principles of Hydrology*, I.K. International, New Delhi, 2013.

CHAPTER

7 Hydrology Statistics and Flood Frequency Analysis

7.1 Introduction

Several variables in hydrology cannot be predicted with certainty. These random hydrological variables, by definition, take different values based on different times/locations. Various hydrological processes are complex and uncertain in nature such that they can be explained only in a probabilistic sense. Hence, we specify the probability that the variable will fall in certain range. The methods of probability and statistics are employed for the analysis of random variables.

7.2 Elements of Statistics

The principal elements of statistics include central tendency, grouping of observations about a central value, variability, dispersion of the observations, skewness and the degree of asymmetry of the distribution. Some of these elements are discussed here.

7.2.1 Central Tendency

Central tendency refers to where in a sample the data are most concentrated. Several measures of central tendency such as arithmetic mean, geometric mean, harmonic mean, median, mode are discussed here.

Arithmetic mean

Arithmetic mean or *average* is the sum of all the elements of a set divided by the number of elements in the set. The formulae for computing mean of sample data may be written as

1. For ungrouped data,

$$\overline{x} = \frac{\sum_{i=1}^{N} x}{N} \tag{7.1}$$

2. For grouped data,

$$\overline{x} = \frac{\sum_{i=1}^{N} fx}{N} \tag{7.2}$$

where $\overline{x}$ is arithmetic mean of the given data, x is the given data, N is the total number of observations, and f is frequency of observation.

Geometric mean

This is defined as

$$\overline{x}_g = (x_1, x_2, x_3, ..., x_N)^{1/N} \tag{7.3}$$

This is useful while working with highly skewed data.

Harmonic mean

It is given by the following formula:

$$\overline{x}_h = \frac{N}{\sum_{i=1}^{N} \frac{1}{x}} \tag{7.4}$$

The last two measures, i.e., the geometric mean and the harmonic mean are less frequently used.

Median

The *median* in a database is the value in the middle of a given set of numbers arranged in order of increasing magnitude.

Mode

The *mode* is defined as the value that is most frequent in a dataset. For a given dataset, there can be more than one mode.

7.2.2 Variability

Variability or *dispersion* refers to how data are distributed around their central location. The measurement of variability or dispersion is done based on the following parameters:

1. Range: The range is simply the difference between the largest and the smallest observation in the sample.

2. Mean deviation: It is the mean of absolute deviation from the mean value.

$$\text{MD} = \frac{\sum |x - \bar{x}|}{N} \tag{7.5}$$

3. Standard deviation: The standard deviation is the square root of the mean squared deviation from the mean. The standard deviation for the sample σ_s is given by

$$\sigma_s = \sqrt{\frac{\sum (x - \bar{x})^2}{(N - 1)}} \tag{7.6}$$

4. Variance: It is a square of standard deviation.

$$\text{Variance} = \sigma^2 \tag{7.7}$$

5. Coefficient of variation (C_v): It is defined as the ratio of standard deviation and mean. This is a dimensionless measure of variability and is useful for comparing different samples or types of data.

$$C_v = \frac{\sigma}{\bar{x}} \tag{7.8}$$

6. Skewness (C_s): Skewness describes the degree of symmetry in the data about the mean. It may be negative skewness or positive skewness, depending on whether data points are skewed to the left (negative skew) or to the right (positive skew) of the data average, respectively.

For univariate data, the formula for skewness is

$$C_s = \frac{N \sum_{i=1}^{N} (x_i - \bar{x})^3}{(N - 1)(N - 2)\sigma^3} \tag{7.9}$$

7. Kurtosis (K_s): Kurtosis describes whether the data are peaked or fiat relative to a normal distribution. The kurtosis of the standard normal distribution is 3. Therefore, any distribution with kurtosis 3 is called *mesokurtic*. A distribution with kurtosis less than 3 is called *platykurtic*. A distribution with kurtosis greater than 3 is called *leptokurtic*. Its central peak is higher and sharper as compared to a normal distribution.

For univariate data, the formula for kurtosis is

$$K_s = \frac{\sum_{i=1}^{N} (x_i - \bar{x})^4}{(N - 1)\sigma^4} \tag{7.10}$$

EXAMPLE 7.1 The following table shows the annual maximum peak flow rate (cumec) records at a gauging site of a stream for 25 years.

Year	1	2	3	4	5
Peak flow rate (cumec)	112	110	116	100	95

(Contd.)

Year	6	7	8	9	10
Peak flow rate (cumec)	114	121	140	135	125
Year	11	12	13	14	15
Peak flow rate (cumec)	125	111	110	109	115
Year	16	17	18	19	20
Peak flow rate (cumec)	99	97	110	123	136
Year	21	22	23	24	25
Peak flow rate (cumec)	139	110	118	124	114

Arrange the peak flow rate and grouped data, and find:
(a) Average (b) Variance (c) Coefficient of skewness (d) Coefficient of kurtosis

Solution Using the given table, calculation are shown below:

(a) Average flow rate $= \frac{\sum x}{N} = \frac{2908}{25} = 116.32$ cumec

(b) Variance $= \sigma^2$, where σ = Standard deviation

$$\text{Standard deviation} = \sqrt{\frac{\sum_{i=1}^{N}(x_i - \overline{x})^2}{(N-1)}} = \sqrt{\frac{3757.44}{24}} = 12.51239$$

So, variance = 156.56

(c) Coefficient of skewness,

$$(S_k) = \frac{N\sum(x_i - \overline{x})^3}{(N-1)(N-2)\sigma^3} = \frac{25 \times 13146.0384}{24 \times 23 \times (12.51239)^3} = 0.30393$$

So, it is positively skewed or skewed to the right.
The required quantities are tabulated below:

Year	*Peak flow rate* (cumec)	$(x_i\text{-}mean)^2$	$(x_i\text{-}mean)^3$	$(x_i\text{-}mean)^4$
1	112	18.6624	–80.621568	348.2851738
2	110	39.9424	–252.435968	1595.395318
3	116	0.1024	–0.032768	0.01048576
4	100	266.3424	–4346.70797	70938.27404
5	95	454.5424	–9690.84397	206608.7934
6	114	5.3824	–12.487168	28.97022976
7	121	21.9024	102.503232	479.7151258
8	140	560.7424	13278.38003	314432.0392

(*Contd.*)

Year	*Peak flow rate* (cumec)	$(x_i\text{-}mean)^2$	$(x_i\text{-}mean)^3$	$(x_i\text{-}mean)^4$
9	135	348.9424	6518.244032	121760.7985
10	125	75.3424	653.972032	5676.477238
11	125	75.3424	653.972032	5676.477238
12	111	28.3024	–150.568768	801.0258458
13	110	39.9424	–252.435968	1595.395318
14	109	53.5824	–392.223168	2871.07359
15	115	1.7424	–2.299968	3.03595776
16	99	299.9824	–5195.69517	89989.44031
17	97	373.2624	–7211.42957	139324.8193
18	110	39.9424	–252.435968	1595.395318
19	123	44.6224	298.077632	1991.158582
20	136	387.3024	7622.111232	150003.149
21	139	514.3824	11666.19283	264589.2534
22	110	39.9424	–252.435968	1595.395318
23	118	2.8224	4.741632	7.96594176
24	124	58.9824	452.984832	3478.92351
25	114	5.3824	–12.487168	28.97022976
Total	2908	3757.44	13146.0384	1385420.238

(d) Coefficient of kurtosis $= \sum_{i=1}^{N} \dfrac{(x_1 - \bar{x})^4}{(N-1)\sigma^4} = \dfrac{1385420.238}{24 \times (12.51239)^4} = 2.355$

7.3 Probability

The probability $P(A)$ of an event A is the relative number of occurrences of event $A(n_a)$ after a very large number of trials (n). It is a measure of how likely that given event will occur. Two trials are independent if the probability of one outcome does not depend on any previous outcomes.

$$P(A) = \frac{n_a}{n} \tag{7.11}$$

EXAMPLE 7.2 The annual precipitation data for 10 years at a rain gauge station were recorded as given below:

Year	1	2	3	4	5
Precipitation (mm)	140.25	130.14	150.5	110.21	180.14
Year	6	7	8	9	10
Precipitation (mm)	190.25	200.21	110.21	150.55	130.23

Find the probability that the annual precipitation in any year will be
(a) Less than 130 mm
(b) Greater than 140 mm
(c) Greater than 140 mm and less than 130 mm.

Solution Here, total number of data points in the dataset $n = 10$

If any particular event occurs m times, then the probability of occurrence of that event $= \dfrac{m}{n}$

(a) For precipitation less than 130 mm/year, $m = 2$

$$\text{Hence, probability} = \frac{2}{10} = 0.2$$

(b) For precipitation greater than 140 mm/year, $m = 6$

$$\text{Hence, probability} = \frac{6}{10} = 0.6$$

(c) For precipitation less than 130 mm and greater than 140 mm, $m = 2$

$$\text{Hence, probability} = \frac{2}{10} = 0.2.$$

7.4 Return Period

Return period is defined as the average time interval between occurrences of a hydrological event of a given or greater magnitude. This is also known as *recurrence interval*. It is inverse of probability of an event. The return period is defined as follows:

$$T = \frac{1}{P} \tag{7.12}$$

An annual maximum event has a return period of T years if its magnitude is equaled or exceeded once, on an average, every T years.

EXAMPLE 7.3 What would be the return period of a design storm to be used for the design of a hydraulic structure? Let there be a probability of 30% that the storm will not occur in the next 30 years.

Solution Given, probability of occurrence of storm = 0.3

$$\text{So, the return period of the storm} = T = \frac{1}{P} = \frac{1}{0.3} = \frac{10}{3} = 3.333 \text{ years}$$

7.5 Probability Distribution

Random variables are characterised by probability distributions. These distributions are approximate description of underlying phenomenon and have proved useful in describing the observed data. A number of probability distributions are commonly used in hydrology. Seven continuous distributions that are widely applied are the normal, log-normal, log-Pearson type-III, gamma, exponential distribution, binomial distribution and Poisson distribution. The use of distribution in a particular situation depends upon the characteristics of the data and problem type. In addition to continuous distribution, two discrete distribution, namely, binomial distribution and the exponential distribution are also used in hydrology. However, the use of these two distributions are restricted generally to those random events in which the outcome can be described either as a success or failure.

7.5.1 Normal Distribution

This is also known as *Gaussian distribution.* This distribution is symmetrical and bell-shaped. In this distribution, mean, median and mode are the same and the area under the curve is unity. The probability density function (PDF) for this distribution with two parameter's μ (mean) and σ^2 (variance) is given by

$$f(x) = \frac{1}{\sigma\sqrt{2\pi}} \exp\left[-\frac{(x-\mu)}{2\sigma^2}\right] \tag{7.13}$$

where μ and σ are the population mean and standard deviation, respectively.

Defining standard normal variate $(z) = \dfrac{(z-\mu)}{\sigma}$, then

$$f(x) = \frac{1}{\sigma\sqrt{2\pi}} \exp\left(-\frac{z^2}{2}\right) \tag{7.14}$$

The standard normal variate (z) is normally distributed with zero mean and unit standard deviation.

EXAMPLE 7.4 From a cumulative monthly rainfall data of 10 years, the average is 75 mm and standard deviation is 7 mm. Calculate the percentage of the magnitude of the rainfall which is:

(a) Between 70 mm and 85 mm
(b) At least 85 mm

Assume that the rainfall data are normally distributed. Use normal distribution table (given in Appendix) to find the value of variate z.

Solution Given, rainfall years, $n = 10$ years, average rainfall, $\overline{P} = 75$ mm, standard deviation, $\sigma = 7$ mm

(a) Percentage of magnitude of rainfall between 70 mm and 85 mm:
Assumption, rainfall data is normally distributed. (Using normal distribution table in given Appendix)
(b) Percentage of magnitude of rainfall at least 85 mm $= P(x \geq 85)$
$= 1 - P(0 \leq x \leq 84)$

$$= 1 - P\left[\frac{(0-75)}{7} \le z \le \frac{(84-75)}{7}\right]$$
$$= 1 - P(-10.7143 \le z \le 1.2857)$$
$$= 1 - [F(1.2857 - F(-10.7143)]$$
$$= 1 - (0.9006 - 8.8597 \times 10^{(-11)}$$
$$= 0.0994$$
$$= 9.94\%$$

7.5.2 Log-Normal Distribution

This is a special case of normal distribution in which the variates are replaced by their logarithmic transformed values with base *e*. The density function of this distribution is given as

$$f(x) = \frac{1}{x\sigma_y\sqrt{2\pi}} \exp\left[-\frac{(y-\mu_y)^2}{2\sigma_y^2}\right] \tag{7.15}$$

where μ_y and σ_y are the mean of y ($y = \ln x$) and standard deviation of y, respectively.

This is used when hydrologic data involves higher skewness towards right side. Following Chow, the statistical parameters for x can be obtained as

$$\mu_x = \exp\left(\mu_y + \frac{\sigma_y^2}{2}\right)$$

$$\sigma = \mu\sqrt{\exp\sigma_y^2 - 1}$$

Examples of variables that have been known to follow a log-normal distribution include streamflow (monthly and annually), daily precipitation and daily peak discharge rates, etc.

7.5.3 Log-Pearson Type-III Distribution

The probability density function of this distribution is given as

$$f(x) = f_0\left(1 - \frac{x}{a}\right)^c \exp\left(-c\frac{x^2}{2}\right) \tag{7.16}$$

where,

$$c = \frac{4}{\beta} - 1,\ a = \frac{c}{2}\frac{\mu_3}{\mu_2},\ \beta = \frac{\mu_3^2}{\mu_2^2}$$

$$f_0 = \frac{n}{ae^c}\frac{c^{c+1}}{\Gamma(c+1)}$$

Here, μ_2 and μ_3 are variance and third moment about mean, respectively, and n is the number of year record.

7.5.4 Gamma Distribution

The density function for this distribution is given as

$$f(x) = \frac{x^a \exp\left(-\frac{x}{b}\right)}{a!b^{a+1}} \quad \text{for } 0 < x < \infty \tag{7.17}$$

$$f(x) = 0 \quad \text{elsewhere} \tag{7.18}$$

Here, a and b are two parameters such that $\mu = b(a + 1)$ and $\sigma_p^2 = b^2(a+1)$.

7.5.5 Exponential Distribution

The exponential distribution is given as

$$f(x) = \lambda \exp(-\lambda x) \tag{7.19}$$

Here, $\lambda > 0$ is the parameter of the distribution.

This distribution is used to obtain the time interval between occurrence of events. The mean and variance of distribution are given as

$$E(t) = \frac{1}{\lambda} \tag{7.20}$$

$$\text{Var}(t) = \frac{1}{\lambda^2} \tag{7.21}$$

7.5.6 Binomial Distribution

In hydrology,it is common to deal with a sequence of independent events for which outcome of each can be either success or failure. Therefore, in such cases, binomial distribution is used. The number of possible ways to choose k events out of n possible events is given by

$$\binom{n}{k} = {}^nC_k = \frac{n!}{k!(n-k)!} \tag{7.22}$$

Therefore, desired probability $\overline{P}$ is the product of probability of any one sequence p and the number of ways in which a sequence can accrue as follows:

$$P(n, k) = \binom{n}{k} p^k(1-p^k) = {}^nC_k p^k(1-p)^k \tag{7.23}$$

EXAMPLE 7.5 Determine the probability of rainfall depth equal to or greater than 1200 mm occurring

(a) Once in five successive years
(b) Twice in ten successive years

(c) At least once in five successive years
(d) Not at all in five years

Assume that rainfall depth of 1200 m has 80 years return period.

Solution $T = 80$ years, $p = \dfrac{1}{80}$, $q = \dfrac{79}{80}$

Let x = rainfall depth $\geq$ 1200 mm.

(a) Occurrence of x once in five successive years,
$n = 5 \quad r = 1$

$$P(5, 1) = {}^5C_1 \left(\frac{1}{80}\right)^1 \left(\frac{79}{80}\right)^4 = 0.0594$$

(b) Occurrence of x twice in ten successive years,
$n = 10 \quad r = 2$

$$P(10, 2) = {}^{10}C_2 \left(\frac{1}{80}\right)^2 \left(\frac{79}{80}\right)^8 = 0.0064$$

(c) Occurrence of x at least, once in five successive years,
$n = 5 \quad r \geq 1$

$$P(r \geq 1) = 1 - P(r = 0) = 1 - {}^5C_0 \left(\frac{1}{80}\right)^0 \left(\frac{79}{80}\right)^5 = 0.0609$$

(d) Occurrence of x not at all in five years,
$n = 5 \quad r = 0$

$$P(5, 0) = {}^5C_0 \left(\frac{1}{80}\right)^0 \left(\frac{79}{80}\right)^5 = 0.939$$

EXAMPLE 7.6 Using Binomial distribution find:

(a) Find the probability that 6 years flood will occur 7 times in a period of 20 years.
(b) Find the probability that it will not occur at all.
(c) How many floods of this magnitude will occur on an average during 20 years? Compare the results with those obtained from binomial distribution.

Solution It is solved using the binomial process only not the Poisson process.

(a) The probability that 6 years flood will occur 7 times in a period of 20 years

$$= {}^nC_r \times P_r \times q_{n-r}$$

$$= 20C_7 \times \left(\frac{1}{6}\right)^7 \times \left(1 - \frac{1}{6}\right)^{20-7}$$

$$= 0.026$$

(b) Probability that it will not occur at all $= \left(1-\frac{1}{T}\right)^n$

$$= \left(1-\frac{1}{6}\right)^{20}$$

$$= 0.026$$

(c) The exceedence probability of 6-year return flood in a year $= \frac{1}{6}$

Therefore, in 20 years, number of 6-year flood that will occur $= 20 \times \frac{1}{6} = 3.33$

7.5.7 Poisson Distribution

The Poisson distribution is used to model the number of events occurring within a given time interval. This distribution deals with mutually independent events. The Poisson distribution is

$$P(x) = \frac{\lambda^x \exp^{-\lambda}}{x!} \tag{7.24}$$

Here, $\lambda(=np)$ is the shape parameter, which indicates the average number of events in the given time interval. Here n is the number of years and p is probability of exceedence.

7.6 Flood Frequency Analysis

Flood frequency analysis deals with the peak flows. In this analysis, first, data series with annual maximum flow is obtained. The annual maximum series should be for as long as the data record allows. Then, frequency histogram of data should be drawn. From this diagram, a relative frequency (divide the number of readings in each class interval by the total number of readings in the data series) curve may be plotted. Then, the following terms can be computed:

1. Probability of exceedence $P(X)$: This is the probability that a flow Q is greater than or equal to a value X.

2. Relative frequency $F(X)$: This is the probability of the flow Q being less than a value X. Then, from relative frequency curve, values of $F(X)$ can be computed.

A general equation for flood frequency analysis of hydrologic events for variate X is written as (Chow, 1964)

$$x_T = \bar{x} + K\sigma$$

where x_T is variate X with return period T, $\bar{x}$ is the average value of the variate, K is a frequency factor and σ is standard deviation of the variate.

The commonly used frequency distribution are discussed here.

7.6.1 Gumbel's Distribution

Gumbel's probability distribution is widely used for extreme value analysis of hydrologic and meteorological data like floods, maximum rainfalls, maximum wind speed and other events. As per the extreme value distribution, the probability of occurrence P of a flood peak equal and larger than a given value is

$$P(X \geq x) = 1 - e^{-e^{-y}} \tag{7.25}$$

Here, X is the event of hydrologic series: x is the desired value of event and y is the reduced variate, which may be written as

$$y = \alpha\,(x - a);\ a = 0.45005\ \sigma_x;\ \alpha = \frac{1.2825}{\sigma_x}.$$

Here, σ_x is the standard deviation of the variate X.

The reduced variate y is determined as follows:

$$y = \frac{1.2825\,(x - \bar{x})}{\sigma_x} + 0.577 \tag{7.26}$$

The value of variate X is required for practical purposes for a given probability. By transposing Eq. (7.25), y may be written in relation to probability as

$$y = \ln\,[-\ln(1 - P)] \tag{7.27}$$

using $T = \dfrac{1}{P}$,

$$y = -\ln\ln\frac{T}{T-1} \tag{7.28}$$

$$y = -0.834 - 2.303\log\log\frac{T}{T-1} \tag{7.29}$$

Using Eq. (7.26), the reduced variate for a return period T may be written as

$$y_T = \frac{1.2825\,(x - \bar{x})}{\sigma_x} + 0.577$$

Rearranging

$$(x - \bar{x}) = (y_T - 0.577)\frac{\sigma_x}{1.2825}$$

or

$$x = \bar{x} + (y_T - 0.577)\frac{\sigma_x}{1.2825}$$

Sample size correction: Frequency factors depends on the sample size.

$$K = \frac{y_T - y_n}{s_n} \tag{7.30}$$

where y_n and s_n are reduced mean and reduced standard deviation, respectively. The value of variate X with recurrence interval T is

$$X_T = \bar{X} + K\sigma_z \tag{7.31}$$

Tables 7.1 and 7.2 provides the value of reduced mean and reduced standard deviation based on the sample size.

Table 7.1 Reduced Mean y_n in Gumbel's Extreme Value Distribution (N = Sample Size)

N	0	1	2	3	4	5	6	7	8	9
10	0.4952	0.4996	0.5035	0.5070	0.5100	0.5128	0.5157	0.5181	0.5202	0.5220
20	0.5236	0.5252	0.5268	0.5283	0.5296	0.5309	0.5320	0.5332	0.5343	0.5353
30	0.5362	0.5371	0.5380	0.5388	0.5396	0.5402	0.5410	0.5418	0.5424	0.5430
40	0.5436	0.5442	0.5448	0.5453	0.5458	0.5463	0.5468	0.5473	0.5477	0.5481
50	0.5485	0.5489	0.5493	0.5497	0.5501	0.5504	0.5508	0.5511	0.5515	0.5518
60	0.5521	0.5524	0.5527	0.5530	0.5533	0.5535	0.5538	0.5540	0.5543	0.5545
70	0.5548	0.5550	0.5552	0.5555	0.5557	0.5559	0.5561	0.5563	0.5565	0.5567
80	0.5569	0.5570	0.5572	0.5574	0.5576	0.5578	0.5580	0.5581	0.5583	0.5585
90	0.5586	0.5587	0.5589	0.5591	0.5592	0.5593	0.5595	0.5596	0.5598	0.5599
100	0.5600									

Table 7.2 Reduced Standard Deviation s_n in Gumbel's Extreme Value Distribution (N = Sample Size)

N	0	1	2	3	4	5	6	7	8	9
10	0.9496	0.9676	0.9833	0.9971	1.0095	1.0206	1.0316	1.0411	1.0493	1.0565
20	1.0628	1.0696	1.0754	1.0811	1.0864	1.0915	1.0916	1.1004	1.1047	1.1086
30	1.1124	1.1159	1.1193	1.1226	1.1255	1.1285	1.1313	1.1339	1.1363	1.1388
40	1.1413	1.1436	1.1458	1.1480	1.1499	1.1519	1.1538	1.1557	1.1574	1.1590
50	1.1607	1.1623	1.1638	1.1667	1.1681	1.1681	1.1696	1.1708	1.1721	1.1734
60	1.1747	1.1759	1.1770	1.1782	1.1793	1.1803	1.1814	1.1824	1.1834	1.1844
70	1.1854	1.1863	1.1873	1.1881	1.1890	1.1898	1.1906	1.1915	1.1923	1.1930
80	1.1938	1.1945	1.1953	1.1959	1.1967	1.1973	1.1980	1.1987	1.1994	1.2001
90	1.2007	1.2013	1.2020	1.2026	1.2032	1.2038	1.2044	1.2049	1.2055	1.2060
100	1.2065									

EXAMPLE 7.7 A hydraulic structure on a stream has been designed for a discharge of 450 cumec. If the available flood data is for 25 years and the mean and standard deviation for annual flood series are 120 and 65 cumec, respectively, estimate the return period for the design flood by using Gumbel's method.

Solution Given,

$$N = 25 \text{ years}$$

$$\text{Mean } \overline{X} = 120 \text{ cumec}$$

$$\text{Standard deviation, } \sigma_{n-1} = 65 \text{ cumec}$$

$$\text{From Gumbel's method, } X_T = \overline{X} + K\sigma_{n-1}$$

$$450 = 120 + K \times 65$$

$$K = 5.0769$$

Using Table 7.1, now, for $N = 25$ years, $y_n = 0.5309$

Using Table 7.2, and $s_n = 1.0915$

$$\text{Again we have, } K = \frac{y_T - y_n}{s_n}$$

$$\Rightarrow \quad 5.0769 = \frac{y_T - 0.5309}{1.0915}$$

$$\Rightarrow \quad y_T = 6.0723$$

$$\Rightarrow \quad \left[\ln \ln\left(\frac{T}{T-1}\right)\right] = 6.072$$

$$\Rightarrow \quad T = 435.78$$

$$\Rightarrow \quad \text{Return period} = 453.78 \text{ years}$$

Log-Pearson Type-III Distribution

This distribution is most commonly used for flood frequency analysis. In this distribution, the natural values of variate are converted into logarithmic form.

If x is variate of hydrologic series, z series is developed by converting the data series to logarithms.

$$z = \log x$$

The procedure developed for applying log-Pearson type-III distribution is used to compute the parameters using the following formula:

1. $$\text{Mean } \bar{z} = \frac{\sum \log x}{n}$$

2. $$\text{Standard deviation, } \sigma_z = \sqrt{\frac{\sum(\log x - \log \bar{x})^2}{n-1}} = \sqrt{\frac{\sum(z - \bar{z})^2}{n-1}}$$

3. $$\text{Skew coefficient, } C_s = \frac{\sum(\log x - \log \bar{x}^3)}{(n-1)(n-2)\sigma^3} = \frac{n\sum(z - \bar{z}^3)}{(n-1)(n-2)\sigma_z^3}$$

The expected values of variate x for various recurrence interval are determined as follows:

$$\log x = \log \bar{x} + K_f \sigma_{\log x}$$

$$z_T = \bar{z} + K_f \sigma_z$$

where K_f is frequency factor for the computed value of C_s and the desired recurrent interval T. After computing z_T, the corresponding value of x_T is calculated as

$$x_T = \text{antilog}\,(z_T)$$

The values of frequency factor K are given in Table 7.3.

Table 7.3 Relationship between c_s and T for log-Pearson Type-III Distribution to Get the Value of K_f

Coefficient of skew, c_s	*K_f for occurrence interval T (years)*							
	2	5	10	25	50	100	200	1000
3.0	–0.396	0.420	1.180	2.278	3.152	4.051	4.970	7.250
2.5	–0.360	0.518	1.250	2.262	3.048	3.845	4.652	6.600
2.2	–0.330	0.574	1.284	2.240	2.970	3.705	4.444	6.200
1.8	–0.282	0.643	1.318	2.193	2.848	3.499	4.147	5.660
1.6	–0.254	0.675	1.329	2.163	2.780	3.388	3.990	5.390
1.4	–0.225	0.705	1.337	2.128	2.706	3.271	3.828	5.110
1.2	–0.195	0.732	1.340	2.087	2.626	3.149	3.661	4.820
1.0	–0.164	0.758	1.340	2.043	2.542	3.022	3.489	4.540
0.9	–0.148	0.769	1.339	2.018	2.498	2.957	3.401	4.395
0.8	–0.132	0.780	1.336	1.998	2.453	2.891	3.312	4.250
0.7	–0.116	0.790	1.333	1.967	2.407	2.824	3.223	4.105
0.6	–0.099	0.800	1.328	1.939	2.359	2.755	3.132	3.960
0.5	–0.083	0.808	1.323	1.910	2.311	2.686	3.041	3.815
0.4	–0.066	0.816	1.317	1.880	2.261	2.615	2.949	3.670
0.3	–0.050	0.824	1.309	1.849	2.211	2.544	2.856	3.525
0.2	–0.033	0.830	1.301	1.818	2.159	2.472	2.763	3.380
0.1	–0.017	0.836	1.292	1.785	2.107	2.400	2.670	3.235
0.0	0.000	0.842	1.282	1.751	2.054	2.326	2.576	3.090
–0.1	0.017	0.836	1.270	1.716	2.000	2.252	2.482	2.950
–0.2	0.033	0.850	1.258	1.680	1.945	2.178	2.388	2.810
–0.3	0.050	0.853	1.245	1.643	1.890	2.104	2.294	2.675
–0.4	0.066	0.855	1.231	1.606	1.834	2.029	2.201	2.504
–0.5	0.083	0.856	1.216	1.567	1.777	1.955	2.108	2.400
–0.6	0.099	0.857	1.200	1.528	1.720	1.880	2.016	2.275
–0.7	0.116	0.857	1.183	1.488	1.663	1.806	1.926	2.150
–0.8	0.132	0.856	1.166	1.448	1.606	1.733	1.837	2.035
–0.9	0.148	0.854	1.147	1.407	1.549	1.660	1.749	1.910
–1.0	0.164	0.852	1.128	1.366	1.492	1.588	1.664	1.880
–1.4	0.225	0.832	1.041	1.198	1.270	1.318	1.351	1.465
–1.8	0.282	0.799	0.945	1.035	1.069	1.087	1.097	1.130
–2.2	0.330	0.752	0.844	0.888	0.900	0.905	0.907	0.910
–3.0	0.396	0.636	0.660	0.666	0.666	0.667	0.667	0.668

EXAMPLE 7.8 The frequency analysis of flood data of the Brahmaputra river using log-Pearson type-III distribution yielded the given data:

Return period T (years)	*Peak discharge* (cumec)
60	16000
250	20000

Given the following data regarding the variation of the frequency factor K with the return period T for Skewness coefficient = 0.5, estimate the flood magnitude in the river with a return period of 1000 years.

Return period T	60	250	1000
Frequency factor K	2.36	3.12	3.98

Solution

We know $x_T = \overline{x} + K \times s$

where x_T is peak discharge at T return period, $\overline{x}$ is mean of the sample observations and s is the standard deviation of the sample observation.

As given,

$$16000 = (\overline{x}) + 2.36 \times s \qquad \text{(i)}$$

$$20000 = (\overline{x}) + 3.12 \times s \qquad \text{(ii)}$$

Following Eq. (ii)-(i), we get

$$4000 = 0.76s$$

$$s = 5263.16$$

Substituting the value of s in Eq. (i),

$$16000 = (\overline{x}) + 2.36 \times 5263.16$$

$$(\overline{x}) = 3578.9 \text{ m}^3/\text{s}$$

Now, for T = 1000 years

$$x_{1000} = 3578.9 + 3.98 \times 5263.16$$

$$x_{1000} = 24526.28 = 24526 \text{ m}^3/\text{s (approx.)}$$

Log-Normal Distribution

In log-Pearson Type-III distribution, when the coefficient of skewness C_s is zero, then this distribution is called *log-normal distribution*.

EXAMPLE 7.9 The given table shows the observed annual flood discharge values in the river at some location. Estimate the flood peaks with return periods 50 and 100 and using

(a) Gumble's extreme value distribution,
(b) Log-Pearson Type-III distribution and
(c) Log-normal distribution.

Year	1980	1981	1982	1983	1984	1985	1986
Discharge (cumec)	3500	3450	3600	3215	3300	3212	4010
Year	1987	1988	1989	1990	1991	1992	–
Discharge (cumec)	3780	3250	3466	3850	3210	3360	–

Solution

(a) Procedure:

(i) Here, the annual flood value is the variate *X*. Find $\overline{X}$ and σ_{n-1} for the given data, where is the σ_{n-1} is the standard deviation of the sample $\left(=\sqrt{\frac{\sum(X-X)^2}{N-1}}\right)$. The calculation is shown in the table below.

S. No.	*Year*	*Observed discharge X* (cumec)	$X-\overline{X}$	$(X-\overline{X})^2$
1	1980	3500	22.8462	521.9489
2	1981	3450	–27.1538	737.3289
3	1982	3600	122.8462	15091.19
4	1983	3215	–262.154	68724.61
5	1984	3300	–177.154	31383.47
6	1985	3212	–265.154	70306.54
7	1986	4010	532.8462	283925.1
8	1987	3780	302.8462	91715.82
9	1988	3250	–227.154	51598.85
10	1989	3466	–11.1538	124.4073
11	1990	3850	372.8462	139014.3
12	1991	3210	–267.154	71371.15
13	1992	3360	–117.154	13725.01
	Mean = 3477.15	3477.1538	Sum	838239.7

(ii) Determine the reduced mean $\overline{(y_n)}$ and reduced standard deviation s_n appropriate to the given $N = 13$ by using Table 7.1 and Table 7.2, respectively.

(iii) Find reduced variate y_T for a given *T* by using $y_T = \left[\ln\ln\left(\frac{T}{(T-1)}\right)\right]$.

(iv) Find *K* using $K = \frac{\left(y_T - \overline{(y_n)}\right)}{s_n}$

(v) Determine the required X_T using $X_T = \overline{X} + K\sigma_{n-1}$.

Here, $$\sigma_{n-1} = \sqrt{\frac{\sum(X-\overline{X})^2}{(N-1)}} = 264.298 \text{ cumec}$$

Return period (years)	*Reduced mean*	*Reduced variate*	*Reduced standard deviation*	*Frequency factor*	*Flood peaks*
T	y_n	$y_T = [\ln \ln (T/(T-1)]$	s_n	K	m^3/s
50	0.5485	3.902	1.1607	2.89	4240.97
100	0.560	4.60	1.2065	3.35	4362.55

(b) Procedure

(i) Here, the annual flood value is the variate X.

(ii) Calculate $z = \log X$.

(iii) Calculate mean of $z, \bar{z}$.

(iv) Calculate standard deviation of z variate sample $\sigma_z = \sqrt{\dfrac{\sum (Z - \bar{Z})^2}{N-1}}$.

(v) Determine, coefficient of skew $C_s = \dfrac{N\sum (z-\bar{z})^3}{(N-1)(N-2)(\sigma)^3}$.

(vi) Determine K_z from standard table. K_z is the frequency factor, which is a function of recurrence interval T and the coefficient of skew C_s.

(vii) For any return period, calculate $z_T = \bar{z} + K_z \sigma_z$, where z_T indicates the peak flood with return period T. All calculations are shown in the table below.

S. No.	*Year*	*Observed Discharge* X (cumec)	$z = \log X$	$z - \bar{z}$	$(z-\bar{z})^2$	$(z-\bar{z})^3$
1	1980	3500	3.544068	0.003967	1.57374E-05	6.24311E-08
2	1981	3450	3.537819	–0.00228	5.20709E-06	–1.1882E-08
3	1982	3600	3.556303	0.016202	0.000262489	4.25271E-06
4	1983	3215	3.507181	–0.03292	0.001083728	–3.5676E-05
5	1984	3300	3.518514	–0.02159	0.000466001	–1.006E-05
6	1985	3212	3.506776	–0.03333	0.001110587	–3.7011E-05
7	1986	4010	3.603144	0.063043	0.003974467	0.000250564
8	1987	3780	3.577492	0.037391	0.001398072	5.2275E-05
9	1988	3250	3.511883	–0.02822	0.000796235	–2.2468E-05
10	1989	3466	3.539829	–0.00027	7.42244E-08	–2.0222E-11
11	1990	3850	3.585461	0.04536	0.002057505	9.33279E-05
12	1991	3210	3.506505	–0.0336	0.001128689	–3.7919E-05
13	1992	3360	3.526339	–0.01376	0.000189385	–2.6063E-06
		Mean	3.540101	Sum	0.012488176	2.5473×10^{-4}

$$\text{Standard deviation } \sigma_z = \sqrt{\frac{\sum (z-\bar{z})^2}{(N-1)}} = 0.03225959$$

Coefficient of skew of z, $$C_s = \frac{N\sum(z-\overline{z})^3}{(N-1)(N-2)(\sigma)^3}$$

$$= \frac{13 \times 2.5473 \times 10^{-4}}{(13-1)(13-2)(0.03225959)^3} = 0.75$$

Return period T (years)	K_z	$z_T = z + K_z\sigma_z$	$X_T = z_T$ (*Peak flood*) (cumec)
50	2.43	3.6185	4154.32
100	2.85	3.6323	4288.44

(c) Procedure: Log-normal distribution is a special case of log-Pearson distribution with $C_s = 0$. Thus, in this case, from the previous calculation,

$$\text{Mean } \overline{z} = 3.540101$$

$$\text{Standard deviation of } z = \sqrt{\frac{\sum(z-\overline{z})^2}{N-1}} = 0.03225959$$

All calculations are shown in the table below:

Return period T (years)	$K_z = f(C_s = 0, T)$	$Z_T = Z + K_z\,\sigma_z$	X_T = antilog z_T (*Peak flood*) (cumec)
50	2.054	3.60636	4039.8
100	2.326	3.615137	4122.274

7.6.2 Limitations of Flood Frequency Analysis

The limitations of flood frequency analysis are as follows:

1. The streamflow records of shorter duration are not worth to carry out frequency analysis.
2. The assumption is made that each storm is independent of another used in the dataset.
3. Hydrological regime has not altered due to several factors such as climate change, land use changes, etc. during the complete period of record.

7.7 Risk and Reliability

Risk is defined as the probability of at least one occurrence of the T-year event in n years. The probability P of an occurrence in one year is

$$P = \frac{1}{T} \tag{7.32}$$

Therefore, the risk is the sum of probabilities of one such flood, two floods, three floods,..., n floods.

$$\text{Risk} = 1 - (1-P)^n = 1 - \left(1 - \frac{1}{T}\right)^n \tag{7.33}$$

Reliability is defined as the probability of non-failure.

$$\text{Reliability} = 1 - \text{Risk} = \left(1 - \frac{1}{T}\right)^n \tag{7.34}$$

EXAMPLE 7.10 Calculate the probability that a 100-year flood will occur at a given site at least once during the next 5, 10, 50, 100 years. What is the chance that a 100-year flood will not occur at this site during the next 100 years?

Solution Let P_n be probability that the flood would occur at least once during the next n years.

So,

$$P_5 = 1 - \left(1 - \frac{1}{100}\right)^5 = 0.049$$

$$P_{10} = 1 - \left(1 - \frac{1}{100}\right)^{10} = 0.096$$

$$P_{50} = 1 - \left(1 - \frac{1}{100}\right)^{50} = 0.395$$

$$P_{100} = 1 - \left(1 - \frac{1}{100}\right)^{100} = 0.634$$

Chance that a 100-year flood would not occur during the next 100 years = $\left(1 - \frac{1}{100}\right)^{100} = 0.366$

EXAMPLE 7.11 What will be the risk involved for a hydraulic structure having a design life of 80 years, if it is designed for

(a) 50 year return period
(b) 800 year return period

Solution (a) Risk involved in 50-year return period = $1 - \left(1 - \frac{1}{T}\right)^n = 1 - \left(1 - \frac{1}{50}\right)^{80}$

$$= 1 - (1 - 0.02)^{80} = 0.801$$

Here, probability of occurrence $\left(\frac{1}{P}\right) = \frac{1}{50} = 0.02$

(b) Risk involved in 800-year return period = $1 - \left(1 - \frac{1}{T}\right)^n = 1 - \left(1 - \frac{1}{800}\right)^{80}$

$$= 1 - (1 - 0.00125)^{80} = 0.095$$

Here, probability of occurrence $\left(\frac{1}{P}\right) = \frac{1}{800} = 0.00125.$

EXAMPLE 7.12 A barrage is proposed to be constructed on the edge of the 60-year flood plain of a river. If the design of the barrage is 30 years, what is the reliability that it will not be flooded during its design life?

Solution Given, return period, $T = 60$ years, Expected life of barrage, $n = 30$
Now, risk of occurring flood

$$(R\sim) = 1 - (1 - P)^n$$

$$= 1 - \left(1 - \frac{1}{T}\right)^n$$

$$= 1 - \left[1 - \left(\frac{1}{60}\right)\right]^{30} = 1 - 0.60 = 0.4$$

Therefore, reliability $= 1 - R\sim = 1 - 0.4 = 0.6$.

7.8 Safety Factor and Safety Margin

The *margin of safety* or *safety margin* (S_m) is defined as the difference between the project capacity-flood magnitude used in practical design, Q_p and hydrological design flood, Q_h.

$$S_m = Q_P - Q_h \tag{7.35}$$

The *factor of safety* or *safety factor* (S_f) is given by the ratio of flood magnitude used in practical design, Q_p and hydrological design flood, Q_h.

$$S_f = \frac{Q_p}{Q_h} \tag{7.36}$$

These both factors are computed by analysing high flood levels in the structures for long periods.

EXAMPLE 7.13 Determine the flood discharge by Gumbel's method for annual high volume discharge data covering the period from 1955 to 1986. The mean and standard deviation estimated for the annual flood discharge are 35000 and 17500 cumec, respectively. A bridge is proposed on this river near this site. It is decided to have an acceptable risk of 15% in its expected life of 60 years. Also, compute factor of safety and margin of safety relating to maximum discharge if the actual flood volume adopted in the design is 130000 cumec.
Solution Given, life period of structures, $n = 60$ years, Risk, $R = 0.15$

Hence, $$R = 0.15 = 1 - \left(1 - \frac{1}{T}\right)^{60}$$

$$\left(1 - \frac{1}{T}\right) = (1 - 0.15)^{1/60}$$

So, $$T = 370 \text{ years}$$

Using Gumbel's method to estimate the flood magnitude for this return period of $T = 370$ years.

Record length, $N = 1982 - 1951 = 31$ years

From Tabled 7.1, $y_n = 0.5371$

$s_n = 1.1159$

$$y_T = -\left[\ln \ln \frac{T}{(T-1)}\right] = -\left[\ln \ln \frac{370}{(370-1)}\right] = 5.9121$$

$$K = \frac{y_T - y_n}{s_n} = \frac{5.9121 - 0.5371}{1.1159} = 4.817$$

$$X_T = \overline{x}_T + K\sigma_{(n-1)}$$

$$= 35000 + 4.817 \times 17500$$

$$= 119297.5 \text{ m}^3/\text{s} \qquad \text{(hydrological design flood magnitude)}$$

Actual flood magnitude adopted in the project = 130000 m^3/s

$$\text{Safety factor, } (S_f)_{\text{flood}} = \frac{130000}{119297.5} = 1.0897$$

Safety margin for flood magnitude, $S_m = 130000 - 119297.5 = 10702.5$ m^3/s.

Summary

- Return period is defined as the average time interval between occurrences of a hydrological event of a given or greater magnitude.
- A plot between the value of random variable as abscissa and the corresponding exceedence probability as the ordinate is called distribution graph or probability plot.
- A normal distribution is bell-shaped and symmetric with respect to mean.
- The log-normal distribution is used in hydrologic event analysis when random variables cannot be negative.
- The three-parameter gamma distribution is used in hydrology for the application of flood flow.
- Risk is the probability of an event occurring per year during a specific time period.
- For estimating the floods of higher return periods and probabilities, Gumbel's distribution is widely used.

Objective Type Questions

1. Probability of a 10-year flood to occur at least once in the next 4 years is (IES, 1997)
(a) 25% (b) 35% (c) 50% (d) 65%

2. The standard project flood is (IES, 1997)
   (a) Derived from the probable maximum precipitation in the region
   (b) Derived from the severe most meteorological conditions anywhere in the country
   (c) The flood with return period of 1000 years
   (d) The same as the probable maximum flood
3. Consider the following statements:
   1. A 100-year flood discharge is greater than a 50-year flood discharge.
   2. 90% dependable flow is greater than 50% dependable flow.
   3. Evaporation from salt water surface is less than that from fresh water surface.

   Which of these statements are correct? (IES, 1999)
   (a) 1 and 2 (b) 2 and 3 (c) 1 and 3 (d) 1, 2 and 3
4. A culvert is designed for a peak flow Q_p on the basis of rational formula. If a storm of the same intensity as used in the design and twice the duration occurs, then the resulting peak discharge will be (IES, 1999)
   (a) Q_p (b) $\frac{Q_p}{2}$ (c) $\sqrt{Q_p}$ (d) $2Q_P$
5. The probability that a 100-year flood is equalled or exceeded, at least once in 100-years is (IES, 2000)
   (a) 99% (b) 64% (c) 36% (d) 1%
6. An effective storage of a flood control reservoir is (IES, 2002)
   (a) The amount of water which can be supplied from it in a particular interval of time
   (b) The storage between the minimum and maximum reservoir levels under ordinary operating conditions
   (c) The useful storage plus the surcharge storage less the valley storage
   (d) The storage volume of floodwater above maximum reservoir level
7. Match List-I (Floods) with List-II (Parameters) and select the correct answer using the codes given below the lists. (IES, 2003)

   List-I
   A. Standard project flood (SPF)
   B. Maximum probable flood (MPF)
   C. Design flood
   D. Maximum flood

   List-II
   1. Includes catastrophic floods
   2. Includes floods of severe conditions
   3. Peak flow obtained from observed data
   4. Flood of desired recurrence interval

   Codes

	A	B	C	D
(a)	2	1	4	3
(b)	1	2	3	4
(c)	2	1	3	4
(d)	1	2	4	3

8. A storm with 14 cm precipitation produced a direct runoff of 8 cm. The time distribution of the storm is shown in the table below: (IES, 2003)

Time from start (h)	*Incremental rainfall in* cm
1	1
2	2
3	2.8
4	3.3
5	2.5
6	1.8
7	0.6

What is the value of ϕ-index of the storm ?

(a) 0.5 cm/h (b) 0 7 cm/h (c) 0.8 cm/h (d) 0.9 cm/h

9. A bridge has an expected life of 50 years and is designed for a flood magnitude of return period 100 years. What is the risk associated with this hydrologic design? (IES, 2003)

(a) $1 - (0.99)^{50}$ (b) $(0.5)^{50}$ (c) $(0.99)^{50}$ (d) $(0.99)^{100}$

10. Which one of the following statements is correct in respect of the two important aspects of flood forecast? (IES, 2003)

1. Reliability of the forecast, and
2. The time available in between the forecast and the occurrence of flood

(a) Meteorological forecast is least reliable and time available is also the least
(b) Hydrological forecast is most reliable, but the time available is the least
(c) River forecast is least reliable and the time available is the maximum
(d) River forecast is most reliable, but the time available is the least

11. By using Gumbel's method, the flood discharge with a return period of 500 years at a particular township neighbourhood was estimated as 18000 m^3/s with a probable error of 2000 m^3/s. What are the 95% confidence probability limits of the 500-year flood at the location? (IES, 2006)

(a) 16100 m^3/s to 19900 m^3/s (b) 17050 m^3/s to 18950 m^3/s
(c) 14080 m^3/s to 21920 m^3/s (d) 13600 m^3/s to 22400 m^3/s

12. In a river carrying a discharge of 142 m^3/s, the stage at a station A was 3.6 m and the water surface slope was 1 in 6000. If during a flood, the stage at A was 3.6 m and the water surface slope was 1 in 3000, what was the flood discharge (approximately)?

(IES, 2006)

(a) 284 m^3/s (b) 200 m^3/s (c) 164 m^3/s (d) 96 m^3/s

13. During a flood in a stream of width 200 m (taken as rectangular section), the gauge reading was found to raise by 18 cm/h. What would be the difference in discharge at two sections, each 250 m on either side of the gauge station? (IES, 2007)

(a) 10 m^3/s less at downstream section
(b) 7.5 m^3/s less at downstream section

(c) 5 m^3/s less at downstream section
(d) 5 m^3/s more at downstream section

14. Kirpich equation is used to determine which one of the following? (IES, 2008)
(a) Runoff from a given rainfall
(b) Base time of a unit hydrograph
(c) Time of concentration in rainoff hydrograph
(d) None of the above

15. A culvert is designed for a flood magnitude of return period 100 years and has expected life of 20 years. The risk in this hydrologic design is (IES, 2010)
(a) $1 - 0.99^{20}$ (b) $1 - 0.01^{20}$ (c) $1 - 0.09^{20}$ (d) $1 - 0.10^{20}$

16. The hydrologic risk of a 100-year flood occurring during the 2-year service life of a project is (IES, 2012)
(a) 9.8% (b) 9.9% (c) 19.9% (d) 1.99%

17. The design flood commonly adopted in India for barrages and minor dams is (IES, 2012)
(a) Probable maximum flood
(b) A flood of 50–100 years return period
(c) Peak flood
(d) Standard protect flood or a 100-year flood, whichever is higher

18. The successive annual rainfall magnitudes at a place for a period of 10 years from 2001 to 2010, both inclusive, are 30.3, 41.0, 33.5, 34.0, 33.3, 36.2, 33.6, 30.2, 35.5 and 36.3 cm. The mean and median values of this annual rainfall series are, respectively, (IES, 2014)
(a) 33.8 cm and 34.39 cm (b) 34.39 cm and 33.8 cm
(c) 34.39 cm and 40.2 cm (d) 33.8 cm and 40.2 cm

19. The probability of a 10-year flood to occur at least once in the next 5 years is (IES, 2014)
(a) 35% (b) 41% (c) 60% (d) 65%

20. It is proposed to design a causeway along a village road. The return period for the annual maximum flood of a given magnitude is found to be 5 years. The probability that this flood magnitude will be exceeded at least once during next 2 years is (IES, 2013)
(a) 0.8 (b) 0.5 (c) 0.45 (d) 0.36

Answers

1. (b) **2.** (b) **3.** (c) **4.** (a) **5.** (b) **6.** (c) **7.** (a) **8.** (d) **9.** (a) **10.** (d)
11. (c) **12.** (b) **13.** (c) **14.** (c) **15.** (a) **16.** (d) **17.** (d) **18.** (b) **19.** (b) **20.** (d)

Explanations

1. The probability of flood occurring at least once in 4 years is

$$1 - (1 - P)^4 = 1 - \left(1 - \frac{1}{10}\right)^4 = 1 - 0.944 = 0.3439 = 35\%$$

8. Amount of infiltration = Rainfall – Direct runoff = 14 – 8 = 6 cm.

Let $t_e = 7$ hours, ϕ-index $= \frac{6}{7} = 0.86$ cm/h

9. Risk $= 1 - \left(1 - \frac{1}{100}\right)^{50} = 1 - (0.99)^{50}$

12.
$$Q_2 = Q_1 \times \sqrt{\frac{S_2}{S_1}}$$

$$Q_2 = 142 \times \sqrt{\frac{1}{3000} \times 6000} = 200.8 \text{ m}^3/\text{s}$$

13.
$$Q_1 - Q_2 = \frac{18 \times 10^{-2}}{60 \times 60} \times (250 + 250) \times 200$$

$\Rightarrow$
$$Q_1 - Q_2 = 5 \text{ m}^3/\text{s}$$

15.
$$R = 1 - \left(1 - \frac{1}{T}\right)^n$$

$$R = 1 - \left(1 - \frac{1}{100}\right)^{20}$$

$$R = 1 - 0.99^{20}$$

16.
$$R = 1 - \left(1 - \frac{1}{100}\right)^2$$

$$R = 0.0199$$

$$R = 1.99\%$$

18.
$$\text{Mean value} = \frac{\sum x}{n} = 34.49 \text{ cm}$$

$$\text{Median} = \frac{(34 + 33.6)}{2} = 33.8 \text{ cm}$$

19. The probability of a 10-year flood to occur at least once in the next 5 years is given as

$$= 1 - (1 - P)^n$$
$$= 1 - (1 - 0.1)^5$$
$$= 41\%$$

20.
$$P = 1 - \left(1 - \frac{1}{T}\right)^n = 1 - \left(1 - \frac{1}{5}\right)^2 = 0.36$$

IES Conventional Questions

PROBLEM 7.1 The annual maximum flood data in a river at a station have been processed to estimate the maximum flood for different return periods using the Gumbel's method. If the estimated maximum flood for return periods of 100 and 50 years are 450 m^3/s and 400 m^3/s, respectively, estimate the flood discharge for return period of 500 years. (IES, 2013)

Solution Given, Maximum flood for return periods 100 years 450 m^3/s

Maximum flood for return periods 50 years 400 m^3/s,

Now, using Gumbel's method,

$$X_T = \bar{X} + K_T \sigma_{n-1}$$

where,

$$K_T = \frac{y_T - y_n}{s_n}$$

$$X_{50} = \bar{X} + K_{50}\sigma_{n-1} \quad \text{(i)}$$

$$\Rightarrow \quad 400 = \bar{X} + K_{50}\sigma_{n-1}$$

$$X_{100} = \bar{X} + K_{100}\sigma_{n-1} \quad \text{(ii)}$$

$$450 = \bar{X} + K_{100}\sigma_{n-1}$$

$$X_{500} = \bar{X} + K_{500}\sigma_{n-1} \quad \text{(iii)}$$

$$\frac{\text{(iii)} - \text{(ii)}}{\text{(ii)} - \text{(i)}} = \frac{X_{500} - X_{100}}{X_{100} - X_{50}} = \frac{K_{500} - K_{100}}{K_{100} - K_{50}}$$

$$\Rightarrow \quad \frac{X_{500} - 450}{450 - 400} = \frac{y_{500} - y_{100}}{y_{100} - y_{50}}$$

$$\Rightarrow \quad \frac{X_{500} - 450}{50} = \frac{\left[\ln\ln\left(\frac{500}{499}\right)\right] + \left[\ln\ln\left(\frac{100}{99}\right)\right]}{\left[\ln\ln\left(\frac{100}{99}\right)\right] + \left[\ln\ln\left(\frac{50}{49}\right)\right]}$$

$$\Rightarrow \quad \frac{X_{500} - 450}{50} = \frac{-6.2136 - 4.6001}{-4.6001 - 3.9019}$$

$$\Rightarrow \quad X_{500} = 513.59 \text{ m}^3/\text{s}$$

Therefore, flood discharge with return period of 500 years = 565.54 m^3/s.

PROBLEM 7.2 Flood frequency computation yields the following results:

Return period (years)	*Peak flood* (m^3/s)
50.0	20500
100.0	25400

Using Gumbel's method, estimate the flood for a return period of 150 years. (IES, 2010)

Solution Using Gumbel's method.

$$x_T = \bar{x} + K\sigma$$

and

$$K = \frac{y_T - y_n}{s_n}$$

where

$$y_T = -\log_e \log_e \left[\frac{T}{T-1}\right]$$

Buy, $\overline{y}_n$ and s_n are same for one analysis because the total number of observations are fixed in one analysis.

$$X_{50} = \overline{X} + \left(\frac{y_{50}}{s_n} - \frac{y_n}{s_n}\right)\sigma$$

$$\Rightarrow \qquad 20500 = \overline{X} + \left(\frac{y_{50}}{s_n} - \frac{y_n}{s_n}\right)\sigma \qquad \text{(i)}$$

Similarly,

$$25400 = \overline{X} + \left(\frac{y_{100}}{s_n} - \frac{y_n}{s_n}\right)\sigma \qquad \text{(ii)}$$

Subtracting Eq. (i) from Eq. (ii), we get

$$\left(\frac{y_{100}}{s_n} - \frac{y_{50}}{s_n}\right)\sigma = 4900 \qquad \text{(iii)}$$

But

$$y_{100} = -\log_e \log_e \left[\frac{100}{100-1}\right] = 4.6$$

$$y_{50} = -\log_e \log_e \left[\frac{50}{50-1}\right] = 3.9$$

Subtracting values of y_{50} and y_{100} in Eq. (iii), we get

$$(4.6 - 3.9) \times \frac{\sigma}{s_n} = 4900$$

$$\Rightarrow \qquad \frac{\sigma}{s_n} = 7000 \qquad \text{(iv)}$$

Now,

$$X_{150} = \overline{X} + \left(\frac{y_{150}}{s_n} - \frac{y_n}{s_n}\right)\sigma \qquad \text{(v)}$$

Subtracting Eq. (ii) from Eq. (v), we get

$$x_{150} - 25400 = (y_{150} - y_{100}) \times \frac{\sigma}{s_n}$$

But,

$$y_{150} = -\log_e \log_e\left(\frac{150}{150-1}\right) = 5$$

Therefore,

$$X_{150} = 25400 + (5 - 4.6) \times 7000 \qquad \text{[from Eq. (iv)]}$$

$$\Rightarrow \qquad X_{150} = 28200 \text{ m}^3/\text{s}$$

PROBLEM 7.3 Analysis of the maximum annual flood over the past 150 years in a small river indicates the following cumulative distribution:

Number	*Q*	$P(X < x_n)$
1	0	0
2	25	0.19
3	50	0.35
4	75	0.52
5	100	0.62
6	125	0.69
7	150	0.88
8	175	0.92
9	200	0.95
10	225	0.98
11	250	1.00

Estimate 10, 50 and 100 years flood. (IES, 2011)

Solution Probability of exceedence $P(X > x_n)$ of

$$\text{10-year flood} = \frac{1}{10} = 0.1 = 0.1$$

$$\text{50-year flood} = \frac{1}{50} = 0.02$$

$$\text{100-year flood} = \frac{1}{100} = 0.01$$

Probability of non-exceedence $P(X < x_n)$ of

$$\text{10-year flood} = 1 - 0.1 = 0.9$$
$$\text{50-year flood} = 1 - 0.02 = 0.98$$
$$\text{100-year flood} = 1 - 0.01 = 0.99$$

Using interpolation from given table data,

$$\text{10-year flood} = 150 + \frac{175-150}{0.92-0.88} \times (0.9 - 0.88) = 162.5 \text{ m}^3/\text{s}$$

50-year flood = 225 m^3/s

$$100\text{-year flood} = 225 + \frac{250-225}{1.00-0.98} \times (0.99-0.98) = 237.5 \text{ m}^3/\text{s}$$

PROBLEM 7.4 The working life of a dam built to store irrigation requirement is expected to be 100 years. The spillway capacity is designed to accommodate the peak flood having a return period of 500 years. Calculate the risk of failure of the dam. (IES, 2012)

Solution Given,

$$n = 100 \text{ years, return period, } T = 500 \text{ years}$$

Risk,

$$\overline{R} = 1 - (1-P)^n$$

$$\overline{R} = 1 - \left(1 - \frac{1}{T}\right)^n$$

where, P is probability $\left(= \frac{1}{T}\right)$ and T is return period.

Therefore,

$$\overline{R} = 1 - \left(1 - \frac{1}{500}\right)^{100} = 0.1814$$

PROBLEM 7.5 For data of maximum recorded flood of a river, the mean and standard deviation are 4200 m^3/s and 1705 m^3/s respectively. Using Gumbel's extreme value distribution, estimate the return period of a design flood of 9550 m^3/s. Assume an infinite sample size.

(IES, 1996)

Solution From the given data we can write the following:

$$X_T = 9550 \text{ m}^3/\text{s}; \ \overline{X} = 4200 \text{ m}^3/\text{s}; \ \sigma = 1705 \text{ m}^3/\text{s}$$

Again, we can write the following relationship as

$$X_T = \overline{X} + K\sigma$$

where X_T is design flood, discharge, $\overline{X}$ is mean of flood discharge, σ is standard deviation of flood discharge and K is frequency factor.

Again, K can be determined from the following expression:

$$K = \frac{y_T - y_n}{s_n}$$

where y_T is reduced variate, $\overline{y}_n$ is reduced mean and s_n is reduced standard deviation when sample size in infinite.

Here,

$$\overline{y}_n = 0.577$$

$$s_n = 1.2825$$

Again, we know

$$X_T = \bar{X} + K\sigma \quad \text{or} \quad X_T = \bar{X} + \left(\frac{y_T - \bar{y}_n}{s_n}\right)\sigma$$

Now, after putting values, we get

$$9550 = 4200 + \left(\frac{y_T - 0.577}{1.2825}\right) \times 1705$$

$$y_T = 4.6013$$

Again, we can write the expression of y_T as a function of recurrence interval.

$$y_T = -\log_e \log_e \left[\frac{T}{T-1}\right]$$

$$4.6013 = -\log_e \log_e \left[\frac{T}{T-1}\right]$$

$$e^{-4.6013} = \log_e \left[\frac{T}{T-1}\right]$$

$$1.0101 = \frac{T}{T-1}$$

$$T = 1.0101 \times (T - 1)$$

$$T = 100.01 \text{ years}$$

Recurrence interval or return period is 100.01 years.

PROBLEM 7.6 Given below the monthly rainfall, P and the corresponding runoff values, R for a period of 10 months for a catchment. Develop a correlation between R and P.

Number of month	P (cm)	R (cms)
1	4	0.2
2	22	7.1
3	28	10.9
4	15	4
5	12	3
6	8	1.3
7	4	0.4
8	15	4.1
9	10	2
10	5	0.3

(IES, 2003)

Solution Interdependency of rainfall and runoff is quite complex due to their dependency on various other topographical and climatic parameters. Here, linear regression technique has been used to relate P with R.

Number of Month	*P*	*R*	P^2	R^2	*PR*
1	4	0.2	16	0.04	0.8
2	22	7.1	484	50.41	156.2
3	28	10.9	784	118.81	305.2
4	15	4	225	16	60
5	12	3	144	9	36
6	8	1.3	64	1.69	10.4
7	4	0.4	16	0.16	1.6
8	15	4.1	225	16.81	61.5
9	10	2	100	4	20
10	5	0.3	25	0.09	1.5
$\sum$	123	33.3	2083	217.01	653.2

For linear regression, we can write

$$R = aP + b$$

Now, the value of a can be calculated as

$$a = \frac{\left(N\sum PR\right) - \left(\sum P \sum R\right)}{\left(N\sum P^2\right) - \left(\sum P\right)^2}$$

where, N is the number of months.

$$a = \frac{(10 \times 653.2) - (123 \times 33.3)}{(10 \times 2083) - (123)^2} = 0.4273$$

Now, the value of b can be calculated as

$$b = \frac{\sum R - \left(a \times \sum P\right)}{N}$$

$$b = \frac{33.3 - (0.4273 \times 123)}{10} = -1.93$$

Again, the coefficient of correlation can be calculated by the following formula:

$$r = \frac{\left(N\sum PR\right) - \left(\sum P \sum R\right)}{\sqrt{\left[N\sum P^2 - \left(\sum P\right)^2\right]\left[N\sum R^2 - \left(\sum R\right)^2\right]}}$$

$$r = 0.99$$

PROBLEM 7.7 Flood frequency computations for a flashy river at a point 50 km upstream of a bund site indicated the following:

Return period T (years)	*Peak flood* (m^3/s)
50	20600
100	22150

Estimate the flood magnitude in the river with a return period of 500 years using Gumbel's method. (IES, 2003)

Solution Using Gumbel's method, we can write that

$$X_T = \overline{X} + K\sigma \tag{i}$$

where X_T is design flood discharge, $\overline{X}$ is mean of flood discharge, σ is standard deviation of flood discharge and K is frequency factor.
Again, K can be determined from the following expression:

$$K = \frac{y_T - \overline{y}_n}{s_n}$$

where y_T is called reduced variate, $\overline{y}_n$ is reduced mean and s_n is reduced standard deviation when sample size is infinite.
Again,

$$y_T = -\log_e \log_e \left[\frac{T}{T-1}\right]$$

From the above given data, we write
$X_{50} = 20600$, $X_{100} = 22150$

$$y_{50} = -\log_e \log_e \left[\frac{50}{50-1}\right] = 3.90194$$

$$y_{100} = -\log_e \log_e \left[\frac{100}{100-1}\right] = 4.60015$$

$$y_{500} = -\log_e \log_e \left[\frac{500}{500-1}\right] = 6.21361$$

Now, putting these values in Eq. (i), we get

$$20600 = \overline{X} + \left(\frac{y_{50}}{s_n} - \frac{\overline{y}_n}{s_n}\right)\sigma \tag{ii}$$

$$22150 = \overline{X} + \left(\frac{y_{100}}{s_n} - \frac{\overline{y}_n}{s_n}\right)\sigma \tag{iii}$$

Now, subtracting Eq. (ii) from Eq. (iii), we can write

$$\left(\frac{y_{100}}{s_n} - \frac{\overline{y}_{50}}{s_n}\right)\sigma = 1550 \tag{iv}$$

$$(4.60015 - 3.90914)\frac{\sigma}{s_n} = 1550$$

$$\frac{\sigma}{s_n} = 2219.9625$$

Again, we know

$$X_{500} = \bar{x} + \left(\frac{y_{500}}{s_n} - \frac{\bar{y}_n}{s_n}\right)\sigma$$

Again, subtracting Eq. (iii) from Eq. (iv), we can write

$$X_{500} - 22150 = (y_{500} - y_{100})\frac{\sigma}{s_n}$$

$$X_{500} = 25731.82 \text{ m}^3/\text{s}$$

PROBLEM 7.8 A hydraulic structure is sized for a 50-year recurrence interval design discharge. What is the risk that the flow capacity will be exceeded during any future 20-year period? What is the probability that the 50-year recurrence interval peak flow rate will be exceeded in the next 50 years? [IES, 2005]

Solution The recurrence interval, T is given as 50 years. Now, the probability of peak flow exceedance

$$P = \frac{1}{T} = \frac{1}{50} = 0.02$$

So, the probability that the peak flow will not be exceeded

$$q = 1 - P = 1 - 0.02 = 0.98$$

The risk that the flow capacity will be exceeded during any future 20-year period

$$R = 1 - q^{20} = 1 - (0.98)^{20} = 0.3324$$

The risk that the flow capacity will be exceeded during any future 50-year period

$$R = 1 - q^{50} = 1 - (0.98)^{50} = 0.6358$$

PROBLEM 7.9 A flood series has mean equal to 300 m^3/s and standard deviation as 50 m^3/s. Compute the magnitude of 50-years flood using Gumbel and Chow methods. (IES, 2008)

Solution Using Gumbel's method we can write that

$$X_T = \bar{X} + K\sigma \quad \text{(i)}$$

where X_T is design flood discharge, $\bar{X}$ is mean of flood discharge, σ is standard deviation of flood discharge and K is frequency factor.
Again, K can be determined from the following expression,

$$K = \frac{y_t - \bar{y}_n}{s_n}$$

where y_T is reduced variate, y_n is reduced mean and s_n is reduced standard deviation when sample size is infinite

$$K = \frac{y_T - 0.577}{1.2825} \tag{ii}$$

Again,

$$y_T = -\log_e \log_e \left[\frac{T}{T-1}\right]$$

Here, $T = 50$ years,

$$y_{50} = -\log_e \log_e \left[\frac{50}{50-1}\right] = 3.90194$$

Using Eq. (ii), we can write

$$K = \frac{y_T - 0.577}{1.2825}$$

$$K = \frac{3.90194 - 0.577}{1.2825} = 2.5925$$

Now, putting these values in Eq. (i), we get

$$X_T = \overline{X} + K\sigma$$

$$X_T = 300 + 2.5925 \times 50 = 429.63 \text{ m}^3/\text{s}.$$

PROBLEM 7.10 The regression analysis of a 30-year flood data at a point on a river yielded sample mean $\overline{x} = 1200 \text{ m}^3/\text{s}$ and a standard deviation $S_X = 650 \text{ m}^3/\text{s}$. For what discharge would you design the structure at this point to provide 95% assurance that the structure would not fail in the next 50 years? Use Gumbel's method. The value of mean and standard deviation of the reduced variate for $n = 30$ are 0.53622 and 1.11238, respectively. (IES, 2000)

Solution From the given data, we can write the following:

$$N = 30$$

where N is the number of years.

Mean of reduced variate, $\overline{y}_n = 0.53622$

Standard deviation of reduced variate, $\overline{s}_n = 1.11238$

Reliability of the structure = 95%

Again, regression analysis of a 30-year flood data at a point on the river yielded the following data:

Sample mean, $\overline{x} = 1200 \text{ m}^3/\text{s}$

Standard deviation, $S_X = 650 \text{ m}^3/\text{s}$

Again, we know

$$\text{Reliability} = \left(1 - \frac{1}{T}\right)^n$$

Here, n = 50 years.

$$\frac{95}{100} = \left(1 - \frac{1}{T}\right)^{50}$$

$$T = 975.29 \text{ years}$$

We can also write that

$$X_T = \bar{X} + KS_X$$

Again, the expression for K is

$$K = \frac{y_T - \bar{y}_n}{S_n} \quad \text{(i)}$$

$$K = \frac{6.88222 - 0.53622}{1.11238} = 5.7049$$

Now, using Eq. (i), we can write

$$y_T = y_{975.29} - \log_e \log_e\left[\frac{975.29}{975.29 - 1}\right]$$

$$= 6.88222$$

So,

$$X_T = 1200 + (5.7049 \times 650) = 4908.17 \text{ m}^3/\text{s}.$$

PROBLEM 7.11 The daily flows in a river for three consecutive years are given in the following table by class intervals along with the number of days the flow belonged to this class. What are the 50% and 75% dependable flows for the river?

Daily mean discharge (m^3/s)	*Number of days in each class interval* 1981	1982	1983
100–90.1	0	6	10
90–80.1	16	19	16
80–70.1	27	25	38
70–60.1	21	60	67
60–50.1	43	51	58
50–40.1	59	38	38
40–30.1	64	29	70
30–20.1	22	48	29
20–10.1	59	63	26
10–negligible	54	26	13

Estimate the average depth of precipitation over the basin. (IES, 2002)

Solution If we multiply the given ranges of daily mean discharges of flow by their corresponding days of flow for a particular year and are added together, then it will provide us with the total amount of annual discharge of the river for that particular year. In this way, annual discharge for all the three years mentioned above are calculated in the table below:

Daily mean discharge (m^3/s)	*Mean value of discharge*	*Number of days* 1981	*Yield* (million m^3)	*Number of days* 1982	*Yield* (million m^3)	*Number of days* 1983	*Yield* (million m^3)
100–90.1	95.05	0	0	6	49.27	10	82.12
90–80.1	85.05	16	117.57	19	139.62	16	117.57
80–70.1	75.05	27	175.08	25	162.11	38	246.4
70–60.1	65.05	21	118.03	60	337.22	67	376.56
60–50.1	55.05	43	204.52	51	242.57	58	275.87
50–40.1	45.05	59	229.65	38	147.91	38	147.91
40–30.1	35.05	64	193.81	29	87.82	70	211.98
30–20.1	25.05	22	47.61	48	103.89	29	62.76
20–10.1	15.05	59	76.72	63	81.92	26	33.81
10–negligible	5.05	54	23.56	26	11.34	13	5.67
	Total	365	1186.55	365	1363.67	365	1560.65

Here, considering N as number of years, we get

$$N = 3$$

Now we know,

$$P = \frac{m}{N} \times 100$$

where m is order number

For 50% dependable flow,

$$50 = \frac{m}{3} \times 100$$

$$m = 1.5$$

50% dependability of water yield is taken as average of water yield for $m = 1$ and $m = 2$, as the value of m, obtained is not an integer.

$$= \frac{1560.65 + 1363.37}{2}$$

$$= 1462.01 \text{ million m}^3$$

Hence, annual flow with 50% dependability = 1462.01 million m^3

$$\text{Daily discharge with 50\%} = \frac{1462.16 \times 10^6}{365 \times 24 \times 60 \times 60}$$

$$= 46.36 \text{ m}^3\text{/s}$$

For 75% dependable flow,

$$75 = \frac{m}{3} \times 100$$

$$m = 2.25$$

75% dependability of water yield is taken as average of water yield for $m = 2$ and $m = 3$, as the value of m obtained is not an integer. So 75% dependability of water yield is given as

$$= \frac{(1363.67 + 1186.55)}{2}$$

$$= 1275.11 \text{ million m}^3$$

$$\text{Daily discharge with 75\% dependability} = \frac{1275.11 \times 10^6}{365 \times 24 \times 60 \times 60}$$

$$= 40.43 \text{ m}^3\text{/s}$$

PROBLEM 7.12 A large sample of peak floods data was available for a river. Flood frequency computation, using Gumbel's method, yields the following results:

Return period T (years)	*Peak period* (m^3/s)
50	30800
100	36300

Estimate the magnitude of a flood for this river with return period of 200 years. [IES, 1997]

Solution Using Gumbel's method, we can write that

$$X_T = \overline{X} + K\sigma \quad \text{(i)}$$

where X_T is design flood discharge, $\overline{X}$ is mean of flood discharge, σ is standard deviation of flood discharge and K is frequency factor.

Again, K can be determined from the following expression,

$$K = \frac{y_T - \overline{y}_n}{s_n}$$

where y_T is called reduced variate, $\overline{y}_n$ is reduced mean and s_n is reduced standard deviation when sample size is infinite.

Again,

$$y_T = -\log_e \log_e \left[\frac{T}{T-1}\right]$$

From the given data, we can write

$$X_{50} = 30800, X_{100} = 36300$$

$$y_{50} = -\log_e \log_e \left[\frac{50}{50-1}\right]$$

$$= 3.90194$$

$$y_{100} = -\log_e \log_e \left[\frac{100}{100-1}\right]$$

$$= 4.60015$$

$$y_{500} = -\log_e \log_e \left[\frac{500}{500-1}\right]$$
$$= 6.21361$$

Now, putting these above values in Eq. (i), we get

$$30800 = \bar{X} + \left(\frac{y_{50}}{s_n} - \frac{\bar{y}_n}{s_n}\right)\sigma \quad \text{(ii)}$$

$$36300 = \bar{X} + \left(\frac{y_{100}}{s_n} - \frac{\bar{y}_n}{s_n}\right)\sigma \quad \text{(iii)}$$

Now, subtracting Eq. (ii) from Eq. (iii), we can write

$$\left(\frac{y_{100}}{s_n} - \frac{y_{50}}{s_n}\right)\sigma = 5500 \quad \text{(iv)}$$

$$(4.60015 - 3.90194)\frac{\sigma}{s_n} = 5500$$

$$\frac{\sigma}{s_n} = 7877.2862$$

Again, we know

$$X_{200} = \bar{x} + \left(\frac{y_{200}}{s_n} - \frac{\bar{y}_n}{s_n}\right)\sigma$$

Again subtracting Eq. (iii) from Eq. (iv), we can write

$$X_{200} - 36300 = (y_{200} - y_{100})\frac{\sigma}{s_n}$$

$$y_{200} = -\log_e \log_e \left[\frac{200}{200-1}\right] = 5.2958$$

$$X_{200} = 41779.83 \text{ m}^3/\text{s}$$

Theoretical Questions

1. What do you understand by standard deviation and skewness?
2. What do you understand by return period?
3. What is standardised normal random variable? What are its mean and variance? Describe the method of determining T-year flood using Gumbel's distribution.
4. What do you understand by frequency factor?
5. Write a short note on the log-normal distribution method for frequency analysis.

Unsolved Problems

Problem 1: The mean monthly rainfall (mm) records at a station for a year are given in the table below. Find the mean rainfall, variance, standard deviation and coefficient of variation.

Month	January	February	March	April	May	June	July	August	September	October	November	December
Rainfall	12	10	5	2	4	8	20	35	27	15	12	12

Problem 2: For a rain gauge station, the annual precipitation data for 12 years were recorded

Year	1	2	3	4	5	6	7	8	9	10	11	12
Rainfall (mm)	142.4	133	147.6	125.2	143.8	152	130.15	150	147	129.8	140.89	137

(a) The annual precipitation in any year will be less than 130 mm
(b) The annual precipitation in any year will be greater than 140 mm
(c) The annual precipitation in any year will be less than 140 mm
(d) There will be 3 successive years of precipitation less than 130 mm

Problem 3: Let the probability of rainfall occurrence on any day in the month of May be 0.3. If you are wishing a no rainfall 5-day period as you are planning for a 7-day study tour
(a) Find the probability that the rainfall will not occur in these 5 days
(b) Find the probability of 1 rainy day during these 5 days

Problem 4: What is the probability that a 5-year flood will occur
(a) In the next year
(b) At least once during the next five years
(c) At least once during the next 50 years

Problem 5: Using 25 years data and Gumbel's method, the flood magnitudes for return periods of 200 and 100 years for a river are found to be 1400 cumec and 1200 cumec, respectively. Determine the mean and standard deviation of the data used, and, also estimate the magnitude of a flood with a return period of 500 years.

Problem 6: A flood control structure has a life of 30 years. Estimate the required return period of a flood if the acceptable risk of failure of the spillway is 10%. (a) in any year, and (b) over its design life.

Problem 7: A temporary dam is to be built to protect a 5-year construction activity for a hydroelectric power plant. If the dam is designed to withstand 25-year flood, determine the risk that the structure will be overtopped
(a) At least once in the 5-year construction period
(b) Not at all during the 5-year period

Problem 8: The data of maximum recorded annual floods of a river, i.e., the mean and the standard deviation are 4500 cumec and 17500 cumec, respectively. Using Gumbel's extreme value distribution, estimate the return period of a design flood of 10000 cumec. Assume an infinite sample size.

Problem 9: A stream has the flood peaks of 1000 m^3/s and 900 m^3/s for the return period of 100 and 50 years respectively, using Gumbel's method for the dataset spanning 30 year. Find the

(a) Mean and standard deviation

(b) Flood peak corresponding to a return period of 500 years

Problem 10: Analysis of annual flood series of a river yielded a sample of 1200 cumec and standard deviation of 600 cumec. Calculate the design flood of a structure on this river to provide 95% assurance that the structure will not fail in the next 50 years. Use Gumbel's method and assume the sample size as too large.

Problem 11: A maximum 1-day rainfall depth of 250 mm at a station shows a return period of 60 years. Find the probability of 1-day rainfall depth equal to or greater than 250 mm occurring

(a) Once in 15 successive years,

(b) Twice in 12 successive years

(c) At least, once in 40 successive years.

Problem 12: Determine the probability of precipitation depth equal to or greater than 1100 mm occurring

(a) Once in 10 successive years

(b) Twice in 5 successive years

(c) At least once in 10 successive years

(d) Not at all in 10 years

Assume that the precipitation depth of 1100 mm has a 100-year return period.

Further Reading

Chow, V.T., Maidment, D.R., and Mays, L.W., *Applied Hydrology*, McGraw-Hill, Singapore, 1988.

Das, Ghanshyam, *Hydrology and Soil Conservation Engineering Including Watershed Management*, PHI Learning, Delhi, 2014.

Davie, Tim, *Fundamentals of Hydrology*, 2nd ed., Routledge, Taylor and Francis Group, London and New York, 2008.

Deodhar, M.J., *Elementary Engineering Hydrology*, Pearson, New Delhi, 2013.

Raghunath, H.M., *Hydrology—Principles, Analysis and Design*, New Age Publishers, New Delhi, 2014.

Reddy, P. Jaya Rami, *A Textbook of Hydrology*, 3rd ed., University Science Press, Delhi, 2014.

Subramanya, K. *Engineering Hydrology*, Tata McGraw-Hill, New Delhi, 2010.

Suresh, R., *Watershed Hydrology*, Standard Publishers Distributors, New Delhi, 2015.

Yadupathi Putty, Mysooru R., *Principles of Hydrology*, I.K. International, New Delhi, 2013.

CHAPTER 8

Flood Routing

8.1 Introduction

A hydrological boundary or watershed or catchment receives precipitation (in different forms such as rainfall, snowfall, etc.) as input and produces a runoff as output. The outflow or runoff hydrograph mostly differs from the input or inflow hydrograph in shape, duration and magnitude. These variations are featured to storage characteristics of the watershed system. The flood hydrograph can be termed as a high intensity wave or flood wave. When the input hydrograph and other physical dimensions of the storage are known, then the method to compute the outflow hydrograph is called *flood routing*. A *rating curve* is a graph of discharge versus stage for a given point on a stream, and it is necessary to determine the flood waves. The passage of waves of the river depth and discharge can be represented by stage and discharge hydrographs, respectively. Flood routing is a procedure to compute the flood hydrograph at a section of a river by applying the data of flood flow at one or more upstream sections. The routed hydrograph is delayed by a time lag known as *translation* and is attenuated. The study of the basic features of these changes in a flood wave passing through a channel system (flood flows from upstream to downstream river channel) forms the subject matter of this chapter.

Flood routing methods are generally used for flood forecasting, design of spillways, reservoirs and flood control structures, etc. Flood routing methods are classified into two broad categories—(a) reservoir routing and (b) channel routing.

In *reservoir routing*, an inflection effect on a flood wave is considered when it passes through a reservoir. In this, the outflow hydrographs can be determined with attenuated peaks and enlarged time bases. In reservoir routing, variations in reservoir elevation and outflow (discharge) can be predicted with respect to time. In this case, the relationships between elevation and volume, and elevation and outflow of the reservoir are known.

In *channel routing*, the changes in the shape of the input hydrograph are mainly studied while the flood wave passes through a channel downstream. Channel routing helps in predicting flood hydrographs at various sections of the channel. In this case, the input hydrograph and channel-reach characteristics are known. The information generated on attenuation of the hydrograph peak and durations of high water levels assists in forecasting floods and taking preventive measures against floods.

A variety of procedures of flood routing is available and they can be broadly classified as hydrologic routing and hydraulic routing. Hydrologic routing methods employ principally, the law of conservation of mass or continuity equation. Hydraulic methods can be employed by continuity equation, together with the equation of motion or mass momentum of unsteady flow (flow varies with time).

The routing methods can also be termed as *lumped and distributed system routing methods*. The difference between lumped and distributed system routing is that in a lumped system model, the flow is calculated as a function of time only alone at a particular location. While, in a distributed system routing, the flow is calculated as a function of time and space both. Routing by lumped system method is referred to as *hydrologic routing*, and routing by distributed system methods is termed as *hydraulic routing*.

8.2 Inflow-Outflow Basic Equations

In case of the open channel hydraulics, the passage of a flood hydrograph through a reservoir is defined as an unsteady flow phenomenon. In open channel hydraulics, the continuity equation is used in all hydrologic routing as the primary equation. The continuity equation states that the difference between the inflow and outflow rate is equal to the rate of change of storage. The differential continuity equation for the reservoir can be written as

$$I - Q = \frac{dS}{dt} \tag{8.1}$$

where, I is inflow rate, Q is outflow rate and S is the storage. Alternatively, in a small time interval (Δt), the difference between the total inflow and total outflow in a reach is equal to the change in storage in that reach.

$$\bar{I}\Delta t - \bar{Q}\Delta t = \Delta S \tag{8.2}$$

where, I is the average inflow in time Δt, $\bar{Q}$ is the average outflow in time Δt and ΔS is the change in storage. Here, $\bar{I} = \frac{(I_1 + I_2)}{2}$, $\bar{Q} = \frac{(Q_1 + Q_2)}{2}$ and $\Delta S = S_2 - S_1$. Here, 1 and 2 are used to denote the starting and end of the time interval Δt. Equation (8.2) can be written as

$$\left(\frac{I_1 + I_2}{2}\right)\Delta t - \left(\frac{Q_1 + Q_2}{2}\right)\Delta t = S_2 - S_1 \tag{8.3}$$

The time interval Δt should be sufficiently short so that the inflow and outflow hydrographs can be assumed to be straight lines in the given time interval. The time of interval Δt must be shorter than the time of transit of the flood wave within the river channel.

8.3 Level Pool Routing (Hydrologic Storage Routing)

Level pool routing is a method for producing the outflow hydrograph from a reservoir with a horizontal water surface. In level pool routing, its inflow hydrograph and storage outflow characteristics are known (8.1). The outflow is a function of the reservoir elevation, which can be written as

$$Q = Q(h) \tag{8.4}$$

The storage in the reservoir is a function of the reservoir elevation, $S = S(h)$. When the flood waves passes through the reservoir, the water level in the storage or reservoir varies with time, $h = h(t)$. Therefore, the storage and discharge change with time. The computation of variations in the values of Q, S and h with respect to time t is called *reservoir routing* as shown in Figure 8.1. This is a genuine case and is applicable for the controlled spillway.

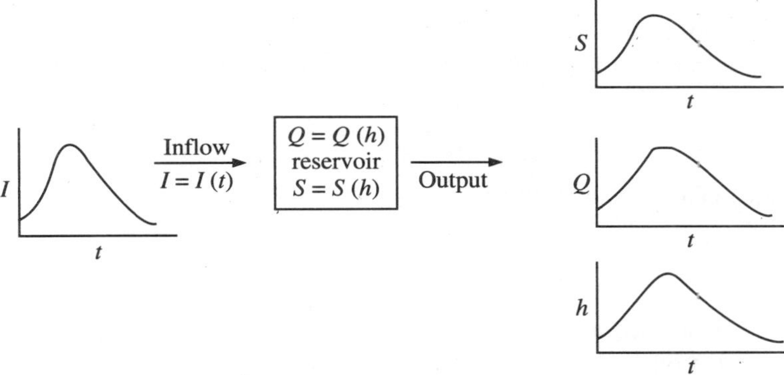

Figure 8.1 Storage routing.

For an uncontrolled spillway, the outflow discharge Q can be computed as

$$Q = \frac{2}{3} C_d \sqrt{2g} L_e H^{3/2} = Q(h) \tag{8.5}$$

where H is the head over the spillway, L_e is the effective length of the spillway crest and C_d is the coefficient of discharge. In case of the reservoir routing, the following information should be collected from the field:

1. Field data, establishing relationship between elevation, outflow and storage of the reservoir
2. Inflow hydrograph, $I(t)$
3. Initial values of storage, outflow and inflow, i.e., recorded values of S, Q and I at the time of start of flow ($t = 0$)

8.3.1 Methods of Hydrologic Storage Routing

There is a variety of routing methods available for flood hydrograph computation in the reservoir. Typical elevation-storage, elevation-discharge and storage-discharge curves for a reservoir are shown in Figure 8.2. In this chapter, two very common methods, i.e., Goodrich method and modified Pul's method are discussed ahead.

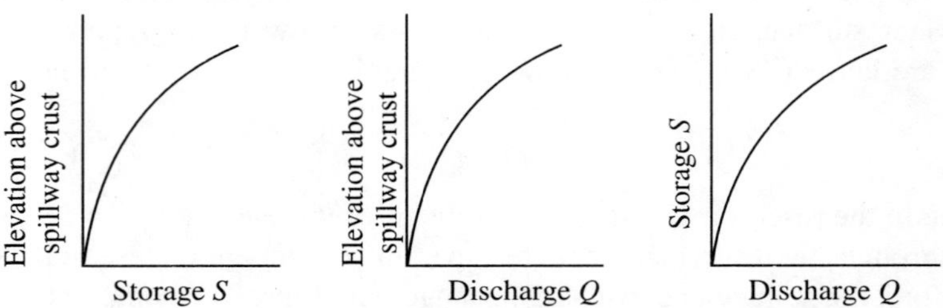

Figure 8.2 Typical elevation-storage, elevation-discharge and storage-discharge curves for a reservoir.

Goodrich method

In the Goodrich method, the inflow-outflow relationships, as per Eq. (8.3), are rearranged in terms of the flow rate and storage rate. The new equation can be written as

$$\frac{2S_2}{\Delta t} - \frac{2S_1}{\Delta t} = I_1 + I_2 - Q_1 - Q_2 \tag{8.6}$$

Equation (8.6) can be expressed as

$$\frac{2S_2}{\Delta t} + Q_2 = I_1 + I_2 + \left(\frac{2S_1}{\Delta t} - Q_1\right) \tag{8.7}$$

Equation (8.7) is termed as *Goodrich equation for reservoir routing*. The stepwise solution of the Goodrich equation is given as follows:

1. Select the time increment Δt, which is equal to about 20%–40% of the time of rise of the inflow hydrograph. Then, calculate $\left(\frac{2S}{\Delta t} + Q\right)$ from the relationship of the field data between S and Q.
2. The two graphical relationships between the $\left(\frac{2S}{\Delta t} + Q\right)$ and reservoir elevation and another between the discharge and reservoir elevation can be plotted.
3. Based on the step 2, for the initial time ($t = 0$), obtain the values for outflow (discharge, Q), reservoir elevation and $\left(\frac{2S}{\Delta t} + Q\right)$. Now, from these values, $\left(\frac{2S_1}{\Delta t} - Q_1\right)$ will be calculated by subtracting $2Q_2$ from $\left(\frac{2S}{\Delta t} + Q\right)$.
4. Now, calculate $\left(\frac{2S_2}{\Delta t} - Q_2\right)$ for the first time interval by using Eq. (8.9), where the values of I_1, I_2, Q_1 (taken from the available data) and $\left(\frac{2S_1}{\Delta t} - Q_1\right)$ (as computed at step 3) are known.

5. Now, from step 2, find out the elevation of the reservoir and Q for $\left(\frac{2S_2}{\Delta t} - Q_2\right)$ and calculate $\left(\frac{2S}{\Delta t} + Q\right)$.
6. Keep repeating the procedure till the entire flood is routed.

Modified Pul's method

In this method, the outflow from the reservoir is to be released as per the requirements of the downstream channel capacity. When using the modified Pul's method, it is assumed that a unique and single-valued stage-storage-outflow relationship exists for each channel, and that changing downstream conditions will not alter this relationship. This method is basically preferred for the flood control measures. However, this method is generally not recommended for channels with gradients less than 3 ft/mile and channels with time varying downstream boundaries such as tidal influences or rapidly rising flood hydrographs such as dam breaks.

In modified Pul's method, Eq. (8.3) is rewritten in terms of inflow, outflow and storage of the reservoir as follows:

$$\frac{1}{2}(I_1 + I_2)\Delta t + \left(S_1 - \frac{Q_1 \Delta t}{2}\right) = S_2 + \frac{Q_2 \Delta t}{2} \tag{8.8}$$

The right term $\left(S_2 + \frac{Q_2 \Delta t}{2}\right)$ of Eq. (8.8) is generally denoted as a storage indication.

For this method, measured or observed discharge (outflow) and initial storage of the reservoir are necessary. In Eq. (8.8), L.H.S. values, at the time step Δt, are computed based on the existed relationship between the reservoir elevation, storage and outflow (discharge). Apart from this, the values of the term $\left(S + \frac{Q\Delta t}{2}\right)$ on the R.H.S., the reservoir elevation, and subsequently, the outflow at the end of the time step Δt are also calculated. The process is repeated for all the values of the inflow hydrograph. The steps to be followed are as follows:

1. Take a time interval Δt equal to about 20%–40% of the time of rise of the inflow hydrograph.
2. Solve the term $\left(S + \frac{Q\Delta t}{2}\right)$ versus reservoir elevation from the respective relationships of the recorded data of elevation versus storage and outflow (discharge) of the reservoir.
3. As per step 2, with the elevation on the ordinate scale, plot the elevation versus discharge relationship from the recorded data.
4. As per the initial step of the routing method, the initial storage, elevation and outflow (discharge) are already known. Hence, for the time interval Δt of the inflow hydrograph, compute values for the L.H.S. of Eq. (8.8), i.e., $\left(\frac{I_1 + I_2}{2}\right)\Delta t$ and $\left(S_1 - \frac{Q_1 \Delta t}{2}\right)$ and measure the value of the R.H.S. term $\left(S_2 + \frac{Q_2 \Delta t}{2}\right)$.

5. Now, calculate the reservoir elevation corresponding to $\left(S_2 + \frac{Q_2 \Delta t}{2}\right)$ by utilising the calculation of step 2 and also calculate discharge, Q at the end of the time step Δt from the same plot.
6. Now, deduct $Q_2 \Delta t$ from $\left(S_2 + \frac{Q_2 \Delta t}{2}\right)$. This gives a new term $\left(S - \frac{Q \Delta t}{2}\right)$ which is the R.H.S. of Eq. (8.8) for the start of the next time step Δt.
7. This process will be repeated till the full inflow hydrograph is routed.

EXAMPLE 8.1 A reservoir has the following elevation, discharge and storage relationships-

Elevation (m)	*Storage* (10^6 m^3)	*Outflow discharge* (m^3/s)
91.0	3.050	0
91.5	3.075	10
92.1	3.375	25
92.5	3.550	45
93.0	3.950	70
93.5	4.150	90
93.75	4.450	105
94.0	4.950	130
94.5	5.250	145
95.0	5.750	160

When the reservoir level was at 91.5 m, the following flood hydrograph entered the reservoir:

Time (h)	0	4	8	12	16	20	24	28	32	36	40	44	48
Discharge (m^3/s)	12	18	28	45	55	60	85	65	40	32	25	15	10

Route the flood and obtain the outflow hydrograph and reservoir elevation curve for the flood discharge.

Solution Graphs depicting Q versus elevation and $\left(\frac{2S}{\Delta t} + Q\right)$ [Figures (a) and (b)] are prepared from the following data given in the table:

Elevation (m)	*Storage* (10^6 m^3)	*Outflow discharge* (m^3/s)	$\left(\frac{2S}{\Delta t} + Q\right)$
91	3.05	0	423.611
91.5	3.075	10	437.083
92.1	3.375	25	493.750

(*Contd.*)

Elevation (m)	*Storage* (10^6 m^3)	*Outflow discharge* (m^3/s)	$\left(\frac{2S}{\Delta t}+Q\right)$
92.5	3.550	45	538.055
93	3.950	70	618.611
93.5	4.150	90	666.388
93.75	4.450	105	723.055
94	4.950	130	817.500
94.5	5.250	145	874.165
95	5.750	160	958.611

At $t = 0$,

$$\text{Elevation} = 91.5 \text{ m}$$

$$Q = 10 \text{ m}^3/\text{s}$$

$$\left(\frac{2S}{\Delta t}+Q\right)_1 = 437.08 \text{ m}^3/\text{s}$$

$$\left(\frac{2S}{\Delta t}-Q\right)_1 = 437.08 - 2 \times 10 = 417.08 \text{ m}^3/\text{s}$$

For the first time interval of 4 h,

$$I_1 = 12, I_2 = 18, Q_1 = 12 \text{ m}^3/\text{s}$$

$$\left(\frac{2S}{\Delta t}+Q\right)_2 = (12 + 18) + 417.08 = 447.08 \text{ m}^3/\text{s}$$

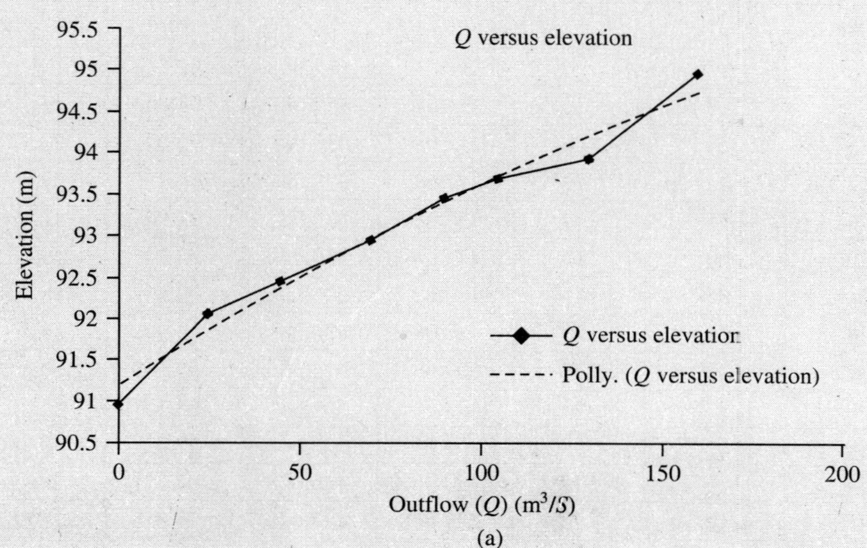

(a)

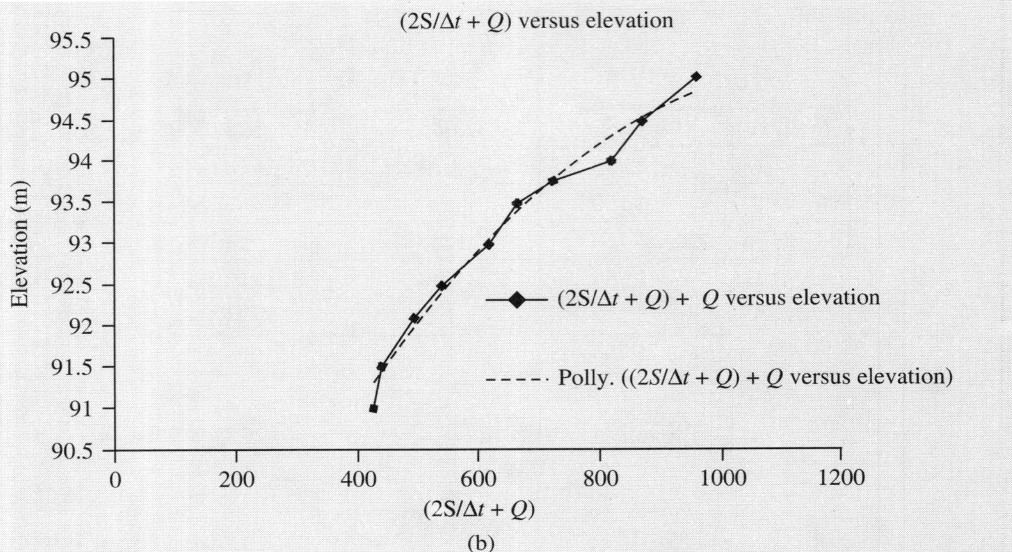

(b)

From the graph, reservoir elevation is calculated to be 91.844 m, and the corresponding discharge is 17 m^3/s.

For the next increment,

$$\left(\frac{2S}{\Delta t} - Q\right)_2 = 447.08 - 2 \times 17 = 413.08 \text{ m}^3\text{/s}$$

This process is repeated and presented in tabular form below. It is the reservoir routing using level pool method.

Time	*I*	$I_1 + I_2$	$2S/\Delta t - Q$	$2S/\Delta t - Q$	*Elevation*	*Discharge*
0	12			437.08	91.5	10
4	18	30	417.08	447.08	91.844	17
8	28	46	413.08	459.08	92.02	20.87
12	45	73	417.34	490.34	92.456	29.83
16	55	100	430.68	530.68	92.94	42.87
20	60	115	444.94	559.94	93.24	49.66
24	85	145	460.62	605.62	93.65	58.77
28	65	150	488.08	638.08	93.9	64.2
32	40	105	509.68	614.68	93.729	60.5
36	32	72	493.68	565.68	93.301	51.01
40	25	57	463.66	520.66	92.827	40.296
44	15	40	440.068	480.068	92.321	28.71
48	10	25	422.648	447.648	91.853	18.08

Using the time, elevation and discharge data, the outflow hydrograph and reservoir elevation curve are plotted with time for the flood discharge [Figure (c)].

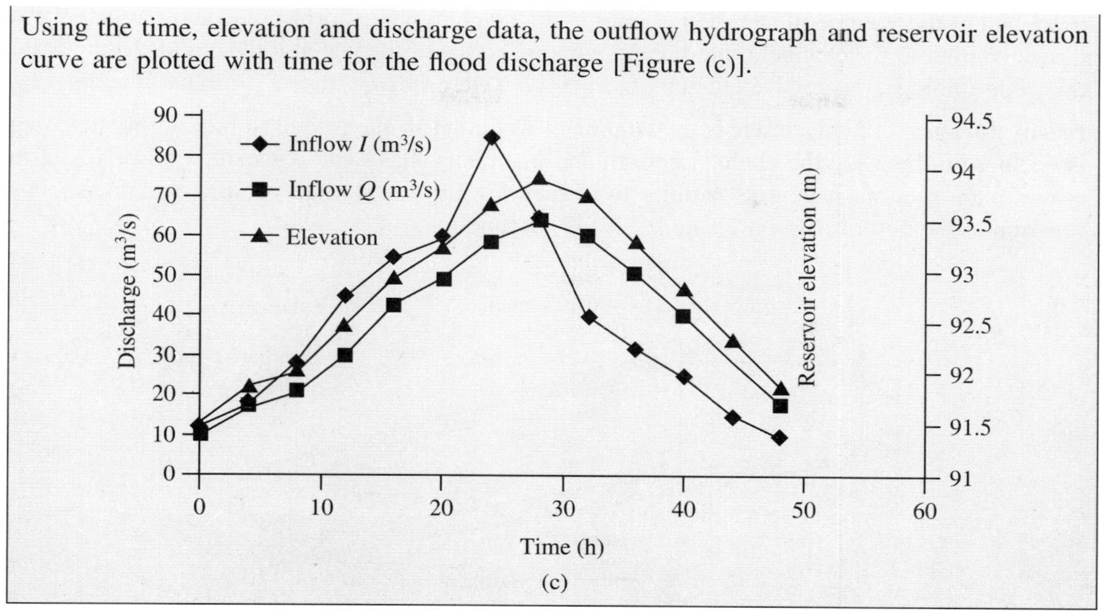

(c)

8.4 Hydrologic Channel Routing

When flood waves enter into the stream channel, it experiences delay and attenuation. If the peak of the outflow hydrograph is smaller than that of the inflow hydrograph, then the reduction in the peak value is called *attenuation*. Further, the peak of the outflow occurs after the peak of the inflow; the time difference between the two peaks is known as *lag*. If the river channel corresponds to a reservoir, then the attenuation and lag of a flood hydrograph are two very important aspects of a reservoir operating under a flood control criterion.

The study of the changes in the flood wave at various sections of the channel, generally expressed by flood hydrographs, is called *channel routing*. As the flood waves move down the river, the shape of the waves gets modified due to various factors such as channel resistance, storage, lateral addition or withdrawal of flows, etc. Flood waves passing down a river have their peaks attenuated due to friction if there is no lateral inflow. The addition of lateral inflows can cause a reduction of attenuation or even amplification of a flood wave.

8.4.1 Channel Storage

In the channel routing, the storage function depends on the outflow and inflow discharges both. Therefore, for channel routing, different routing methods have been formulated. In a river channel, a flood wave may be reduced in magnitude and lengthened in travel time, by storage in the reach between two sections, i.e., prism storage and wedge storage, as shown in Figure 8.3.

Wedge storage: When flood wave enters into the channel, the inflow exceeds the outflow and creates a wedge of storage, which is also called *positive wedge storage*. While during recession, the outflow exceeds the inflow, which is called *negative wedge storage*. *Wedge storage* can be described as the volume of storage in between the plane of water surface and an imaginary plane drawn parallel to the channel bottom from the top surface of the water at the outlet. At the outlet, the wedge storage gets reduced to zero. If S_w is the wedge storage, then $S_w = f(I)$, a function of I.

Prism storage: Prism storage is the volume of water below the wedge storage. A prism storage is enclosed in between the channel bottom and imaginary plane and constant at all points of the downstream section. It is, thus, similar to a reservoir and the storage volume is expressed as a function of the outflow discharge, i.e., $S_p = f(Q)$, a function of Q.

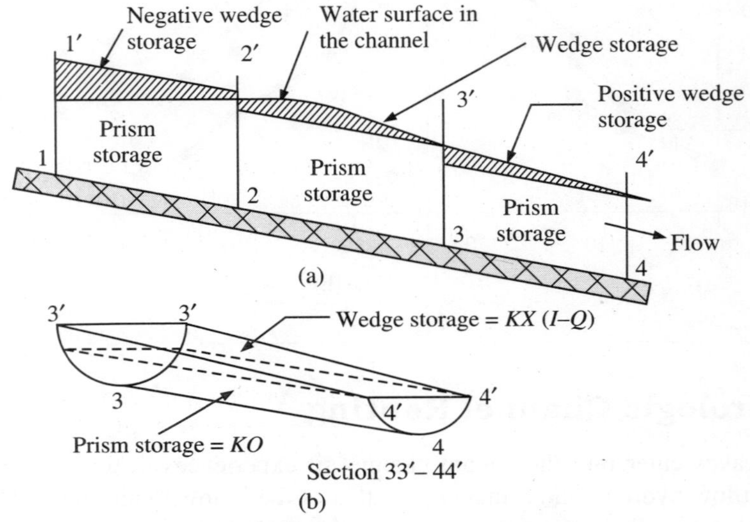

Figure 8.3 Storage in channel reach.

So, the total storage S in the channel is a function of the sum of the above two storages, which may be written as

$$S = f(S_w + S_p) \tag{8.9}$$

Here, $S_w = f(I)$, i.e., $S_w = xI^m$, where x is a coefficient.
$S_p = f(Q)$, i.e., $S_p = zQ^m$, where z is another coefficient.
S_w and S_p are added together in the proportion of x and z such that, $x + z = 1$, $z = (1 - x)$.
Therefore, Eq. (8.9) may be written as

$$S = K[xI^m + (1 - x)Q^m] \tag{8.10}$$

where k and x are coefficients and m is an exponent. The value of the exponent m varies from 0.6 for rectangular channels to 1.0 for natural channels.

Muskingum method

The Muskingum storage routing equation was developed by G.T. McCarthey under the Muskingum Conservancy District Flood Control Project of the US Army Corps of Engineers for natural river channels. This method is mostly used hydrologic routing method for handling discharge-storage relationships, where the value of $m = 1.0$.

Now, substituting the value of m in Eq. (8.10),

$$S = K[xI + (1-x)Q] \tag{8.11}$$

Equation (8.11) is known as the *Muskingum equation* and this equation represents a linear model for routing flow in river channels. In this, the parameter x is denoted as weighting factor and corresponds a value between 0 and 0.5. When $x = 0$, then the storage is a function of discharge only and this can be written as

$$S = KQ \tag{8.12}$$

When $x = 0.5$, then both the inflow and outflow are equally important in determining the storage. The coefficient K is known as *storage-time constant* and approximately equal to the time of travel of a flood wave through the channel reach or the basin lag.

Computation of K and x

By the use of continuity Eq. (8.1), the storage increment at any time t and time element Δt can be estimated. Figure 8.4 shows inflow and outflow hydrograph through a channel reach. Here, it should be noted that the outflow peak does not occur at the point of intersection of the inflow

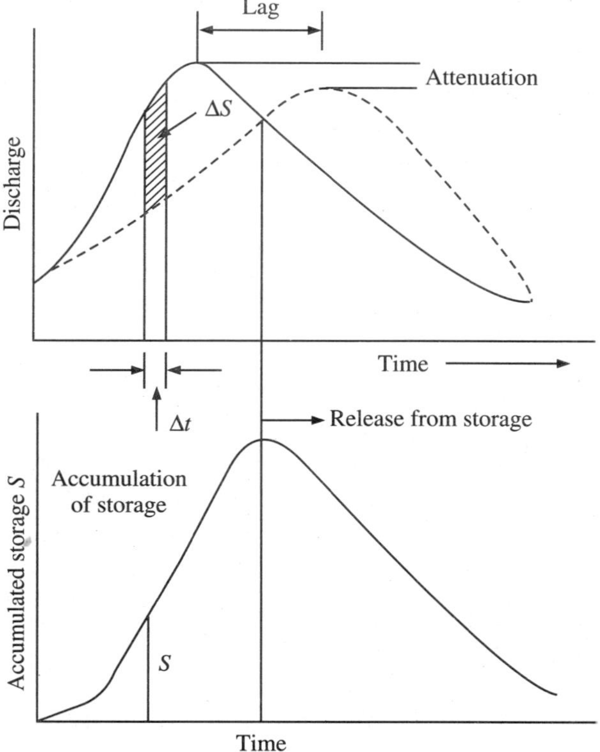

Figure 8.4 Hydrograph.

and outflow hydrographs. Figure 8.4 also shows the summation of the various incremental values, which represents the channel storage S versus time t relationship.

The values of S at various time intervals can be determined using the inflow and outflow hydrographs for a given reach. By selecting a trial value of x, values of S at any time t are plotted against the corresponding $[xI + (1 - x)Q]$ values. If the value of x is selected correctly, a straight line relationship will be occurred as per Eq. (8.12). If the value is incorrect, the plotted points will trace a looping curve. By trial and error, a value of x is so chosen that the data very nearly describe a straight line (Figure 8.4). The inverse slope of this straight line will give the value of K, i.e., $\dfrac{dS}{d[xI + (1 - x)Q]}$. Normally, for natural channels, the value of x falls between 0 to 0.3. For a given channel reach, the values of x and K are taken as constant.

Determination of K by empirical methods

1. Clark's method:

$$K = \frac{cL}{\sqrt{s}} \tag{8.13}$$

where, L is length of the main stream, c is constant (0.8 to 2.2, when the distance L is taken in miles) and s is mean slope of the channel.

2. Linsley's method:

$$K = \frac{bL\sqrt{A}}{\sqrt{s}} \tag{8.14}$$

where, L is length of the main stream (m), A is drainage area (km^2), b is constant (0.04 to 0.08, when the distance L is taken in miles) and s is mean slope of the channel.

As per the Himalayan catchment of Ramganga river, Agarwal and Das (1987) proposed the following values of constants for the above Eqs. (8.13) and (8.14):

$$c = 0.00427$$
$$b = 0.000133$$

Muskingum routing model

For any channel flow, if the routing interval between any two points (say 1 and 2) on the channel reach is Δt, then the difference in the storage of channel between the points 1 and 2 can be calculated as [by using the Muskingum Eq. (8.11)]

$$S_2 - S_1 = K[x(I_2 - I_1) + (1 - x)(Q_2 - Q_1)] \tag{8.15}$$

where the subscripts 1 and 2 are the parameters at points 1 and 2 after time Δt.

The continuity equation of the reach is

$$S_2 - S_1 = \left(\frac{I_2 + I_1}{2}\right)\Delta t - \left(\frac{Q_2 + Q_1}{2}\right)\Delta t \tag{8.16}$$

From Eqs. (8.15 and 8.16), Q_2 can be evaluated as

$$Q_2 = C_0 I_2 + C_1 I_1 + C_2 Q_1 \tag{8.17}$$

where

$$C_0 = \frac{-Kx + 0.5\Delta t}{K - Kx + 0.5\Delta t} \tag{8.18}$$

$$C_1 = \frac{Kx + 0.5\Delta t}{K - Kx + 0.5\Delta t} \tag{8.19}$$

$$C_2 = \frac{K - Kx - 0.5\Delta t}{K - Kx + 0.5\Delta t} \tag{8.20}$$

and $C_0 + C_1 + C_2 = 1.0$.

Equation (8.17) can be written in general form for the nth time step as

$$Q_n = C_0 I_n + C_1 I_{n-1} + C_2 Q_{n-1} \tag{8.21}$$

Equation (8.21) is known as *Muskingum Routing Equtaion* and provides a simple linear equation for channel routing. It is noticed that for the best results, the routine interval Δt should be so chosen that

$K > \Delta t > 2Kx$. If $\Delta t < 2Kx$, the coefficient C_0 will be negative. Normally, negative values of coefficients are avoided by choosing appropriate values of Δt.

To determine the inflow hydrograph using Muskingum equation through a river channel, the values of x and K for the river channel and the value of outflow Q_1 from the channel are required. Following is the procedure:

1. Knowing the values of K and x, select an appropriate value of Δt.
2. Calculate C_0, C_1 and C_2.
3. Starting from the initial conditions I_1 and Q_1 calculate I_2 at the end of the first time step Δt; and then, compute Q_2.
4. The outflow calculated in the above step becomes the known initial outflow for the next time step. Now, repeat the calculations for the entire inflow hydrograph.

Muskingum–Cunge method

Most of the flood waves experience attenuation and are better approximated by diffusion waves rather than by kinematic waves. The theoretical development of the Muskingum–Cunge method (Cunge, 1969) is based on the simplification of the convective diffusion equation. In Muskingum–Cunge method, the routing parameters are based on readily measurable hydraulic data (stage-discharge rating, channel slope, channel width and reach length), instead of being based on historic flood discharge data. This method allows routing in ungauged streams with a reasonable expectation of accuracy. The modified Muskingum–Cunge method (Ponce and Yevjevich, 1978) expresses the routing parameters as the dimensionless Courant and cell Reynolds numbers. The equations are given as:

$$C = \frac{c\Delta t}{\Delta x} \tag{8.22}$$

$$D = \frac{q_0}{S_0 c\Delta x} \tag{8.23}$$

where C is Courant number D is cell Reynolds number, c is flood wave celerity, Δx is space step, Δt is time step, q_0 is reference unit-width discharge, and S_0 is bed slope. With the Muskingum routing parameters defined in this way, the routing coefficients are

$$C_0 = \frac{-1 + C + D}{1 + C + D} \quad (8.24)$$

$$C_1 = \frac{1 + C + D}{1 + C + D} \quad (8.25)$$

$$C_2 = \frac{1 - C + D}{1 + C + D} \quad (8.26)$$

and the routing equation is

$$Q_{j+1}^{n+1} = C_0 Q_j^{n+1} + C_1 Q_j^n + C_2 Q_{j+1}^n \quad (8.27)$$

where j is a spatial index and n is a temporal index.

In this method, optimal results are obtained if C is close to and not greater than unity, and if the sum of C and D is greater than 1. Unlike the Muskingum method, the Muskingum–Cunge requires minimal streamflow data. The kinematic wave method is identical to the Muskingum–Cunge method when $D = 0$.

EXAMPLE 8.2 Assuming $K = 24$ h and $x = 0.2$ route hypothetical flood having a constant flow rate of 1000 units and lasting one day, through the reservoir whose storage is simulated by the Muskingarn equation. Plot the inflow and outflow hydrograph considering initial outflow as zero.

Solution $$Q_{(j+1)}^{(n+1)} = C_1 Q_j^n + C_2 Q_j^{(n+1)} + C_3 Q_{(j+1)}^n$$

The relevant calculations are shown in the table below:

t	*Inflow*	$C_1 Q_j^n$	$C_2 Q_j^{(n+1)}$	$C_3 Q_{(j+1)}^n$	*Outflow*
0	1000				0
2	1000	287.129	–188.119	0	99.01
4	1000	287.129	–188.119	89.207	188.217
6	1000	287.129	–188.119	169.581	258.591
8	1000	287.129	–188.119	241.998	341.008
10	1000	287.129	–188.119	307.215	406.252
12	1000	287.129	–188.119	366.031	465.041.
14	1000	287.129	–188.119	418.998	518.006
16	1000	287.129	–188.119	466.720	565.730
18	1000	287.129	–188.119	509.717	608.727
20	1000	287.129	–188.119	548.457	647.467
22	1000	287.129	–188.119	583.361	682.371
24	1000	287.129	–188.119	614.809	713.810

The inflow and outflow hydrograph is shown below:

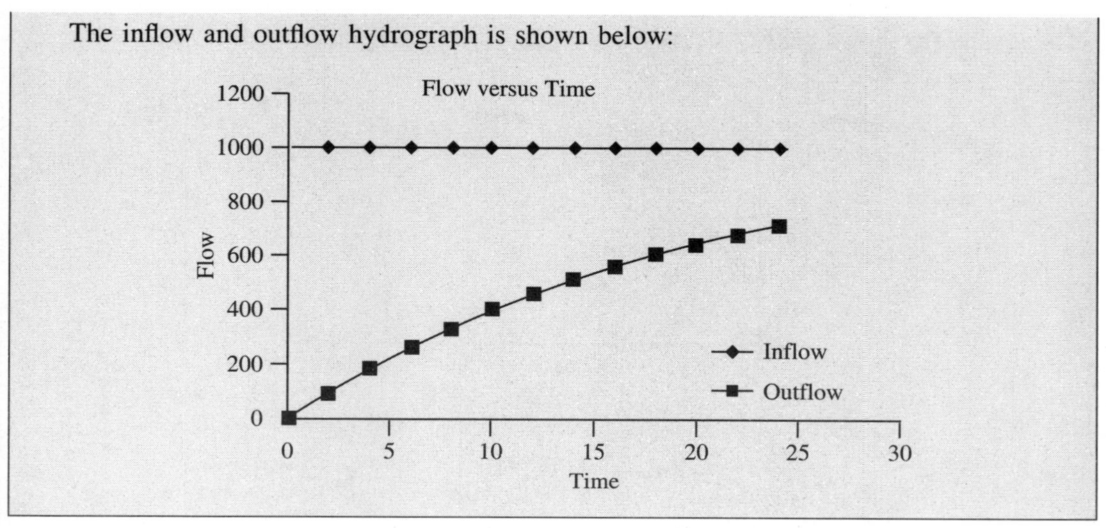

EXAMPLE 8.3 The following inflow and outflow hydrographs were observed in a river reach. Estimate the values of x and K applicable to this reach.

Time (h)	0	4	8	12	16	20	24	28
Inflow (m^3/s)	10	20	40	70	90	60	30	16
Outflow (m^3/s)	8	15	20	30	40	35	20	10

Solution The relevant calculations are shown in the table below:

Time	I	Q	$I - Q$	*Average* $I - Q$	ΔS = Col. 5 × Δt	S (m^3/s × h)	$x = 0.2$	$[xI + (1 - x)Q]$ $x = 0.3$	$x = 0.5$
(h)	(m^3/s)	(m^3/s)	(m^3/s)	(m^3/s)	(m^3/s)				
1	2	3	4	5	6	7	8	9	10
0	10	8	2			0	8.4	8.6	9
4	20	15	5	3.5	14	14	16	16.5	17.5
8	40	20	20	12.5	50	64	24	26	30
12	70	30	40	30	120	184	38	42	50
16	90	40	50	45	180	364	50	55	65
20	60	35	25	37.5	150	514	40	42.5	47.5
24	30	20	10	17.5	70	584	22	23	25
28	16	10	6	8	32	616	11.2	11.8	13

The graphs are shown below:

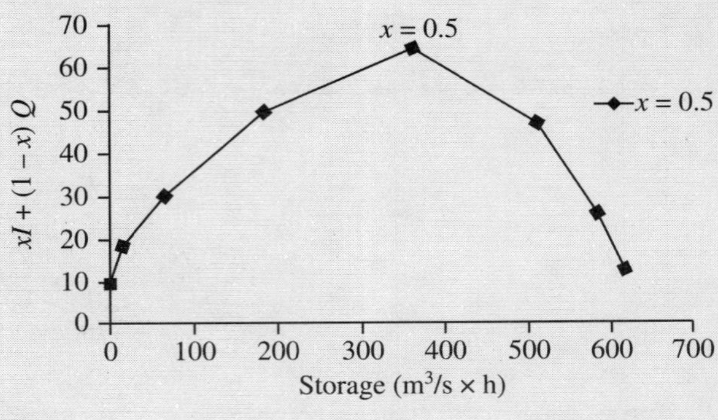

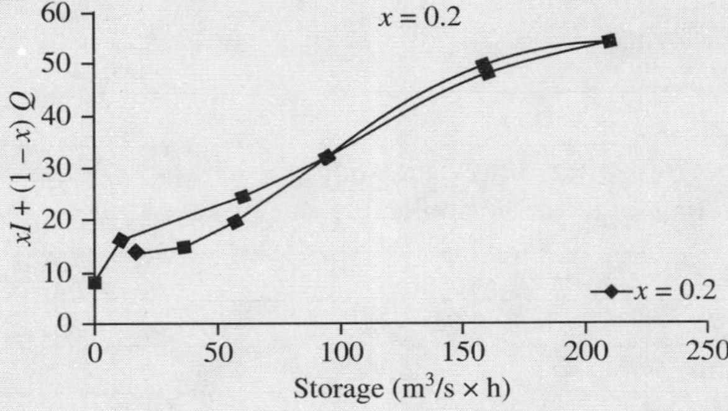

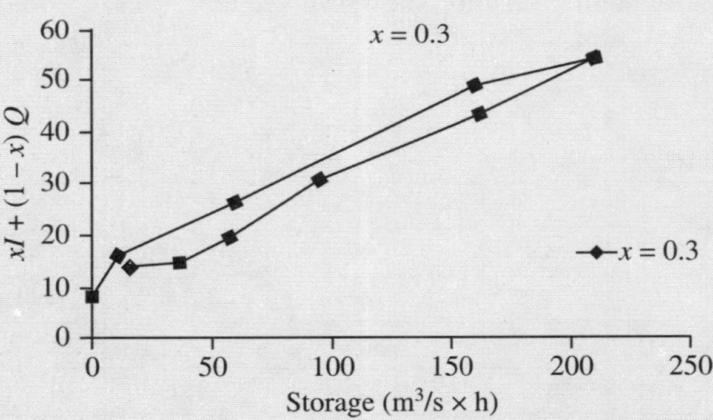

Thus, we can see that when S is plotted against $xI + (1 - x)Q$, there is no straight line relationship. Now, modifying the data.
Therefore, it is seen that for $x = 0.20$, the data nearly describes a straight line.

Thus, at $x = 0.20$, $K = \dfrac{100}{23} = 4.35$ h.

EXAMPLE 8.4 The inflow and outflow hydrograph ordinates of a particular channel are given in the table below. Compute the Muskingum parameter K and x for the channel reach.

Times (h)	0	4	8	12	16	20	24	28	32	36	40	44
Inflow discharge (m^3/s)	20	115	170	290	320	440	365	230	155	90	70	45
Outflow discharge (m^3/s)	20	56	88	130	200	310	450	250	190	150	80	50

Solution Routing period t = 4 h is taken.

Time (h)	*Inflow* (*I*) cumec	*Outflow* (*O*) cumec	*I* – *O* (cumec)	*Mean storage* (cumec-h)	*Cumulative storage* (cumec-h)
0	20	20	0	0	0
4	115	56	59	118	118
8	170	88	82	282	400
12	290	130	160	484	884
16	320	200	120	560	1444
20	440	310	130	500	1944
24	365	450	–85	90	2034
28	230	250	–20	–210	1824
32	155	190	–35	–110	1714
36	90	150	–60	–190	1524
40	70	80	–10	–140	1384
44	45	50	–5	–30	1354

The mean storage is determined from

$$\left(\frac{I_1 + I_2}{2}\right)\Delta t - \left(\frac{Q_1 + Q_2}{2}\right)\Delta t = S_2 - S_1$$

Then, the cumulative storage S is tabulated. For trial values of x = 0.2, 0.25 and 0.3, the values of $[xI + (1 - x)Q]$ are computed. Curves of S versus $[xI + (1 - x)Q]$ for each trial value of x are plotted as shown in the figure.

	$x = 0.2$			$x = 0.25$			$x = 0.3$	
0.2 *I*	0.8 *Q*	*Total*	0.25 *I*	0.75 *Q*	*Total*	0.3 *I*	0.7 *Q*	*Total*
4	16	20	5	15	20	6	14	20
23	44.8	67.8	28.75	42	70.75	34.5	39.2	73.7
34	70.4	104.4	42.5	66	108.5	51	51.6	112.6
58	104	162	72.5	97.5	170	87	91	178
64	160	224	80	150	230	96	140	236

(*Contd.*)

	$x = 0.2$			$x = 0.25$			$x = 0.3$	
88	248	336	110	232.5	342.5	132	217	349
73	360	433	91.25	337.5	428.75	109.5	315	424.5
46	200	246	57.5	187.5	245	69	175	244
31	152	183	38.75	142.5	181.25	46.5	133	179.5
18	120	138	22.5	112.5	135	27	105	132
14	64	78	17.5	60	77.5	21	56	77
9	40	49	11.25	37.5	48.75	13.5	35	48.5

By inspection, the value of $x = 0.3$ approximates a straight line, and, hence, this value of x is chosen. K is determined by measuring the slope of the median straight line which is found to be 1.43 h. Hence, for the given reach of the river, the values of the Muskingum coefficients are $x = 0.3$ and $K = 1.43$ h.

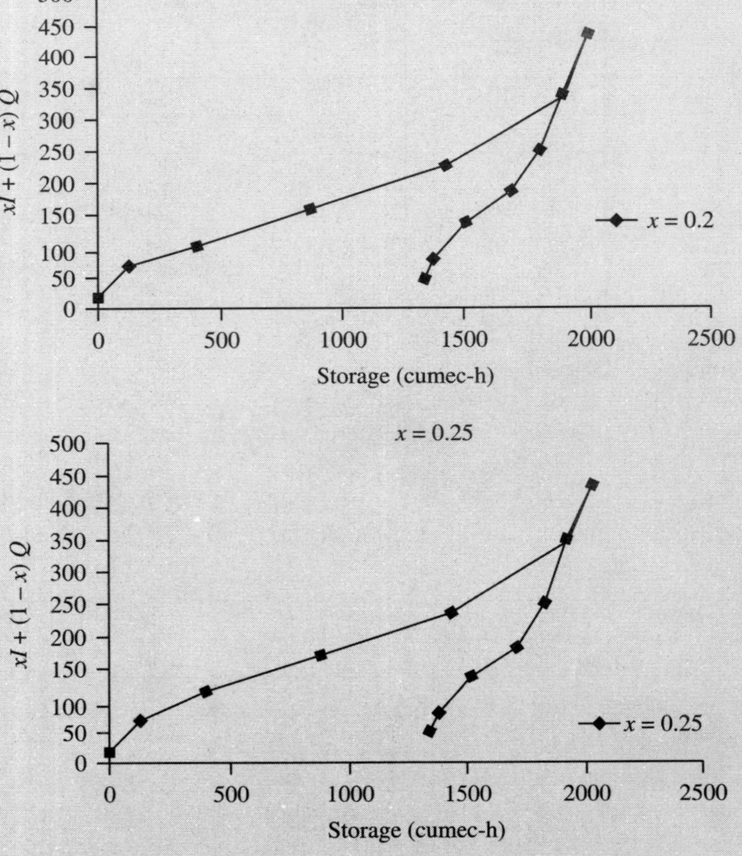

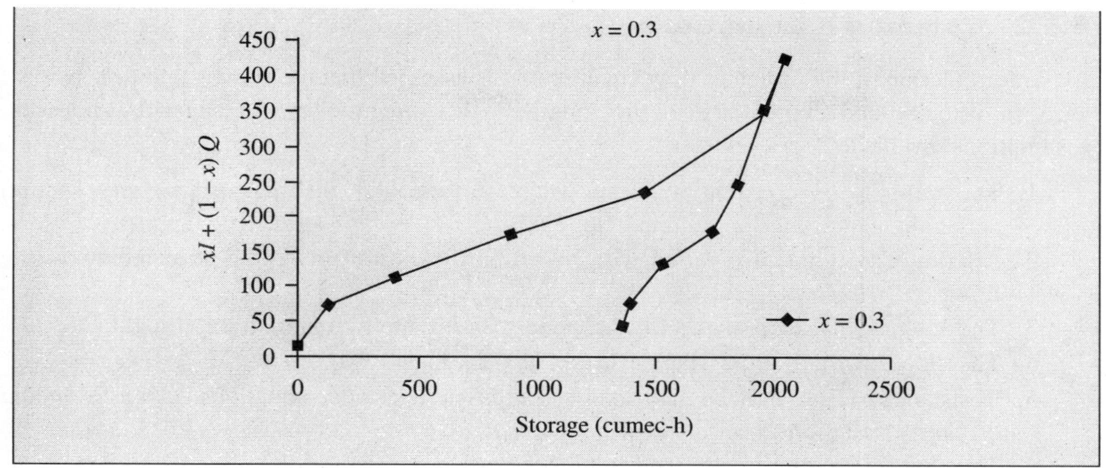

8.5 Hydraulic Routing

Hydraulic routing employs the full dynamic wave (St. Venant equations). These are the continuity equation and the momentum equation, which takes the place of the storage-discharge relationship used in hydrologic routing. The three most common approximations or simplifications to the full dynamic equations are referred to as kinematic, diffusion and quasi-steady models. Neglecting the certain terms of the momentum equation due to their relative orders of magnitude. Kinematic and Diffusion models have found wide application and acceptance in the engineering hydrology. Thus, in this chapter, a kinematic wave model for hydraulic flood routing is discussed in detail.

8.5.1 Kinematic Waves

A kinematic wave method predicts that the motion of the hydrograph along the channel is controlled mostly by friction and gravity forces. Then, the momentum equation becomes a wave equation

$$\frac{\delta Q}{\delta t} + c\frac{\delta Q}{\delta x} = 0 \tag{8.28}$$

where, Q is the discharge, t is the time, x is the distance along the channel and c is the celerity of the wave (speed). A kinematic wave travels downstream with speed c without any attenuation. Thus, diffusion is generally absent. Now, the kinematic behaviour can be assumed if

$$\frac{t_r S_0 V}{d_0} \geq 85 \tag{8.29}$$

where, t_r is the time-to-peak of the inflow hydrograph, S_0 is the channel bed slope and d_0 is the average flow graph.

8.5.2 Common Assumptions

In the formulation of the unsteady flow equations, it is assumed that the flow is one-dimensional such as velocity and depth vary in the longitudinal x-direction of the channel. Following assumptions are made:

1. The velocity is constant and the water surface is horizontal across any section perpendicular to the longitudinal flow axis.
2. All flows are gradually varied with hydrostatic pressure prevailing at all points in the flow.
3. The longitudinal axis of the flow channel can be approximated by a straight line.
4. The slope of the bottom channel is small (less than 1:10).
5. Resistance to flow may be described by empirical resistance equations such as Manning or Chezy equations.
6. The flow is incompressible and homogeneous in density.

8.5.3 Kinematic Waveform as a Stage-Discharge Relationship

Kinematic waves occur when the dynamic terms in momentum equations are negligible. The bed slope is approximately equal to the friction slope ($S_0 = S_f$), Under these circumstances, the discharge can be defined as a function of depth flow only for all x and t.

$$Q = \alpha y^m \tag{8.30}$$

where Q is discharge (ft^3/s) and α and m are kinematic wave routing parameters. These parameters are directly related to flow and catchment characteristics. St. Venant Momentum equation may be written as

$$Q = Q_n \left[1 - \frac{1}{S_0} \left(\frac{\delta y}{\delta x} + \frac{u}{g} \frac{\delta u}{\delta x} + \frac{1}{g} \frac{\delta u}{\delta t} + \frac{qLu}{gy} \right) \right]^{\frac{1}{2}} \tag{8.31}$$

The sum of the terms to the right of the minus sign is much less than one (i.e., pressure, inertia and local inflow are relatively small compared to S_0); the unsteady flows are nearly uniform and may be approximated by a normal flow, e.g.

$$Q \approx Q_n \tag{8.32}$$

This describes kinematic flow and gives a simple method for calculating flows from storm water runoff.

8.5.4 Development of α and m for Cross Section Shapes

Overland flow relationship

The kinematic wave equation for an overland flow with very shallow flow conditions can be derived from Manning's equation and Eq. (8.30). Then, we consider the flow down an overland flow strip of unit breadth. Now, the steady discharge from a unit strip can be determined with the Manning's relationship.

$$Q = \frac{1.486}{n} R^{2/3} S_o^{1/2} A \tag{8.33}$$

where S_0 is average slope of overland flow, y_0 is mean depth of overland flow, and for very shallow flows that are a depth of y_Q, R (the hydraulic radius) and A are simply [($y_o \times 1$)/1 and $y_o \times 1$], respectively. Substitution of these values for R and A in Eq. (8.33) yields

$$Q = \frac{1.486}{N} S_o^{1/2} y_o^{5/3} \tag{8.34}$$

Here, Manning's n has been replaced by N that determines the properties of the runoff surface being modelled. The values of N are generally greater than Manning's n.

Although the discharge represented by Eq. (8.34) is per unit breadth, one can substitute the previously defined q for Q to obtain discharge in terms of flow per unit breadth.

$$q = \left(\frac{1.486}{N} S_o^{1/2}\right) y_o^{5/3} \tag{8.35}$$

Now, rewrite Eq. (8.30) in terms of discharge per unit width, where the subscripts 'o' indicates variables associated with the overland flows.

$$q = \alpha_o y_o^{m_o} \tag{8.36}$$

where $\left[\alpha_o = \left(\frac{1.486}{N} S_o^{1/2}\right)\right]$ and $m_o = \frac{5}{3}$. α_o is conveyance for particular runoff surface, slope and roughness.

Here, two unknowns are observed in Eq. (8.36). Thus, another relationship is needed for mathematical solution. As per the continuity equation, it provides the second important equation to complete the solution. The equation can be written as

$$\frac{\delta y_o}{\delta t} + \frac{\delta q_o}{\delta x} = i - f \tag{8.37}$$

where, $(i - f)$ is the rate of excess rainfall (ft/s), q_o is discharge per unit width (cfs/ft), t is time (s) and y is mean depth of overland flow (ft).

Equations (8.36) and (8.37) together form the complete kinematic wave equations for overland flow. The final equation can be written as

$$\frac{\delta y_o}{\delta t} + \alpha_o m_o y_o^{m_o - 1} \frac{\delta y_o}{\delta x} = i - f \tag{8.38}$$

This equation has only one dependent variable so that it can be solved to give a relationship y_o in terms of x, t and the excess rainfall intensity $(i - f)$. This method provides the necessary information to be able to determine the time-dependent discharge from the overland flow elements.

Main channel routing relationships

Flows entering the rivulets, storm drains (collectors) and stream channels can consist both flows from upstream sections and lateral inflows from adjacent catchment surfaces. Their slopes, lengths, cross-sectional dimensions, shapes and Manning's *n* values, describe these representative channels. The basic form of the equations for kinematic wave routing of collector and streamflows is similar to those developed for shallow overland flow. The equations can be written as

$$\frac{\delta A_c}{\delta t} + \frac{\delta Q_c}{\delta x} = q_o \tag{8.39}$$

$$Q_c = \alpha_c A_c^{m_c} \tag{8.40}$$

where, A_c is cross-sectional area of flow (ft^2), Q_c is discharge (cfs), q_o is lateral inflow per unit length (cfs/ft) from overland flow strips, t is time (s), x is distance along the stream (ft) and $\alpha_c m_c$ are kinematic wave parameters for a particular cross-sectional shape, slope and roughness.

One should notice that the subscripted variables used above are for a typical collector channel and are indicated as such with the subscript c. Identical relationships would be used for routing in the main channel.

Values of α_c and m_c will be different for each differently-shaped cross section. They also vary with Manning's *n* and channel slope as well. The different shapes are presented in Figure 8.5.

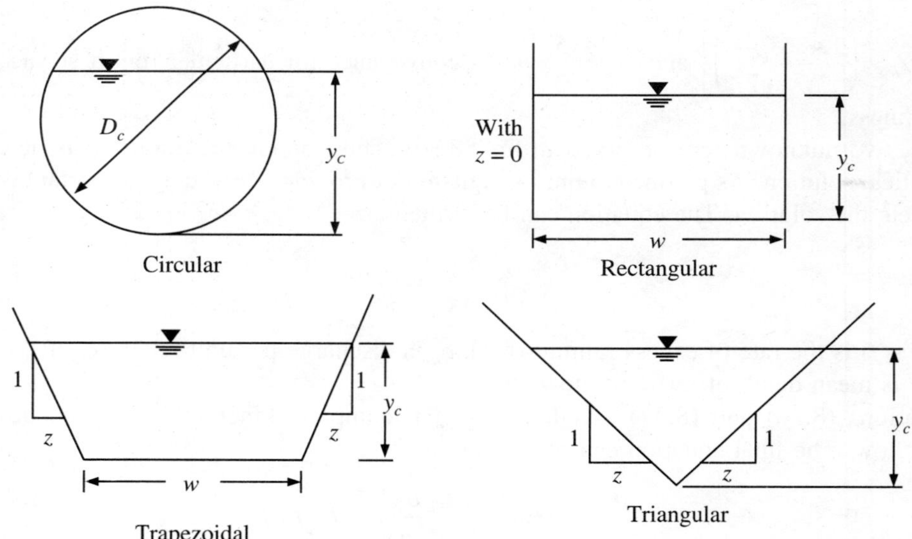

Here, D_c = the diameter of the circular section, w = width of channel, y_c = height of flow, and z = slope as shown in figure

Figure 8.5 Channel shapes and their variations.

Triangular section

Derivation of α_c and m_c for a triangular channel begins with the description of R_c and A_c in Eq. (8.40). As per Figure 8.5 and Eq. (8.39) and (8.40), new equations can be written as

Area of triangle-shaped cross section/wetted perimeter = Hydraulic radius

$$R_c = \frac{A_c}{P_c}$$

where P_c is wetted perimeter and A_c is cross-sectional area.

Also, $$P_c = 2\left(\sqrt{1+z^2}\right) \times y_c$$

and $$A_c = \frac{1}{2}(2y_c \times z) \times y_c = zy_c^2$$

where, z is side slope ratio and y_c is height of flow.

Now, the values of A_c and R_c may be substituted into Eq. (8.40) as

$$Q_c = \frac{1.486}{n} S_c^{1/2} \frac{A^{2/3}}{P_c^{2/3}} A_c = \frac{1.486 S_c^{1/2}}{n} \frac{A_c^{5/3}}{P_c^{2/3}} \tag{8.41}$$

Defining ϕ as $\dfrac{1.486 S_c^{1/2}}{n}$ and substituting the appropriate value for A_c and P_c into Eq. (8.43) and simplifying, we get

$$Q_c = \phi \frac{z^{5/3} y_c^{10/3}}{(2^{2/3})(1+z^2)^{1/3} y_c^{2/3}} = \phi \frac{z^{1/3}(z^{4/3} y_c^{8/3})}{(2)^{2/3}(1+z^2)^{1/3}} \tag{8.42}$$

$$Q_c = \frac{1.486}{1.587} \frac{S_c^{1/2}}{n} \left(\frac{z}{1+z^2}\right)^{1/3} (z \times y_c^2)^{4/3} \tag{8.43}$$

or $$Q_c = \frac{0.94 S_c^{1/2}}{n} \left(\frac{z}{1+z^2}\right)^{1/3} A_c^{4/3} \tag{8.44}$$

Now, Eq. (8.34) can be written in the form of Eq. (8.40) as

$$Q_c = \alpha_c A_c^{m_c} \tag{8.45}$$

where $\alpha_c = \dfrac{0.94 S_c^{1/2}}{n} \left(\dfrac{z}{1+z^2}\right)^{1/3}$ and $m_c = \dfrac{4}{3}$.

Rectangular section

A rectangular shape is obtained by stipulating that z is zero in Figure 8.5. This shape may represent manmade channels and concrete drain sections. For the wide shallow channel case, using Eqs. (8.44) and (8.45), the new equation can be written as

$$\alpha_c = \frac{1.486S_c^{1/2}}{n} \times w^{-2/3} \tag{8.46}$$

and $$m_c = \frac{5}{3}$$

Now, for rectangular channel, where $w \approx y_c$,

$$\alpha_c = \frac{0.72S_c^{1/2}}{n} \tag{8.47}$$

and $m_c = \dfrac{4}{3}$

Trapezoidal section

In trapezoidal section, it is important to describe the side slope z and the channel bottom width w accurately (Figure 8.5). The kinematic wave equation for the trapezoidal section can be written as

$$Q_c = \alpha_c A_c^{m_c} \tag{8.48}$$

In this case, it is not possible to derive a simple relationship for α_c and m_c from the geometric properties alone. Thus, it is necessary to fit α_c and m_c to the Manning's equation at two or more depths y_c and use numerical techniques of fitting the kinematic equation to these values to obtain values for α_c and m_c for various flow conditions.

The Manning's equation for a trapezoid can be written as

$$Q_c = \frac{1.486S_c^{1/2}}{n}(A_c)^{5/3}\left[\frac{1}{w + zy_c(1+z^2)^{1/2}}\right]^{2/3} \tag{8.49}$$

where, A_c is the area of the effective cross section at depth y_c. Here, values of m_c vary from 4/3 for a triangular section to 5/3 for a wide rectangular section.

Circular sections

This type of sections can be used to model storm or sewer pipes in urban areas. The ac and mc relationship for the typical circular sections (Figure 8.5) can be derived as

$$\alpha_c = \frac{0.804S_c^{1/2}}{n} D_c^{1/6} \tag{8.50}$$

and $$m_c = 1.25$$

Here, D_c is the diameter of the circular section (ft).

8.5.5 Numerical Solution of the Kinematic Wave Equations

The kinematic wave equations can be solved using finite difference numerical solutions. The details of these techniques, which have been perfected after years of development and testing, are available from several resources (Harley, 1975). In the standard procedures, time is discretised in constant steps of Δt and distance in steps of Δx, The rainfall excess $(i - f)$ is assumed constant within each time step Δt, but it varies from time step to time step to simulate the variability occurring within a storm event.

Recall that Eqs. (8.39) and (8.40) are the kinematic wave equations for collector and streamflow routing. If Eq. (8.40) is substituted into Eq. (8.39), the following relationship, which has A_c as the only dependent variable, will be computed as

$$\frac{\delta A_c}{\delta t} + \alpha_c m_c A_c^{(m_c - 1)} \frac{\delta A_c}{\delta x} = q_o \tag{8.51}$$

Numerical solution of Eq. (8.51) gives a relationship for A_c in terms of x, t and q_o. These values of A_c may then be substituted into Eq. (8.40) to solve for Q_c. This simple procedure (described previously for overland flows) allows the computation of Q_c as function of the segment length L_c and time t. Thus, one can determine the discharge hydrograph from each of the segments that are of length L_c, This gives a simple procedure for determining first the overland flows, then the flows through the channel system, and finally, the discharges in the main stream.

The movement of a flood wave down a stream or along an overland flow surface can be followed by monitoring the times and locations. Suppose, if lateral inflow is zero for a moment, then a flood wave can be followed in time by noting the times when a flow for a given magnitude occurs at successive downstream stations along the channel.

Flood routing methods such as kinematic wave method depend on several numerical techniques to solve the governing equations, which determines the movement, stage and discharge characteristics of a flood wave as it travels downstream. Several solutions are provided by the various authors. In this chapter, the concepts and basic fundamentals of finite difference method (FDM) are discussed. However, the whole numerical computation is not provided.

Finite difference method

An FDM provides a point-wise approximation to the governing partial difference equations. It uses simple difference equations, which replace the partial difference equations for an array of stationary grid points, located in the space-time $(x - t)$ plane (Figure 8.6).

The intersections of the lines in Figure 8.6 define the time and space points at which the discharge and water surface elevation are calculated. Lines parallel to x-axis are called *time lines* and lines parallel to t-axis are defined as the *space lines*. The regular pattern formed by the intersections of time and space lines is called *computational network*. The points that are marked by solid dots represent computation points, called *nodes*. Solutions to the governing equations via the FDM can be computed using *explicit* and *implicit* methods.

The temporal derivative [Eq. (8.52)] and spatial derivative [Eq. (8.53)] of explicit method can be written as (for variable u).

$$\frac{\delta u}{\delta t} \approx \frac{u_i^{j+1} - u_i^j}{\Delta t} \tag{8.52}$$

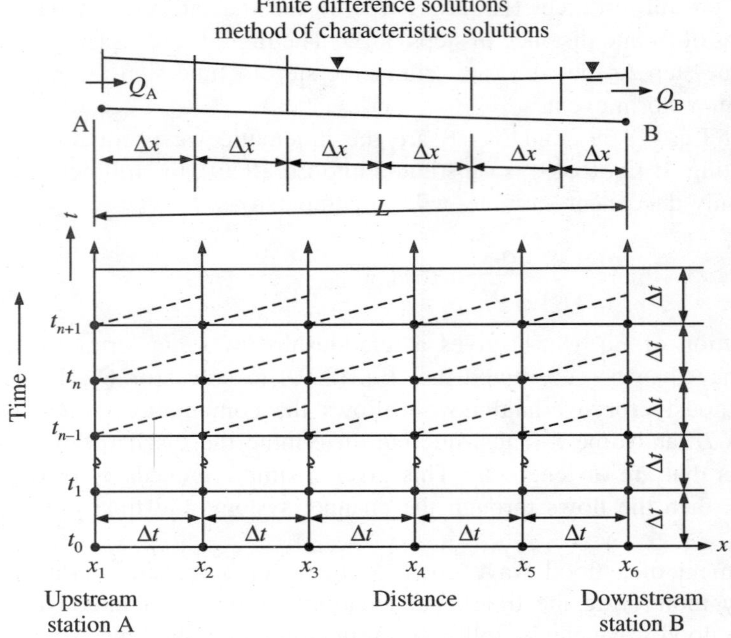

Figure 8.6 Characteristics curves on a fixed Δσ-Δt grid.

$$\frac{\delta u}{\delta x} \approx \frac{u_{i+1}^{j} - u_{i+1}^{j}}{2\Delta x} \tag{8.53}$$

Likewise, the temporal derivative [Eq. (8.54)] and spatial derivative [Eq. (8.55)] of implicit method can be written as

$$\frac{\delta u}{\delta t} \approx \frac{u_{i}^{j+1} u_{i+1}^{j+1} - u_{i+1}^{j}}{2\Delta t} \tag{8.54}$$

$$\frac{\delta u}{\delta x} \approx \frac{u_{i+1}^{j+1} - u_{i}^{j+1}}{\Delta x} + (1-\theta)\frac{u_{i+1}^{j} - u_{i}^{j}}{\Delta x} \tag{8.55}$$

There are detailed methods available for the approximation of derivatives by the FDM. By simply combining the appropriate finite difference approximations for first and second order derivates, complete partial differential equations such as Eqs. (8.54) and (8.55) can be reformulated in terms of finite differences instead of partial derivatives. As an example of the method, the first order derivative $\delta Q/\delta x$ is approximated using a backward finite difference method.

$$\frac{\delta Q}{\delta x} \approx \frac{\Delta Q}{\Delta x} \approx \frac{Q_{i,j} - Q_{i-1,j}}{\Delta x} \tag{8.56}$$

The governing equations developed in the above sections consider a pair of equations for each of the different kinds of flow elements like overland flow elements, collector elements and for the main channel routing elements. Rather than treating each of these three pairs of equations separately, the solution details are developed for one set only because basically, they are all the same.

8.5.6 Standard Form of the Kinematic Wave Equations

As per Eq. (8.39), which is the continuity equation and Eq. (8.40), which relates flow Q_c within a collector element to the collector channel cross-sectional area A_c. It is assumed that the kinematic wave coefficients α_c and m_c are constant for any given system of channel elements. Differentiation of the flow in Eq. (8.40) with respect to x and substitution of this into Eq. (8.39) provide

$$\frac{\delta A_c}{\delta t} + \alpha_c m_c A_c^{(m_c-1)} \frac{\delta A_C}{\delta x} = q_o \tag{8.57}$$

Equation (8.57) is the same as the previous derived Eq. (8.55). Referring to Figure 8.9, the area A_c is known at points on the space-time grid for times prior to the current time (given by index j) and points in space prior to the current location (indicated by index i). Thus, index (i, j) corresponds to the current time and space coordinates. Future times and space locations that are advanced by one Δt and Δx are indicated as $j + 1$ and $i + 1$, respectively. Similarly, one previous time and space location would correspond to a point on the space-time grid indexed by $(i - 1, j - 1)$.

This allows one to rewrite partial differential equations in terms of finite difference approximations of known quantities (located at previous times and space points) and unknown quantities (current time and space points). The area A_c at point B in Figure 8.7 is designated as $A_{c(i-1,\, i)}$ at point C, $A_{c(1-1,\, j-1)}$ at point D, $A_{c(i,\, j-1)}$ and so forth.

Based on the above governing equations in terms of finite differences using the previously defined indexing scheme, solve for the values of A_c and Q_c at the current point A as per Figure 8.8 as

$$\alpha m A^{m-1} \frac{\delta A}{\delta x} \approx \alpha m A^{m-1} \frac{\Delta A}{\Delta x} = \alpha m \left[\frac{A_{(i,j-1)} + A_{(i-1,j-1)}}{2}\right]\left[\frac{A_{(i,j-1)} - A_{(i-1,j-1)}}{\Delta x}\right] \tag{8.58}$$

where the differential of the area A in x-direction is taken as the difference between the values known at points C and D. The area term, which is raised to the $(m - 1)$ power in Eq. (8.58) is taken to be an average area between points C and D.

Now, considering the time derivative term $(\delta A/\delta T)$ in Eq. (8.59) and evaluating between points A and D,

$$\frac{\delta A}{\delta t} = \frac{\Delta A}{\Delta t} = \left[\frac{A_{(i,j)} - A_{(i,j-1)}}{\Delta t}\right] \tag{8.59}$$

The lateral inflow term q is handled as an average lateral inflow, which occurs within a time step Δt and is defined as q here to simplify the final form of the equation.

$$q \approx \frac{q_{(i,j)} + q_{(i,j-1)}}{2} = \bar{q} \tag{8.60}$$

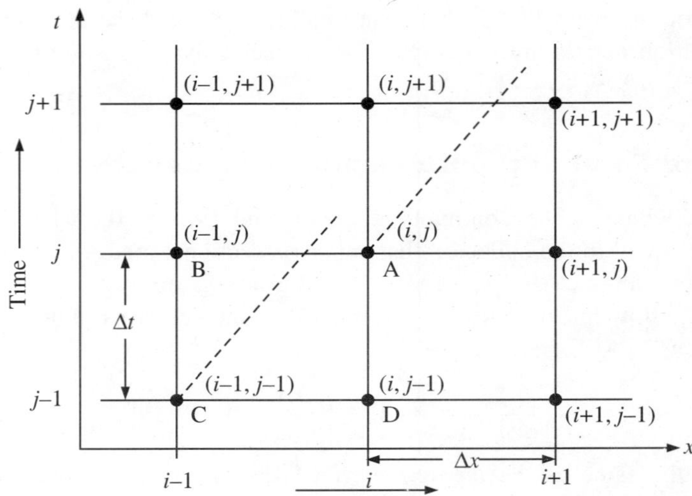

Figure 8.7 Space-time grid used for FDM.

Combining Eqs. (8.58), (8.59) and (8.60) gives the final finite differences form of the original partial differential Eq. (8.57).

$$\left[\frac{A_{(i,j)} - A_{(i,j-1)}}{\Delta t}\right] + \alpha m \frac{\Delta A}{\Delta x} = \alpha m \left[\frac{A_{(i,j-1)} + A_{(i-1,j-1)}}{2}\right]^{m-1} \left[\frac{A_{(i,j-1)} - A_{(i-1,j-1)}}{\Delta x}\right] = \bar{q} \quad (8.61)$$

$A_{i,j}$ is the only unknown term in Eq. (8.61). Thus, this can be solved for

$$A_{i,j} = \bar{q}\Delta t + A_{i,j-1} - \alpha m \frac{\Delta t}{\Delta x}\left[\frac{A_{(i,j-1)} + A_{(i-1,j-1)}}{2}\right]^{m-1} \left[A_{(i,j-1)} - A_{(i-1,j-1)}\right] \quad (8.62)$$

$A_{i,j}$ is computed, the corresponding flow $Q_{i,j}$ can also be calculated from Eq. (8.40). This gives a straightforward method of computing the time-varying discharges along the chanel.

8.6 Flood Control Measures

A *flood* can be defined as the river overflowing its banks and inundating the surrounding marginal area. Floods can occur because of heavy rainfall, glacial lake burst, cloud burst or breaching of river flood banks. Most parts of India are severely affected by flood, which causes loss of economy, property and lives. The Ganges plains, Brahmaputra river basin, and Narmada basin are some of the examples of severely affected flood regions. Severity of flood hazards can be reduced by taking necessary actions before the occurrence of flood. These flood control measures can be grouped in two categories, i.e., structural measures and non-structural measures.

Design magnitude of floods are required for flood control sturctures. The maximum flood that a control structure can safely bear is known as *design flood*. Further, the damage due to floods can be minimised by the following flood control measures (single or in combination):

Flood control structures

1. Reservoirs: Reservoir temporarily stores a portion of excess volume of incoming flow which can cause flood situation in downstream portion of the stream, as shown in Figure 8.8. Reservoirs are constructed upstream of the area to be protected and excess water is discharged to downstream at safe capacity limit. Reservoirs are generally more costly structure if the only purpose is to control the flood. So, generally, some other benefits also taken into consideration such as hydropower and irrigation projects.

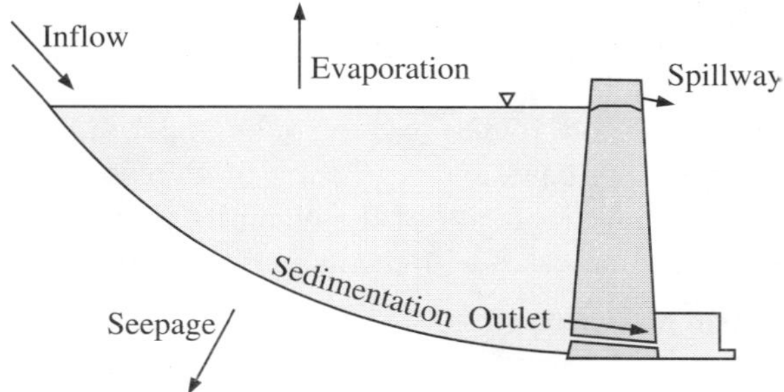

Figure 8.8 Flood control operation of a reservoir.

2. Retarding basins: Retarding basins are the structures made specially for controlling the flood. Retarding basins are provided with large spillways and sluice without providing any control gate, as shown in Figure 8.9. In flood time, discharge from retarding basin will be the safe discharging capacity of the channel downstream. Capacity of the retarding basin should be equal to the volume of design flood minus volume of water released during the flood.

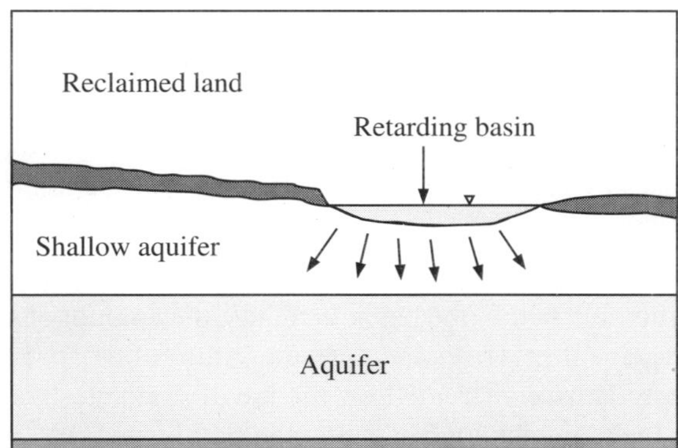

Figure 8.9 Retarding basin.

3. Levees: Levees are the most economical structure for direct mitigation of flood impacts. Levees are constructed along the bank of river beyond the meandering limit of river to stop the river to change its course, as shown in Figure 8.10. During levees construction, there should be least curves in the alignment.

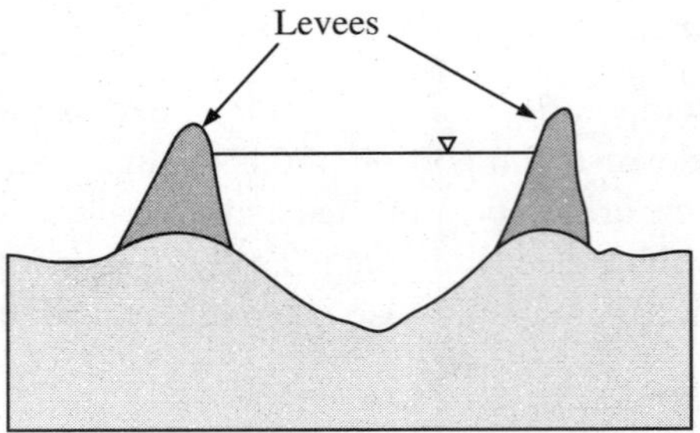

Figure 8.10 Levees.

4. Channel improvement: Flood effects can also be reduced by improving the channel discharging capacity. This capacity could be in the form of increased velocity or increased width of the river. Generally, increasing the depth of the channels preferred to widening the channel width, as hydraulic mean radius increases with the increase in depth.

Limitation of flood control structures: Flood control structure need regular monitoring and maintenance. If it is not done, failure in flood control structures will cause even more devastation as it could have without the structures.

Hydrologic system of any area gets altered significantly on construction of structural projects, which can affect it adversely in long time.

Non-structural flood control measures

Non-structural measures include agronomic practices, reforestation flood plain management, flood plain regulation, removing flood plain development, storm water management, watershed management.

1. Afforestation: Afforestation is the most inexpensive method to reduce the impact of flood as the roots of the plants and trees keep the soil particles firm at their places and reduce the erosion tendency. Well-grown trees reduce the velocity of streamflow and cause reduction in erosion.

2. Flood-prone area away from human development: Keeping flood-prone area away from human development causes less hazards to lives and property, as the unplanned development leads to more vulnerability to flood hazards.

3. Watershed management programme: It includes afforestation, adoption of cropping system, flood control structures within the basin to reduce the amount of generated runoff from the storm such as soil management practices increase the infiltration capacity of soil and less runoff is produced. Small check dams constructed on the small drainage line can stop the significant amount of water to join the main channel, carrying the floodwater.

Summary

- Flood routing is a procedure to compute the flood hydrograph at a section of a river by applying the data of flood flow at one or more upstream sections.
- In Muskingum method of flow routing, storage is a function of both inflow and outflow in the reach.

- When flood wave enters into the channel, the inflow exceeds the outflow and creates a wedge of storage. Prism storage is the volume of water below the wedge storage.
- Flood control measures refer to reduce or minimise damage of lives and property by floods.

Objective Type Questions

1. In a linear reservoir, the (IES, 1999)
   (a) Volume varies linearly with elevation
   (b) Outflow rate varies linearly with storage
   (c) Storage varies linearly with time
   (d) Storage varies linearly with inflow rate
2. The life of a reservoir is determined by its capacity C, volume of annual inflow into the reservoir I and concentration of sediment in the incoming flow Cs. Life will be more if (IES, 2000)
   (a) C, I and C_s are high (b) C and I are high but C_s is low
   (c) C is high but I and C_s are low (d) C, I and C_s are low
3. As a flood wave passes a given section of a river, the time of occurrence of the maximum stage and that of the maximum discharge will be such that
   (a) The maximum discharge passes down before the maximum stage is attained
   (b) The maximum stage is attained before the maximum discharge passes down
   (c) The two events occur simultaneously
   (d) No specific sequence would be universally assignable
4. Clark's method aims at which one of the following? (IES, 2003)
   (a) Developing an IUH due to an instantaneous rainfall excess over catchment
   (b) Developing stage-discharge relationship
   (c) Measurement of infiltration
   (d) Flood routing through channels
5. Which one of the following flood routing methods involve the concepts of wedge and prism storages? (IES, 2005)
   (a) Coefficient method (b) Muskingum method
   (c) Pul's method (d) Lag method
6. The basic equation of flood routing through a reservoir can be modified for discrete successive intervals delta t by which one of the following? (IES, 2006)

   (a) $\frac{(L_1 + L_2)}{2}\nabla t + S_1 + \frac{(Q_1 \nabla t)}{2} = S_2 + \frac{(Q_2 \nabla t)}{2}$

   (b) $\frac{(L_1 + L_2)}{2}\nabla t + S_1 - \frac{(Q_1 \nabla t)}{2} = S_2 + \frac{(Q_2 \nabla t)}{2}$

   (c) $\frac{(L_1 + L_2)}{2}\nabla t + S_1 + \frac{(Q_1 \nabla t)}{2} = S_2 - \frac{(Q_2 \nabla t)}{2}$

   (d) $\frac{(L_1 + L_2)}{2}\nabla t + S_1 - \frac{(Q_1 \nabla t)}{2} = S_2 - \frac{(Q_2 \nabla t)}{2}$

7. Consider the following statements:

In case of flood routing in a river channel by Muskingum method, the coefficient x represents:

1. A dimensionless constant indicating the relative importance of inflow and outflow in determining storage
2. A storage constant having the dimension of time
3. In natural channels, x usually varies between 0.1 and 0.3
4. When the values of x equals 0.5, there exists the influence of both inflow and outflow on storage

Which of these statements are correct? (IES, 2010)

(a) 1, 2, 3 and 4 (b) 1, 3 and 4 only
(c) 1, 2 and 3 only (d) 2, 3 and 4 only

8. The Muskingum method of flood routing is a (IES, 2011)

(a) Form of hydraulic routing of a flood
(b) Form of reservoir routing
(c) Complete numerical solution of St. Venant equations
(d) Hydrological channel routing method

9. A flood wave with a known inflow hydrograph is routed through a large reservoir. The outflow hydrograph will have (GATE, 2008)

(a) Attenuated peak with reduced time-base
(b) Attenuated peak with increased time-base
(c) Increased peak with increased time-base
(d) Increased peak with reduced time-base

10. The type of flood routing (Group I) and the equation(s) used for the purpose (Group II) are given below: (GATE, 2016)

Group I	Group II
P. Hydrologic flood routing	1. Continuity equation
Q. Hydraulic flood routing	2. Momentum equation
	3. Energy equation

The correct match is

(A) P-1; Q-1, 2 and 3
(B) P-1; Q-1, 2 only
(C) P-1 and 2; Q-1 only
(D) P-1 and 2; Q-1, and 2

11. The Muskingum model of routing a flood through a stream reach is expressed as $O_2 = K_0I_2 + K_1I_1 + K_2O_1$, where K_0, K_2 and K_2 are the routing coefficients for the concerned reach, I_1 and I_2 are the inflows to reach, and O_1 and O_2 are the outflows from the reach corresponding to time steps 1 and 2, respectively. The sum of K_0, K_1 and K_2 of the model is

(a) –1 (b) –0.5 (c) 0.5 (d) 1

Answers

1. (b) **2.** (d) **3.** (a) **4.** (a) **5.** (b) **6.** (b) **7.** (b) **8.** (d) **9.** (b) **10.** (B)
11. (d)

Explanations

1. The storage of a linear reservoir varies linearly with the outflow discharge or the outflow rate varies linearly with storage.

6. The basic continuity equation is

$$\left(\frac{L_1 + L_2}{2}\right)\nabla t - \left(\frac{Q_1 + Q_2}{2}\right) = S_2 - S_1$$

$$\left(\frac{L_1 + L_2}{2}\right)\nabla t + \left(S_1 - \frac{Q_1 \nabla t}{2}\right) = S_2 - \frac{Q_2 \nabla t}{2}$$

7. Muskingum equation is given as

$$S = K[xI + (1 - x)Q]$$

The coefficient K is known as storage-time constant and the dimensions of time. Normally, for natural channels, the value of x lies between 0 to 0.3. For a given reach, the values of x and K are assumed to be constant. When $x = 0$, then $S = KQ$ (valid for linear reservoir). When $x = 0.5$, then $S = K[0.5\ 1 + 0.5\ Q)$. There exists the influence of both inflow and outflow on storage.

11. $K_0 + K_1 + K_2 = 1$

IES Conventional Type Questions

PROBLEM 8.1 The inflow hydrograph readings for a stream reach (m^3/s) at 12 hourly intervals are as follows:
42, 45, 88, 272, 342, 288, 240, 198, 162, 133, 110, 90, 79, 68, 61, 56, 54, 51, 45, 45 and 42.

The Muskingum coefficients for the stream reach are $K = 36$ h and $x = 0.15$. Determine the attenuation in peak flow discharge and the time of peak outflow. (IES, 2007)

Solution From the given data, we can write that: $K = 36$ h, $x = 0.15$ and $\Delta t = 12$ h
Again, using Muskingum theorem, we can write

$$O_n = C_0 I_n + C_1 I_{(n-1)} + C_2 O_{(n-1)}$$

where,

$$C_0 = \frac{(-Kx + 0.5\Delta t)}{(K - Kx + 0.5\Delta t)}$$

$$C_0 = \frac{(-36 \times 0.15 + 0.5 \times 12)}{(36 - 36 \times 0.15 + 0.5 \times 12)} = 0.0164$$

$$C_1 = \frac{(Kx + 0.5\Delta t)}{(K - Kx + 0.5\Delta t)}$$

$$C_1 = \frac{(36 \times 0.15 + 0.5 \times 12)}{(36 - 36 \times 0.15 + 0.5 \times 12)} = 0.3115$$

$$C_2 = \frac{(K - Kx - 0.5\Delta t)}{(K - Kx + 0.5\Delta t)}$$

$$C_2 = \frac{(36 - 36 \times 0.15 - 0.5 \times 12)}{(36 - 36 \times 0.15 + 0.5 \times 12)} = 0.6721$$

For first time interval,

$$I_1 = 42,\ C_1I_1 = 0.3155 \times 42 = 13.08$$
$$I_2 = 45,\ C_0I_2 = 0.0164 \times 45 = 0.74$$
$$O_1 = 42,\ C_2O_1 = 0.6721 \times 42 = 28.23$$

Therefore, $O_2 = C_0I_2 + C_1I_1 + C_2O_1 = 0.74 + 13.08 + 28.23 = 42.05$

Time (h)	I (m^3/s)	$0.0164I_2$ ($C_0\,I_2$)	$0.3115I_1$ (C_0, I)	$0.6721O_1$ ($C_2\,O_1$)	O (m^3/s)
0	42				42
		0.74	13.08	28.23	
12	45				42.05
		1.44	14.0175	28.26	
24	88				43.7175
		4.46	27.412	29.38	
36	272				61.252
		5.61	84.728	41.17	
48	342				131.508
		4.72	106.533	88.39	
60	288				199.643
		3.94	89.712	134.18	
72	240				227.832
		3.25	74.76	153.12	
84	198				231.13
		2.66	61.677	155.34	
96	162				219.677
		2.18	50.463	147.65	
108	133				200.28
		1.8	41.4295	134.61	
120	110				177.84
		1.48	34.265	119.53	
132	90				155.27

(*Contd.*)

Time (h)	I (m³/s)	$0.0164I_2$ (C_0 I_2)	$0.3115I_1$ (C_0, I)	$0.6721O_1$ (C_2 O_1)	O (m³/s)
		1.3	28.035	104.36	
144	79				133.69
		1.12	24.6085	89.85	
156	68				115.58
		1	21.182	77.68	
168	61				99.86
		0.92	19.0015	67.12	
180	56				87.04
		0.89	17.444	58.5	
192	54				76.83
		0.84	16.821	51.64	
204	51				69.30
		0.74	15.88	46.58	
216	45				63.19
		0.74	14.0175	42.46	
228	45				57.22
		0.69	14.0175	38.46	
240	42				53.17

From the above table, we can determine that

Peak of inflow hydrograph = 342 m³/s, [t = 48 h]

Peak of outflow hydrograph = 341.1 m³/s, [t = 84 h]

Therefore, attenuation in peak flow discharge = 342 – 341.1 = 0.9 m³/s

PROBLEM 8.2 The amounts of water flowing from a certain catchment area at the proposed dam site are tabulated below: (IES, 1999)

Month	*Inflow* × 10^5 m³
January	2.83
February	4.25
March	5.66
April	18.4
May	22.64
June	22.64
July	19.81
August	8.49
September	7.1
October	7.1
November	5.66
December	5.66

Determine

(a) The minimum capacity of the reservoir if the water is to be used to feed the turbines of hydropower plant at uniform rate and no water is to be spilled over

(b) The initial storage required to maintain the uniform demand as above.

Solution Total amount of inflow to the dam over a year = (2.83 + 4.25 + 5.66 + 18.40 + 22.64 + 22.64 + 19.81 + 8.49 + 7.10 + 7.10 + 5.66 + 5.66) × 105 = 130.24 × 105 m^3

Since the water is to be used to feed the turbines of hydropower plant at uniform rate, therefore the maximum average monthly rate

$$= \frac{\text{Total amount of inflow to the dam over a year}}{12}$$

$$= \frac{(130.24 \times 10^5)}{12} = 10.8533 \times 10^5 \text{ m}^3$$

The calculations for determination of storage capacity are provided in the table below:

Month	*Monthly inflow* (× 10^5 m^3)	*Monthly outflow* (× 10^5 m^3)	*Monthly deficit* (× 10^5 m^3) Col. 3–Col. 2	*Monthly surplus* (× 10^5 m^3) Col. 2–Col. 3	*Cumulative deficit* (× 10^5 m^3)	*Cumulative surplus* (× 10^5 m^3)	*Net water available in reservoir* (× 10^5 m^3)
Col. 1	Col. 2	Col. 3	Col. 4	Col. 5	Col. 6	Col. 7	Col. 8
January	2.83	10.8533	8.0233	–	–	–	–
February	4.25	10.8533	6.6033	–	–	–	–
March	5.66	10.8533	5.1933	–	19.82	–	(–)19.82
April	18.4	10.8533	–	7.5467	–	–	–
May	22.64	10.8533	–	11.7867	–	–	–
June	22.64	10.8533	–	11.7867	–	–	–
July	19.81	10.8533	–	8.9567	–	40.08	(+)20.26
August	8.49	10.8533	2.3633	–	–	–	–
September	7.1	10.8533	3.7533	–	–	–	–
October	7.1	10.8533	3.7533	–	–	–	–
November	5.66	10.8533	5.1933	–	–	–	–
December	5.66	10.8533	5.1933	–	20.26	–	0

(a) Now from the above table the highest value in Col. 6 and Col. 7 is representing the minimun capacity of the reservoir if the water is to be used to feed the turbines of hydropower plant at an uniform rate and no water is to be spilled over, which is 40.08 × 10^5 m^3.

(b) The minimum initial storage in the reservoir is equal to the maximum negative storage in the reservoir which is 19.82 × 10^5 m^3.

Theoretical Questions

1. What do you understand by flood routing?
2. Differentiate between hydrologic and hydraulic method of routing.
3. Describe the Muskingum method of flood routing.
4. Discuss hydraulic routing.
5. Discuss the various flood control measures.

Unsolved Questions

Problem 1: Develop an expression for the output $Q(t)$ for a linear hydrologic system in terms of the input $I(t)$ and the storage constant k. At the initial condition, output is zero.

Problem 2: Route the flood hydrograph given below through a channel reach and derive the outflow hydrograph. The value of K and x for the channel reach may be taken as 12 h and 0.25, respectively.

Times (h)	0	4	8	12	16	20	24	28	32	36	40	44	48
Discharge (m^3/s)	20	45	60	90	120	140	185	140	95	70	60	45	30

Problem 3: The inflow and outflow hydrograph ordinates of a particular channel are given in table below. Compute the Muskingum parameters K and x for the channel reach:

Times (h)	0	6	12	18	24	30	36	42	48	54	60	66	72
Inflow Discharge (m^3/s)	30	145	170	190	220	340	365	230	155	90	70	45	35
Outflow Discharge (m^3/s)	30	76	98	120	210	350	410	250	180	100	80	50	40

Problem 4: A reservoir has the following elevation, discharge and storage relationships:

Elevation (m)	*Storage* 10^6 (m^3)	*Outflow discharge* (m^3/s)
101	4.050	0
101.5	4.075	10
102.1	4.375	25
102.5	4.550	45
103	4.950	70
103.5	5.150	90
103.75	5.450	105
104	5.950	130
104.5	6.250	145
105	6.750	160

When the reservoir level was at 101.5 rn, the following flood hydrograph entered the reservoir:

Time (h)	0	4	8	12	16	20	24	28	32	36	40	44	48
Discharge (m^3/s)	12	18	28	45	55	60	85	65	40	32	25	15	10

Route the flood using modified Pul's method. Also obtain ordinate of the outflow hydrograph attenuation in the peak flow rate and the lag in peak flow time.

Further Reading

Chow, V.T., Maidment, D.R., and Mays, L.W., *Applied Hydrology*, McGraw-Hill, Singapore, 1988.

Das, Ghanshyam, *Hydrology and Soil Conservation Engineering Including Watershed Management*, PHI Learning, Delhi, 2014.

Davie, Tim, *Fundamentals of Hydrology*, 2nd ed., Routledge, Taylor and Francis Group, London and New York, 2008.

Deodhar, M.J., *Elementary Engineering Hydrology*, Pearson, New Delhi, 2013.

Raghunath, H.M., *Hydrology—Principles, Analysis and Design*, New Age Publishers, New Delhi, 2014.

Subramanya, K., *Engineering Hydrology*, Tata McGraw-Hill, New Delhi, 2010.

Suresh, R., *Watershed Hydrology*, Standard Publishers Distributors, New Delhi, 2015.

Yadupathi Putty, Mysooru R., *Principles of Hydrology*, I.K. International, New Delhi, 2013.

CHAPTER

9

Groundwater

9.1 Introduction

Groundwater is the water which occurs below the surface of the earth. Groundwater is easy to extract, often cheaper, and less vulnerable to pollution than surface water. Therefore, it is a popular resource of water in many countries of the world. In arid regions, groundwater is a reliable source of water for domestic and irrigation purposes. Groundwater is an important part of hydrological cycle and it is replenished naturally by surface water from precipitation, streams and rivers. *Groundwater hydrology* deals with the study of occurrence, movement, and quality of underground water. Groundwater exploitation and contamination attracted the attention of several disciplines, including physics, geology, chemistry, soil science, and plant physiology.

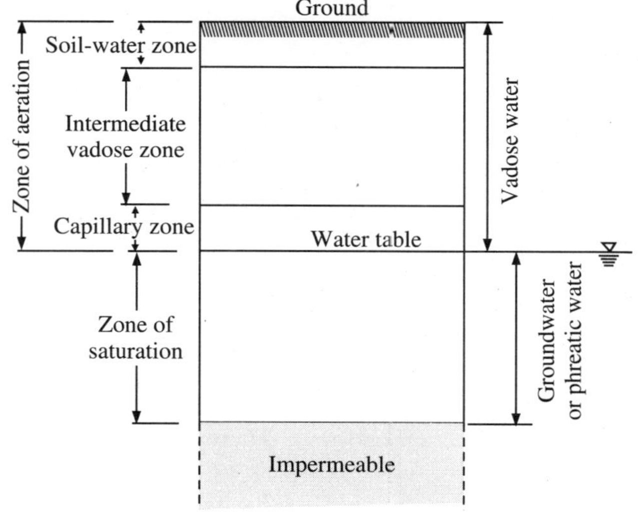

Figure 9.1 Zones of saturation and aeration.

Groundwater available below the ground surface is stored under two different zones, namely, saturated zone and unsaturated zone (aeration zone), as shown in Figure 9.1. These two zones are divided by an irregular surface called *water table*.

Saturated zone

In this zone, all the interstices (pores) are completely filled with water. Hence, water table forms its upper limit and marks a free surface. Since water in this zone is under pressure which is more than atmospheric pressure, water may flow if a hole is dug in this zone.

Unsaturated (aeration zone) zone

In this zone, interstices (pores) are occupied partially by water and partially by air. Hence, the space between the land surface and the water table marks the extent of this zone. This is also known as *vadose zone*. The differences between zone of saturation and zone of aeration are shown in Table 9.1. It is further classified into three parts namely,

1. Soil water zone: This zone extends from the ground surface to a maximum depth of a metre or two. Water in the soil is used by plants, and also, water is lost to the atmosphere by evapotranspiration.

2. Capillary fringe/zone: The boundary between the unsaturated zone and the saturated zone is known as the *capillary fringe*. In this lowest part of unsaturated zones, water is held by capillary action.

3. Intermediate zone: This zone lies in between the soil-water zone and capillary fringe.

Table 9.1 Difference between Zone of Saturation and Zone of Aeration

Sl. No.	*Zone of Saturation*	*Zone of Aeration*
1	The soil pores are completely filled with water	The soil pores are partly filled with water and partly with air
2	The zone of saturation lies beneath the water table	The zone of aeration lies above water table.
3	Piezometers are used to measure the hydraulic head	Tensiometers are used to measure the hydraulic head.
4	The moisture content θ equals the porosity η	The moisture content θ is less than the porosity η.

9.2 Occurrence of Groundwater

Groundwater occurs below the Earth's surface at depths and occupies innumerable pores and cracks found in soil and rock. There are basically four types of geological formations as given below:

1. **Aquifer:** A reservoir of groundwater that contains and transmits significant quantity of water through its pores under normal conditions is called *aquifer*. For example, gravel, sand, vesicular basalts, glacial deposits and limestone.
2. **Aquitard:** A geologic formation that transmits water at a slow rate compared to an aquifer is called *aquitard*. For example, clay lenses interbedded with sand.
3. **Aquiclude:** An impermeable geological formation which may contain water, but does not transmit significant quantity of water is known as *aquiclude*. For example, clays, shales, etc.

4. **Aquifuge:** An impermeable geological formation that does not contain or transmit water is termed as *aquifuge*. For example, solid granites.

A comparison is performed in geological formations based on water and permeability as shown in Table 9.2.

Table 9.2 Differences between Geological Formations

Formation	*Contains water*	*Permeability*
Aquifer	Yes	High
Aquitard	Yes	Low
Aquiclude	Yes	Very low
Aquifuge	No	Negligible

9.3 Types of Aquifer

Aquifers are generally classified as unconfined or confined, depending on the presence or absence of water table, while a leaky aquifer represents a combination of the two types.

9.3.1 Unconfined Aquifer

This type of aquifer is bounded below by a relatively impermeable strata and above by the free water table. The water in such aquifers is said to be unconfined, and the aquifers are referred to as unconfined aquifers. At free water table, groundwater is at atmospheric pressure. Unconfined aquifers are also widely referred to as *water table aquifers*.

9.3.2 Confined Aquifer

In this type of aquifer, groundwater is confined by overlying relatively impermeable strata. The pressure of water in confined aquifers is usually higher than the atmospheric pressure. Such aquifers are also referred to as *artesian aquifers*.

Wells open to unconfined aquifers are referred to as *water table wells*. The water level in these wells indicates the position of the water table in the surrounding aquifer. While, wells drilled into confined aquifers are referred to as *artesian wells* as shown below in Figure 9.2.

The characteristics of confined and unconfined aquifer are summarised in Table 9.3.

Table 9.3 Characteristics of Confined and Unconfined Aquifer

Unconfined Aquifer	*Confined Aquifer*
Free water table	No water table
Greater fluctuations with season	Smaller fluctuations with season
Hydraulic gradient changes rapidly	Hydraulic gradient more uniform
Recharge area is around borehole	Recharge area is away from borehole

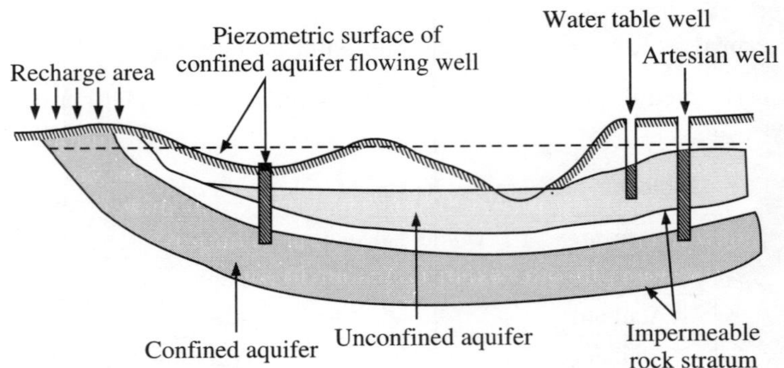

Figure 9.2 Confined and unconfined aquifers.

9.3.3 Leaky Aquifer

This type of aquifer is bounded below by an aquiclude and above by an aquitard. This is also called *semi-confined aquifer*.

9.4 Geological Formations as Aquifers

The geologic formations are classified into two categories, namely, unconsolidated and consolidated deposits.

Unconsolidated deposits

These deposits of sand and gravel form the most important aquifers. The yield of these deposits is good and may be of the order of 50–100 m^3/h. In India, the alluvium deposit of Gangetic areas and the coastal alluvium of Andhra Pradesh are some examples.

Consolidated deposits

Consolidated sediments are sedimentary rocks such as sandstone, shale, and limestone. Groundwater flows through fracture networks and openings in these consolidated sediments. The yield has a value of 20–50 m^3/h. In India, sandstones of Kathiawar and Kutch areas of Gujarat are examples of this kind.

9.5 Properties of Aquifer

Some important properties of aquifers are porosity, specific yield, coefficient of permeability, storage coefficient, transmissivity, hydraulic diffusivity, etc.

These properties are described here.

9.5.1 Porosity

The ratio of volume of openings (voids) to the total volume of a soil or rock is referred to as its *porosity*. The larger the pore space implies the higher the porosity, and thus, the larger the water-holding capacity, as shown in Figure 9.3. Therefore, porosity gives a measure of the water-storage capability of a formation. Thus, it is expressed as

$$n = \frac{V_t - V_s}{V_t} = \frac{V_v}{V_t} \tag{9.1}$$

where, n is the porosity as a decimal fraction or percentage, V_t is the total volume of a soil or rock sample, V_s is the volume of solids in the sample, and V_v is the volume of openings (voids).

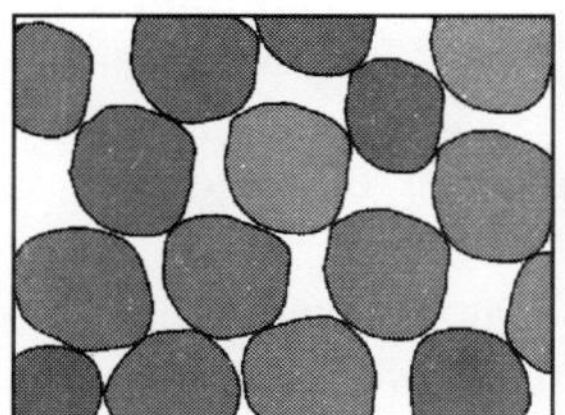

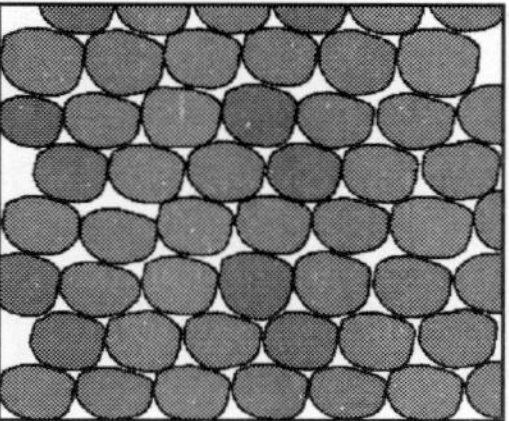

Figure 9.3 Porosity.

Porosity of unconsolidated deposits depends on the range of grain size (sorting) and on the shape of the rock particles but not on their size. In qualitative terms, porosity greater than 20% is considered as large, between 5% and 20% considered as medium, and less than 5% as small.

Effective porosity is the ratio of interconnected pore space to the bulk volume of the soil or rock sample.

Aquifer properties such as porosity, specific yield and specific retention for different geological materials are shown in Table 9.4.

Table 9.4 Aquifer Properties for Geological Materials (Values are in percent of volume)

Material	*Porosity*	*Specific yield*	*Specific retention*
Soil	55	40	15
Clay	50	2	48
Sand	25	22	3
Gravel	20	19	1
Limestone	20	18	2
Sandstone (semi-consolidated)	11	6	5
Granite	0.1	0.09	0.01
Basalt (young)	11	08	03

9.5.2 Specific Yield

Specific yield, S_y is the ratio of the volume of water that drains from a saturated rock (due to gravity) to the total volume of the rock. It is defined as

$$S_y = \frac{V_d}{V_t} \tag{9.2}$$

where, S_y is specific yield, V_d is the volume of water that drains from a total volume, and V_t is the total volume of a soil or rock sample.

The *specific retention*, S_r is the ratio of the volume of water a rock can retain against gravity as a film on rock surfaces and in very small openings to the total volume of the rock. It is defined as

$$S_r = \frac{V_r}{V_t} \tag{9.3}$$

where, S_r is specific retention and V_r is the volume of water retained in a total volume.

The sum of specific yield and specific retention is equal to porosity.

$$\eta = S_y + S_r \tag{9.4}$$

The specific yield and specific retention depend on the size and shape of particles, voids distribution, and compaction of the formation.

In general, specific yield tells how much water is available for man's use, and specific retention tells how much water remains in the rock after it is drained by gravity.

EXAMPLE 9.1 Determine the specific yield of an unconfined aquifer covering an area of 7.6 km^2. It was observed that original water level is dropped by 3 m due to pumping of 4 million m^3. Also, determine the specific retention if the porosity of aquifer is 24%.

Solution Volume of water pumped out from the aquifer

= Area × Change in water table × Specific yield = $7.6 \times 10^6 \times 3 \times S_y$

It is given that volume of water pumped out = 4×10^6 m^3

Therefore, specific yield of the aquifer $\dfrac{4 \times 10^6}{7.6 \times 10^6 \times 3}$ = 0.175 = 17.5%

Specific retention = Porosity – Specific yield = 24 – 17.5 = 6.5%

9.5.3 Coefficient of Permeability

Coefficient of permeability K of a medium, also known as *hydraulic conductivity*, is equal to the volume of water which flows in unit time through unit cross-sectional area of a medium under unit hydraulic gradient at the prevailing temperature. The common unit for K is m/day or cm/s. This involves the properties of both medium and fluid. Table 9.5 shows representative values of hydraulic conductivity for various types of rocks.

Table 9.5 Various Types of Rock Materials

Material	*Hydraulic Conductivity* (m/s)
Gravel	3×10^{-4} to 3×10^{-2}
Coarse sand	9×10^{-7} to 6×10^{-3}
Fine sand	2×10^{-7} to 2×10^{-4}
Silt, loess	1×10^{-9} to 2×10^{-5}
Clay	1×10^{-11} to 4.7×10^{-9}
Siltstone	1×10^{-11} to 1.4×10^{-8}
Shale	1×10^{-13} to 2×10^{-9}

9.5.4 Storage Coefficient

The volume of water given out by a unit prism of an aquifer when the piezometric surface or the water table drops by unit depth is called *storage coefficient* of the aquifer or *storativity*, and it is denoted by S, which is dimensionless. Storage coefficient of an aquifer is given by the following relation:

$$S = \gamma_w nb\left(\frac{1}{K_w} + \frac{1}{nE_s}\right) \tag{9.5}$$

where, S is storage coefficient (decimal), γ_w is specific weight of water, n is porosity of soil (decimal), b is thickness of the confined aquifer, K_w is bulk modulus of elasticity of water and E_s is modulus of compressibility (elasticity) of the soil grains of the aquifer.

Specific capacity

The *specific capacity* Q/S_w of a well is the discharge per unit drawdown in the well and is usually expressed as *l*/min/m or *l*pm/m.

Note:

1. The specific capacity is a measure of the effectiveness of the well.
2. Specific capacity decreases with the increase in the pumping rate Q and prolonged pumping time (t).

9.5.5 Transmissivity

Transmissivity, T is defined as the product of average hydraulic conductivity and the saturated thickness of the aquifer. It is the flow capacity of an aquifer per unit width under unit hydraulic gradient. Transmissivity of most of the formations lies between $1 \times 10^4 - 1 \times 10^6$ l/d/m, with an average value of 1×10^5 (l/d/m).

Note:

1. In a confined aquifer, $T = Kb$ and is independent of the piezometric surface.
2. In a water table aquifer, $T = KH$, where H is the saturated thickness.
3. As the water table drops, H decreases and the transmissivity is reduced.
4. Transmissivity of an unconfined aquifer depends on the depth of groundwater table (GWT).

9.5.6 Hydraulic Diffusivity

The *hydraulic diffusivity*, D is defined as the ratio of transmissivity and storavity.

$$D = \frac{T}{S} \tag{9.6}$$

9.6 Darcy's Law

Henry Darcy (1856) conducted a series of experiments in vertical homogeneous sand filter and based on the experimental observations, he described the flow of a fluid through a porous medium. Darcy's law states that the velocity of flow in a porous medium is proportional to the hydraulic gradient. Darcy's law may be written as

$$V \propto i = Ki = -K\frac{dh}{dl} \tag{9.7}$$

where, V is velocity of flow through the aquifer, K is hydraulic conductivity or coefficient of permeability of aquifer soil, i is hydraulic gradient and h is head lost in a length of flow path.

The negative sign indicates that the flow takes place in the direction of falling head.

The Darcy's law is very similar to the Ohm's law for electrical circuits where $I = \frac{1}{R} \times V$ $\left(\text{or Current} = \frac{\text{Voltage}}{\text{Resistance}}\right)$.

In Figure 9.4, an inclined cylinder filled with sand of cross-section area A is shown. Applying Bernoulli's energy equation for the two sections where piezometers are inserted,

$$\frac{p_1}{\gamma} + z_1 + \frac{V_1^2}{2g} = \frac{p_2}{\gamma} + z_2 + \frac{V_2^2}{2g} + h_L$$

Here, p_1, V_1 are pressure and velocity of flow at one section and p_2, V_2 are pressure and velocity of flow at another section, respectively.

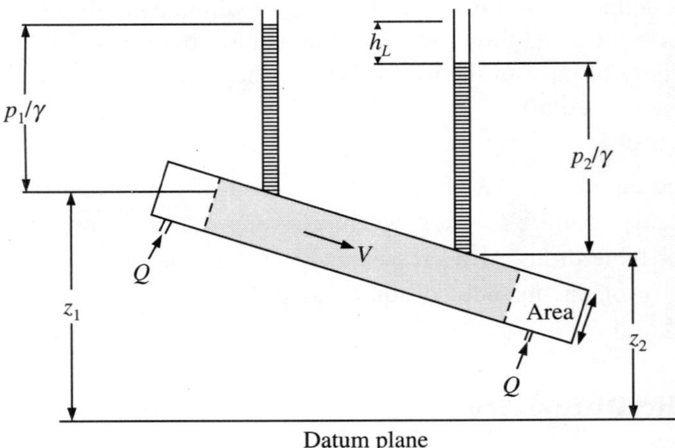

Figure 9.4 Experimental demonstration—Darcy's law.

Here, z_1 and z_2 are elevations of intake of piezometer from the datum.

Neglecting velocity heads as they will be very small compared to other terms, therefore

$$h_L = \left(\frac{p_1}{\gamma} + z_1\right) - \left(\frac{p_2}{\gamma} + z_2\right)$$

From the experiment, Darcy found that rate of flow Q is directly proportional to the area of flow A and head loss and inversely proportional to the length L. Therefore,

$$Q = AK\frac{h_L}{L} \tag{9.8}$$

where, K is hydraulic conductivity and is defined as the flow rate per unit cross-sectional area under the influence of a unit gradient. In differential form,

$$V = -K\frac{dh}{dl}$$

This is based on the assumption that the flow occurs through the entire cross-sectional area. However, the part of cross-sectional area is occupied by solids and the remaining part is void. Thus, the actual velocity of flow in the porous media (V_{act}) is expressed as

$$V_{act} = \frac{V}{\eta} \tag{9.9}$$

The hydraulic conductivity depends on the properties of the porous medium and the fluid. The hydraulic conductivity can be written as

$$K = \frac{cd^2 \rho g}{\mu} \tag{9.10}$$

where cd^2 is a function of medium and μ is coefficient of dynamic viscosity of fluid.

The intrinsic permeability k, which is a property of medium only, can be written as

$$k = \frac{K\mu}{\rho g} = cd^2 \tag{9.11}$$

The unit of intrinsic permeability k is the unit of area, written as 'Darcy', which is equal to 9.87×10^{-9} cm^2.

Note: Darcy's law is valid for laminar flow only. From experiments, it was found that this law is valid up to the Reynolds number $Re < 1$, and does not create much error up to $Re < 10$.

EXAMPLE 9.2 Following information is given regarding an aquifer:

Cross-sectional area of the aquifer = 2×10^4 m^2

Length of the aquifer = 1600 m

Hydraulic head at entry = 350 m

Hydraulic head at exit = 270 m

Rate of groundwater discharge = 1600 m^3/day

Porosity of the soil = 0.4

Determine the (a) hydraulic conductivity of an aquifer, (b) pore velocity of water.

Solution Given, cross-sectional area of the aquifer = 2×10^4 m^2

Length of the aquifer = 600 m

Hydraulic head at entry = 350 m

Hydraulic head at exit = 270 m

Rate of groundwater discharge = 1600 m^3/day

Porosity of the soil = 0.4

$$\text{Darcy's velocity, } V = \frac{Q}{A} = \frac{1600}{2 \times 10^4} = 0.08 \text{ m/day}$$

$$\text{Hydraulic gradient, } -\frac{dh}{dl} = \frac{350 - 270}{1600} = 0.05$$

$$\text{Hydraulic conductivity, } K = -\frac{V}{dh/dl} = \frac{0.08}{0.05} = 1.6 \text{ m/day}$$

$$\text{Pore velocity of water, } V_{\text{act}} = -\frac{V}{\eta} = \frac{0.08}{0.4} = 0.2 \text{ m/day}$$

EXAMPLE 9.3 In a Darcy's experiment, water flows through soil with hydraulic conductivity of 0.02 cm/s. If the viscosity of water is 0.9 × 10^{-3} Ns/m^2, then what will be the intrinsic permeability of the soil?

Solution Given, specific weight of water g = 9810 N/m^3 = 9.81 × 10^{-3} N/cm^3

Hydraulic conductivity, K = 0.02 cm/s

Viscosity of water, (μ) = 0.9 × 10^{-3} Ns/m^2 = 9 × 10^{-8} Ns/cm^2

$$\text{Intrinsic permeability of the soil, } k = \frac{K\mu}{\rho g} = \frac{0.02 \times 9\,10^{-8}}{9.81 \times 10^{-3}}$$

$$k = 18.349 \times 10^{-8} \text{ Ns/cm}^2$$

9.6.1 Stratification

In some cases, it is possible that the aquifer consists of different strata with different hydraulic conductivities. If the flow occurs parallel to the strata, then the equivalent hydrualic conductivity may be given as

$$K_e = \frac{\sum_1^n K_i Z_i}{\sum_1^n Z_i} \tag{9.12}$$

If the flow occurs normal to the strata, then the equivalent hydrualic conductivity may be given as

$$K_e = \frac{\sum_1^n L_i}{\sum_1^n L_i / K_i} \tag{9.13}$$

where Z_i and L_i is width and length of ith stratum.

EXAMPLE 9.4 Compute equivalent hydraulic conductivity of the confined aquifer (a) along the horizontal direction of the alignment of the layers, (b) along the vertical direction of the alignment of the layers, if the following information is given regarding a confined aquifer having three layers.

Hydraulic conductivity of top layer = 3 cm/s and thickness = 4.5 m

Hydraulic conductivity of middle layer = 0.002 cm/s and thickness = 3.5 m

Hydraulic conductivity of bottom layer = 0.02 cm/s and thickness = 5 m

Solution (a) When the flow is horizontal, then the equivalent hydraulic conductivity of the confined aquifer can be determined as

$$K_e = \frac{K_i Z_i}{Z_i} = \frac{(5 \times 100) \times 0.02 + (3.5 \times 100) \times 0.002 + (4.5 \times 100) \times 3}{500 + 350 + 450}$$

$$= 1.047 \text{ cm/s}$$

(b) If the flow is vertical to the alignment of the layers, then the hydraulic conductivity will be

$$K_e = \frac{Z_i}{Z_i / K_i} = \frac{(5 \times 100) + (3.5 \times 100) + (4.5 \times 100) \times 3}{(500/0.02) + (350/0.002) + (450/3)}$$

$$= 6.495 \times 10^{-3} \text{ cm/s}$$

EXAMPLE 9.5 Determine the transmissivity of the confined aquifer along the horizontal alignment of the layer in the aquifer in Example 9.4.

Solution Transmissivity of the confined aquifer

$$= (5 \times 100) \times 0.02 + (3.5 \times 100) \times 0.002 + (4.5 \times 100) \times 3$$

$$= 1360.7 \text{ cm}^2\text{/s}$$

9.7 Anisotropy and Heterogeneity

For a given geological formation, if the hydraulic conductivity depends on the position within the formation, then, such formation is called *heterogeneous*. If the hydraulic conductivity also varies in the direction of measurement, the formation is called *anisotropic*.

9.8 Groundwater Flow Problems

In steady flow, flow properties such as velocity flow do not change with time.

9.8.1 Steady Flow in Unconfined Aquifer

When two streams with difference in the elevation of their water surfaces are separated by unconfined aquifer, then such a situation arises as shown in Figure 9.5.

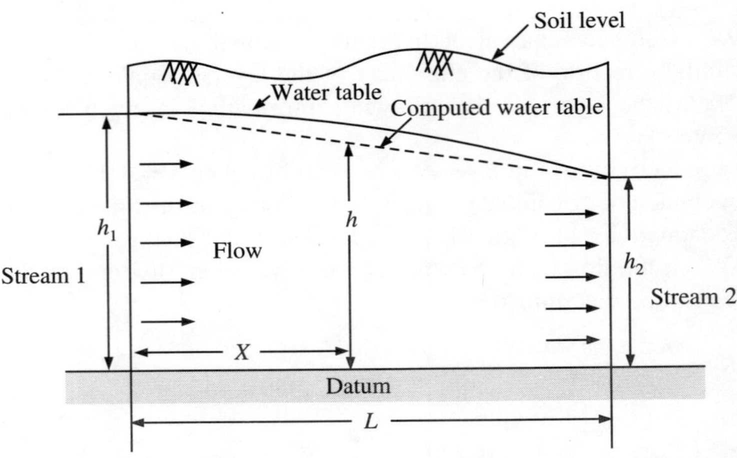

Figure 9.5 Steady flow in unconfined aquifer.

Assume flow is horizontal and uniform at every section. For undirectional flow, discharge per unit width may be written as

$$q = -Kh\frac{dh}{dx} \tag{9.14}$$

Integrating Eq. (9.14),

$$\int q dx = \int -Khdh$$

$$qdx = -K\frac{h^2}{2} + C$$

where C is a constant of integration.

To find C, applying boundary conditions for $x = 0$, $h = h_i$, hence,

$$C = \frac{Kh_1^2}{2}$$

Therefore,

$$q = \frac{K}{2x}(h_1^2 - h^2) \tag{9.15}$$

This equation is known as *Dupuit's equation*, which indicates that the phreatic line is parabola. Now, substituting $x = L$, $h = h_2$,

$$q = \frac{K}{2L}(h_1^2 - h_2^2) \tag{9.16}$$

From Eqs. (9.15) and (9.16),

$$h = \left(h_1^2 - (h_1^2 - h_2^2)\frac{x}{L}\right)^{1/2} \tag{9.17}$$

It is observed that value of h from Dupuit's equation is always lower than the actual water table.

EXAMPLE 9.6 Two observation wells are located on the banks of a river and a canal under which an unconfined aquifer is situated. Find out the seepage into the canal if the following data are given:

Hydraulic conductivity of unconfined aquifer = 40 m/day

Distance between the observation wells = 4 km

Heads in the two observation wells are 70 and 55 m, respectively, above impermeable strata.

Solution Given, L = 4000 m, K = 40 m/day, H_1 = 70 m, H_2 = 55 m

The quantity of seepage is given as

$$q = \frac{K}{2L}(H_1^2 - H_2^2)$$

$$= \frac{40}{2 \times 4000}(70^2 - 55^2) = 9.375 \text{ m}^3\text{/day/m}$$

9.8.2 Steady Flow in Confined Aquifer

Consider a confined aquifer of depth B, as shown in Figure 9.6.

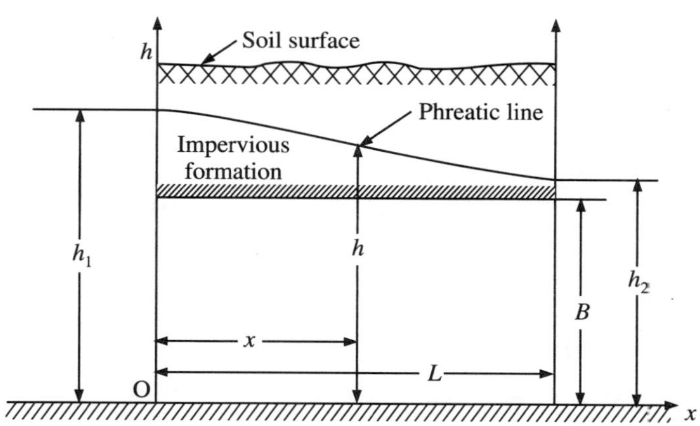

Figure 9.6 Steady flow in confined aquifer.

For undirectional flow, discharge per unit width may be written as

$$q = \text{Area} \times \text{Velocity} = B \times -K\frac{dh}{dx} = -KB\frac{dh}{dx} \tag{9.18}$$

Integrating Eq. (9.18),

$$\int q dx = \int -KB dh$$

$$qx = -KBh + C$$

where C is a constant of integration. To find C, applying boundary conditions for $x = 0$, $h = h_1$, hence,

$$C = KBh_1$$

Therefore,

$$qx = -KBh + KBh_1 = KB(h_1 - h)$$

or

$$h = h_1 - \frac{qz}{KB} \tag{9.19}$$

Equation (9.19) indicates that the phreatic line is a straight line in this case.

Now substituting $x = L$, $h = h_2$,

$$q = \frac{KB}{L}(h_1 - h_2) = \frac{T}{L}(h_1 - h_2) \tag{9.20}$$

Here T is transmissivity.

Discharge per unit width q can be computed by knowing piezometric heights h_1, h_2 and L distance.

EXAMPLE 9.7 If two canals A and B are separated by a soil formation of two layers, which consists of an impervious layer on the top, then compute the discharge from canal A to B if the following data are given:

Width of formation = 7.2 km

Thickness of the bottom pervious layer = 10 m

Hydraulic conductivity of the pervious layer = 45 m/day

Water level at canal A = 20 m

Water level at canal B = 16 m

Solution Given, K = 45 m/day, L = 7.2 km = 7200 m, h_A = 20 m, h_B = 16 m

$$q = \frac{KB(h_A - h_B)}{L} = \frac{45 \times 10(20 - 16)}{7200} = 0.25 \text{ m}^3/\text{day}$$

9.9 Well Hydraulics

Groundwater is usually extracted from well and finds extensive use in water supply and irrigation engineering practices. Hence, wells are used for pumping as well as for recharge purpose.

When water is pumped from the well, water flows towards the well from storage. So, a gradient in the water table towards the well is created, and thus, the head declines, forming a conical shape. This is called *cone of depression*. The amount of decline in water level in the well is called *drawdown*. The areal extent of the cone of depression is known as *area of influence* and its radial extent is called *radius of influence*. After sufficiently long time, an equilibrium stage is reached when contribution from the aquifer to the well is equal to the rate of pumping. Then, the shape of the cone of depression takes a constant position in space, and at this condition, the flow is said to be steady.

Expressions for the steady-state radial flow into a well under both confined and unconfined aquifer conditions are presented here.

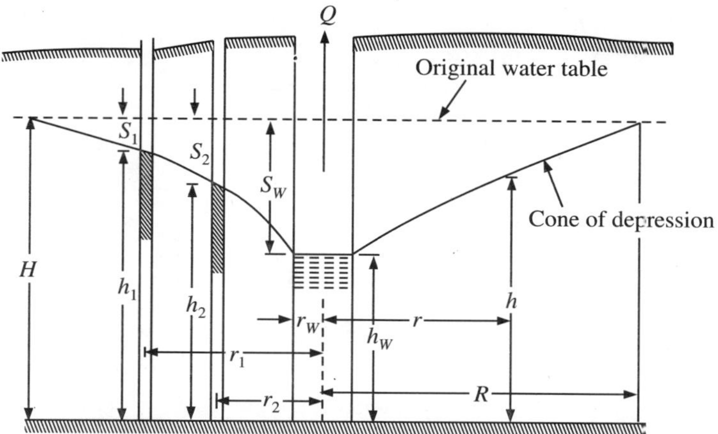

Figure 9.7 Steady radial flow to a well in unconfined aquifer.

9.9.1 Steady Radial Flow to a Well in Unconfined Aquifer

Assuming that the well is pumped at a constant rate Q for a long time and flow is radial even though, near the pumping well, as shown in Figure 9.7, discharge is given by
Using Darcy's law,

$$Q = \text{Area} \times \text{Velocity} = 2\pi rh \times K\frac{dh}{dr} \tag{9.21}$$

Here, the depth of flow at any radial distance r from the well is h.

$$hdh = \frac{Q}{2\pi K}\frac{dr}{r}$$

Integrating between the limits r_1 and r_2 where water table depths are h_1 and h_2.

$$\int_{h_1}^{h_2} hdh = \frac{Q}{2\pi K}\int_{r_1}^{r_2}\frac{dr}{r}$$

$$Q = \frac{\pi K(h_1^2 - h_2^2)}{\ln(r_1/r_2)} \tag{9.22}$$

Equation (9.21) is an equilibrium equation for the steady flow in an unconfined aquifer.

EXAMPLE 9.8 Find the hydraulic conductivity of a fully penetrating well in an unconfined aquifer under steady state condition if the following data are given:

Diameter of the well = 0.5 m
Rate of discharge from the well = 1500 m^3/day
Saturated depth of the aquifer = 55 m
Amount of drawdown in the well = 20 m
Radius of influence = 400 m

Solution Given, $h_1 = 55$ m, $h_2 = (55 - 20) = 35$ m, $Q = 1500$ m^3/day, $r_1 = 400$ m, $r_2 = 0.25$ m

$$Q = \frac{\pi K(h_1^2 - h_2^2)}{\ln(r_1/r_2)}$$

$$h_1^2 - h_2^2 = \frac{Q}{\pi K}\ln\left(\frac{r_1}{r_2}\right)$$

$$K = \frac{Q}{\pi(55^2 - 35^2)}\ln\left(\frac{r_1}{r_2}\right)$$

$$K = \frac{1500}{\pi(55^2 - 35^2)}\ln\left(\frac{400}{0.25}\right)$$

$$K = 1.958 \text{ m/day}$$

9.9.2 Steady Radial Flow to Confined Aquifer

It is assumed that the well completely penetrates the confined aquifer of thickness b and the flow is radial. Under steady-state condition, consider well to be discharging a steady flow Q. At this state, the piezometric head at the pumping well is h_w and drawdown is s_w. Hence, discharge past the section which is situated at radial distance r from the well, as shown in Figure 9.8, becomes

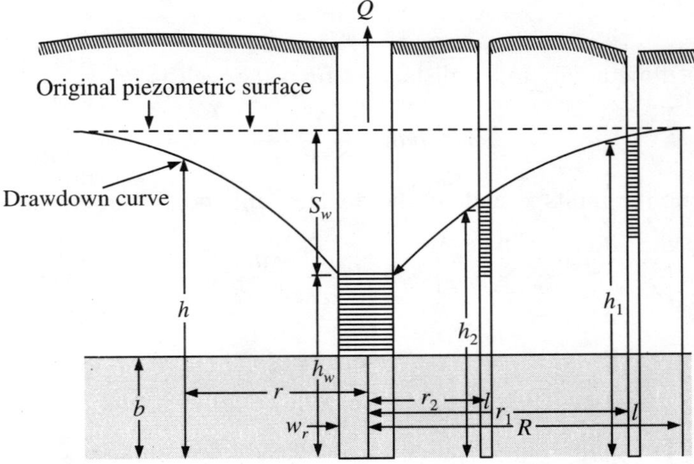

Figure 9.8 Steady radial flow to a well in confined aquifer.

Using Darcy's law,

$$Q = \text{Area} \times \text{Velocity} = 2\pi rb \times K\frac{dh}{dr} \tag{9.23}$$

$$dh = \frac{Q}{2\pi bK}\frac{dr}{r} \tag{9.24}$$

Integrating between the limits r_1 and r_2 where water table depths are h_1 and h_2, respectively.

$$\int_{h_1}^{h_2} dh = \frac{Q}{2\pi bK}\int_{r_1}^{r_2}\frac{dr}{r}$$

$$Q = \frac{2\pi Kb(h_1 - h_2)}{\ln(r_1/r_2)} \tag{9.25}$$

Equation (9.25) is an equilibrium equation for the steady flow in a confined aquifer. This is also known as *Thiem's equation.*

Simplifying and putting $T = Kb$, where T is Transmissivity of the aquifer,

$$Q = \frac{2\pi T(h_1 - h_2)}{\ln(r_1/r_2)} \tag{9.26}$$

If drawdowns in the wells are s_1 and s_2.

$$Q = \frac{2\pi T(s_1 - s_2)}{\ln(r_1/r_2)} \tag{9.27}$$

Note: To avoid incrustation and corrosion at the openings, the length of the screen provided should be half to three-fourths of the thickness of the aquifer for obtaining a suitable velocity (2.5 cm/s). Percentage open area provided in the screen should be usually 15 to 18%.

EXAMPLE 9.9 Calculate the transmissivity and hydraulic conductivity of a confined aquifer if a well is fully penetrated in the aquifer and is being pumped at a constant rate. The following information is given below:

Distances of two observation wells A and B from the centre of the main well = 15 m and 65 m, respectively.

Head drawdowns in the observation wells = 4 m and 1 m, respectively

Rate of pumping from the well = 60 l/s

Diameter of the well = 0.2 m

Thickness of the aquifer = 10 m

Solution Given, $Q = 0.06$ m^3/s, $r_A = 15$ m, $r_B = 65$ m, $s_A - s_B = (4 - 1) = 3$ m

$$\text{Transmissivity, } T = \left(\frac{Q\ln(r_A/r_B)}{2\pi(s_A - s_B)}\right) = \left(\frac{0.06 \times \ln(65/15)}{2\pi \times 3}\right) = 4.67 \times 10^{-3} \text{ m}^2/\text{s}$$

$$\text{Hydraulic conductivity, } K = \frac{T}{b} = \frac{4.67 \times 10^{-3}}{10} = 4.67 \times 10^{-4} \text{ m}^2/\text{day}$$

EXAMPLE 9.10 Determine the radius of influence of a fully penetrating artesian well under steady state condition if the following data are given,

Diameter of the well = 0.5 m

Rate of discharge from the well = 2000 m^3/day

Transmissivity of the aquifer = 0.167 m^2/mm

Amount of drawdown in the well = 10.5 m

Solution Given, diameter of the well 0.5 m

$$Q = 2000 \text{ m}^3/\text{day} = \frac{2000}{60 \times 24} = 1.3889 \text{ m}^3/\text{min}$$

$$T = 0.167 \text{ m}^2/\text{min}$$

$$s_1 - s_2 = \frac{Q}{2\pi T} \ln\left(\frac{r}{r_w}\right)$$

$$10.5 = \frac{1.3889}{2\pi \times 0.167} \ln\left(\frac{r}{0.5}\right)$$

$$r = 1387.72 \text{ m}$$

EXAMPLE 9.11 Three tube wells of equal diameter are situated in such a way that if they are joined by three straight lines, an equilateral triangle will be formed with side length of 150 m. All the three tube wells of 50 cm diameter are penetrated in a confined aquifer of thickness 30 m. Determine the discharge when only one tube well is in operation with a depression head of 4 m. Also, determine the percentage change in discharge of the well if all the three tube wells are in operation with equal depression head 4 m.

Assume the radius of influence for wells and the coefficient of permeability of aquifer to be 350 m and 50 m/day.

Solution Given, diameter of each tube well = 0.5 m

Thickness of the aquifer = 30 m

Radius of each tube well = 350 m

Coefficient of permeability of the aquifer = 50 m/day

When only one well is pumping at steady state condition,

$$s = 4 \text{ m}$$

$$r_w = 0.25 \text{ m}$$

$$R = 350 \text{ m}$$

$$K = 50 \text{ m/day}$$

$$b = 30 \text{ m}$$

$$T = Kb = 1500 \text{ m}^2/\text{day}$$

$$s = \frac{Q}{2\pi T} \ln\left(\frac{R}{r_w}\right)$$

$$Q = \frac{2\pi s T}{\ln(R/r_w)}$$

$$Q = \frac{2 \times \pi \times 4 \times 1500}{\ln(350/0.25)} = 5201.4 \text{ m}^3/\text{day}$$

When all the three wells are pumping at steady-state condition,

$$s_{\text{Total}} = \sum_{i=1}^{3} s_i = \sum \frac{Q_j}{2\pi T} F(r_{ij})$$

Now, $Q_1 = Q_2 = Q_3$

$$s_t = \frac{3Q}{2\pi T}\ln\left(\frac{R}{r^*}\right)$$

$$r^* = \sqrt[3]{r_{11} \times r_{12} \times r_{13}} = 0.25$$

$$s_t = 3 \text{ m},\ r_{11} = 0.25 \text{ m},\ r_{12} = r_{13} = 150 \text{ m}$$

$$Q = \frac{4 \times 2\pi \times 1500}{3 \times \ln\left(\dfrac{350}{\sqrt[3]{0.25 \times 150 \times 150}}\right)} = 4215.32 \text{ m}^3\text{/day}$$

$$\text{Percentage change in discharge} = \left(\frac{5201.4 - 4215.32}{5201.4}\right) \times 100 = 18.96\%$$

9.10 Dupuit's Equations Assumptions

1. For equilibrium stage, pumping continues at constant rate.
2. For constant permeability, aquifer is homogeneous, isotropic and of infinite areal extent.
3. Darcy's law is applicable when the flow is laminar, i.e., flow lines are radial and horizontal.

9.11 Yield Tests

Open wells are popular source of daily needs of domestic water in many parts of the country. Under normal conditions (i.e., no pumping), water level in a well approximately indicates the level of water table in the surrounding of the well. As the water is pumped out, the water level in the well falls more quickly. The difference between the level of water table and the water level in the well is called *depression head*. The rate of flow into the well Q is proportional to the depression head h, $Q \propto h$. This may be expressed as

$$Q = Kh$$

Here, K is constant that depends on the characteristics of the aquifer and the area of well.

Due to continued pumping of water, depression head goes on increasing. Consequently, it may dislodge the soil particles due to higher velocities. The depression head at which this happens is called *critical depression head*. The discharge corresponding to this head is called *critical* or *maximum yield*. Therefore, *maximum yield* of well is the rate at which water percolates into the well under the critical depression head. To avoid critical depression yield, a factor of safety (2.5–4) should be provided. Therefore, a working head is specified and the yield corresponding to this working head is known as the *maximum safe yield*. The following tests can be performed to get an idea of the probable yield of the well:

1. Pumping test
2. Recuperation test

9.11.1 Pumping Test

Pumping test is field experiment to understand the response of aquifers when they are subjected to abstraction or recharge. In this test, stress is applied on the aquifer by extracting groundwater and the aquifer response to this stress is measured by monitoring drawdown as a function of time. These tests help in determining how much groundwater can be extracted from well on long-term yield and hydraulic properties of an aquifer.

Here, water is withdrawn from the well freely till a critical depression head is created. Then, water level is kept constant by adjusting the rate of pumping equal to the percolation into the well. This rate is considered as the yield of the well.

9.11.2 Recuperation Test

In this test, well is pumped at constant rate Q till the water level in the well is depressed below the static level. Then, pumping is stopped and the well is allowed to recuperate. Hence, the depth of recuperation in the well is noted at various time intervals.

- Let the water level inside the well rise from h_1 to h_2 in time T, as shown in Figure 9.9.

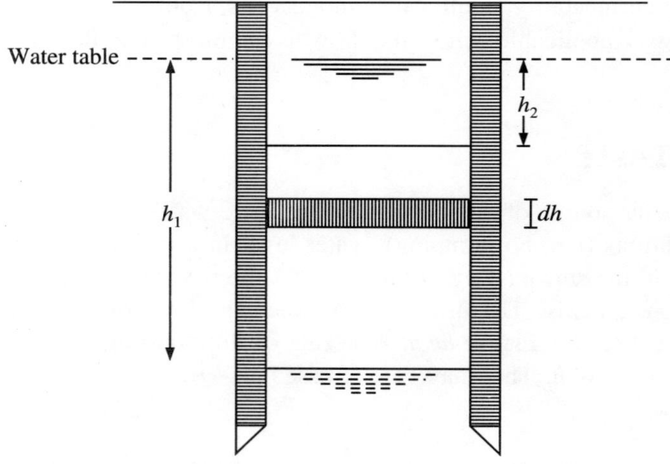

Figure 9.9 Recuperation test.

If head h is lost in a length L of seepage path, then from Darcy's law,

$$Q = \frac{KAh}{L} = CAh$$

where, $C = K/L$.

Assuming the level of water to rise by a value dh in a small time interval dt, then volume of recuperation,

$$dV = Adh$$

The flow in the well is equal to $Q.dt$. Therefore,

$$Qdt = -Adh$$

Here, negative sign indicates head decrease as the time increases,
By using $Q = CAh$,

$$CAhdt = -Adh$$

If h increases from h_1 to h_2 in time interval from 0 to T, then

$$C\int_0^T dt = \int_{h_1}^{h_2} -\frac{dh}{h}$$

$$C = \frac{2.303}{T}\log_{10}\left(\frac{h_1}{h_2}\right)$$

From the available data on h for a range of t, C can be determined. The usual values are given below:

Note:

1. $C = 0.25\ h^{-1}$ for clayey soil
2. $C = 0.50\ h^{-1}$ for fine sand
3. $C = 1.00\ h^{-1}$ for coarse sand

EXAMPLE 9.12 Calculate the diameter of an open well which has clay as subsoil, if the following information are given:

Safe maximum yield from the well = 5.2 l/s
Working head of the well = 4.5 m
Take value of specific yield $C = 0.25\ h^{-1}$

Solution Given, Q = 5.2 l/s = 0.0052 m^3/s = 18.72 m^3/h, H = 4.5 m
We know that

$$Q = CAH$$

$$A = \frac{Q}{CH}$$

$$= \frac{18.72}{0.25 \times 4.5} = 16.64\ \text{m}^2$$

$$\text{Diameter of the well} = \sqrt{\frac{4A}{\pi}} = \sqrt{\frac{4 \times 16.64}{\pi}} = 4.6\ \text{m}$$

EXAMPLE 9.13 Calculate the safe maximum yield from an open well if the following data are obtained from recuperation test:

Reduced level of water level in the well = 155 m
Reduced level of water level when pumping is just stopped in the well = 150 m
Reduced level of water level in the well after 3.5 h of pumping = 152.5 m

Working head in the well = 3.5 m
Diameter of the open well = 5 m

Solution Given, h_1 = 155 – 150 = 5 m, h_2 = 155 – 152.5 = 2.5 m, T = 3.5 h, H = 3.5 m

$$C = \frac{2.303}{T}\log\left(\frac{h_1}{h_2}\right) = \frac{2.303}{3.5}\log\left(\frac{5}{2.5}\right) = 0.19808/\text{h}$$

$$Q = CAH = 0.19808 \times \left(\frac{\pi}{4} \times 5^2\right) \times 3.5 = 13.6055\ \text{m}^3/\text{h} = 3.78\ \text{l/s}$$

9.12 Groundwater Resources in India

In India, several attempts have been made to estimate the groundwater resources of the country. In 1976, the National Commission of Agriculture estimated the total groundwater resources of the country as 670 km^3. The usable ground water was determined as 350 km^3, out of which 260 km^3 was considered available for irrigation. The Central Ground Water Board (2004) reported that the total annual replenishable groundwater resources in the country are about 433 km^3. In 2009, the annual replenishable groundwater resources were 431 km^3.

In some parts of the country, utilisation of groundwater has increased exponentially due to increased population growth and technological advances. The Central Groundwater Board, which monitors the groundwater levels from a network of about 15000 stations in the county, reported that out of 6607 assessment units in the country, 1071 units in various states have been categorised as overexploited. Overexplioted refers to the annual groundwater extraction exceeding the net annual groundwater availability. Measurements of water levels are taken at these stations four times in a year in the months of January, April/May, August and November. The collected data are used to prepare maps of groundwater level depths, water level contours and changes in water levels during different periods and years.

Summary

- About 2/3rd of the fresh water resources exist in the form of groundwater.
- An aquifer is geological formation which contains, transmits and yields significant quantity of water. The permeability of aquifer is high.
- A confined aquifer is called artesian, in which water pressure is greater than the atmospheric pressure.
- The actual volume of water drained by gravity force from a unit volume of aquifer is called specific yield.
- The hydraulic conductivity in Darcy's law is a function of both fluid property and property of porous medium.
- Any geological formation is called anisotropic at a point if the hydraulic conductivity varies with the direction of measurement at that point.

- The ability of movement of water through an aquifer is known as transmissivity.
- Darcy's law states that the velocity of water flow through porous media is directly proportional to the hydraulic gradient The law is valid only for laminar flow.
- During prolonged pumping when there is no further lowering of water level from a well, the condition of flow is called as steady-state condition.
- The radius of influence is defined as the radial distance from the centre of well to the limit where the drawdown is zero.

Objective Type Questions

1. The yield of a well depends on
(a) Permeability of soil (b) Area of aquifer opening into the wells
(c) Actual flow velocity (d) All of the above (IES, 1996)

2. Match List-I with List-II and select the correct answer using the codes given below the lists:

List-I	List-II
A. Specific yield	1. Volume of water retained per unit volume of aquifer
B. Specific capacity	2. Volume of water drained by gravity per unit volume of aquifer
C. Specific retention	3. Difference of porosity and specific storage
D. Specific storage	4. Well yield per unit drawdown
	5. Volume of water released from unit volume of aquifer for unit decline in piezometric head

Codes (IES, 1999)

	A	B	C	D
(a)	2	4	1	5
(b)	4	2	3	5
(c)	2	5	1	4
(d)	4	2	3	1

3. Water present in an artesian aquifer is usually (IES, 1999)
(a) At subatmospheric pressure
(b) At atmospheric pressure
(c) At 0.5 times of the atmospheric pressure
(d) Above atmospheric pressure

4. An aquifer confined at top and bottom by impermeable layers is stratified into three layers as follows:

Layer	Thickness (m)	Permeability (m/day)
Top layer	4	30
Middle layer	2	10
Bottom layer	6	20

The transmissivity (m^2/day) of the aquifer is (IES, 1999)

(a) 260 (b) 227 (c) 80 (d) 23

5. Specific capacity of a well is the (IES, 2000)

(a) Volume of water that can be extracted by the force of gravity from a unit volume of aquifer

(b) Discharge per unit drawdown of the well

(c) Drawdown per unit discharge of the well

(d) Rate of flow through a unit width and entire thickness of aquifer

6. Consider the following statements: (IES, 2001)

Assertion (A): The available yield of a tubewell can be doubled by doubling the diameter of the well.

Reason (R): The yield of a tubewell varies inversely with the logarithm of the reciprocal of the diameter of the well. Of these statements

(a) Both A and R. are true and R is the correct explanation of A

(b) Both A and R are true but R is not a correct explanation of A

(c) A is true but R is false

(d) A is false but R is true

7. Match List-I (Equation) with List-II (Applicability or principle of equation) and select the correct answer using the codes given below the lists: (IES, 2002)

List-I	**List-II**
A. Theim's equation	1. Is based on energy conservation principle
B. Dupuit's assumption	2. Is based on mass conservation principle
C. Bernoulli's equation	3. Is applicable to steady flow towards a well in a confined aquifer
D. Continuity equation	4. Is applicable to steady flow in an unconfined aquifer

Codes

	A	B	C	D
(a)	4	3	2	1
(b)	3	4	2	1
(c)	4	3	1	2
(d)	3	4	1	2

8. The performance of a well is measured by its (IES, 2002)

(a) Specific capacity

(b) Specific yield

(c) Storage coefficient

(d) Permeability coefficient

9. Consider the following statements: (IES, 2003)

A well development

1. Involves reversal of flow through the well screen
2. Increases permeability towards the well
3. Decreases permeability towards the well
4. Is continued till sand/silt-free water is pumped out

Which of these statements is/are correct?

(a) 1, 3 and 4 (b) 1, 2 and 4 (c) 3 only (d) 1 and 4

10. Match List-I (well hydraulics parameter) with List-II (definition) and select the correct answer using the codes given below the lists. (IES, 2003)

List-I		List-II	
A.	Specific yield	1.	Discharge per unit drawdown of well
B.	Safe yield	2.	Same as specific retention
C.	Specific capacity	3.	Measure of water that can be removed by pumping
D.	Field capacity	4.	Limit of withdrawal from well without depletion of the aquifer
			Water-bearing capacity of aquifer

Codes

	A	B	C	D
(a)	4	3	2	5
(b)	3	4	1	2
(c)	4	3	1	2
(d)	3	4	2	5

11. Consider the following statements: (IES, 2004)

Assertion (A): The yield of a well varied from 10 m^3/day to 20 m^3/day when the aquifer area changed from 50 m^2 to 75 m^2

Reason (R): The yield is found to be directly proportional to the area of an aquifer opening into a well.

Of these statements

(a) Both A and R. are true and R is the correct explanation of A
(b) Both A and R are true but R is not a correct explanation of A
(c) A is true but R is false
(d) A is false but R is true

12. Which one of the following correctly defines aquiclude? (IES, 2004)

(a) A saturated formation of earth material which not only stores water but also yields it in sufficient quantity
(b) A formation through which only seepage is possible and thus the yield is insignificant compared to an aquifer
(c) A geological formation which is neither porous nor permeable
(d) A geological formation which is essentially impermeable to the flow of water

13. The vertical hydraulic conductivity of the top soil at certain stage is 0.2 cm/h. A storm of intensity 0.5 cm/h occurs over the soil for an indefinite period. Assuming the surface drainage to be adequate, the infiltration rate after the storm has lasted for a very long time, shall be (GATE, 2003)

(a) Smaller than 0.2 cm/h (b) 0.2 cm/h
(c) Between 0.2 and 0.5 cm/h (d) 0.5 cm/h

14. The discharge per unit drawdown at the well is known as (IES, 2005)

(a) Specific yield (b) Specific storage
(c) Specific retention (d) Specific capacity

15. How is the determination of aquifer parameters S (storage coefficient) and T (transmissivity coefficient) done? (IES, 2007)

1. By recording the drawdown in a pumped well at different time intervals.
2. By recording the drawdown in installed observation wells at different time intervals.

Select the correct answer using the codes given below

(a) 1 only (b) 2 only
(c) Both 1 and 2 (d) Neither 1 nor 2

16. In an aquifer extending over 150 ha, the water table was 20 m below ground level. Over a period of time, the water table dropped to 23 m below the ground level. If the porosity of aquifer is 0.40 and the specific retention is 0.15, what is the change in groundwater storage of the aquifer? (GATE, 2011)

(a) 67.5 ha-m (b) 112.5 ha-m (c) 180.0 ha-m (d) 450.0 ha-m

17. What would be the volume of water stored in a saturated column with a porosity of 0.35, a cross-sectional area of 1 m^2 and depth of 3 m? (IES, 2012)

(a) 2.0 m^3 (b) 0.105 m^3 (c) 1.05 m^3 (d) 3.0 m^3

18. The surface joining the static levels in several non-pumping wells penetrating a continuous confined aquifer represents (IES, 2012)

(a) Water table surface
(b) Capillary fringe
(c) Piezometric surface of the aquifer
(d) Physical top surface of the aquifer

19. Due to continuous pumping for 7 h from a catchment of 25 ha, having porosity of 30% and specific retention of 10%, groundwater level dropped by 1 m. The corresponding change in storage is: (IES, 2013)

(a) 7.5 ha-m (b) 0.75 ha-m (c) 1.87 ha-m (d) 0.187 ha-m

20. As recommended by Sichardt, the radius of influence is (IES, 2013)

(a) Inversely proportional to drawdown
(b) Linearly proportional to drawdown
(c) Independent of drawdown
(d) Proportional to square root of drawdown

21. A tube-well of 30 cm diameter penetrates fully into an artesian aquifer, the strainer length is 15 m. The yield from the well under a drawdown of 3 m through the aquifer consisting of sand of effective size of 0.2 mm and coefficient of permeability of 50 m/day, with radius of drawdown of 150 m, is nearly: (IES, 2014)

(a) 240 l/s (b) 120 l/s (c) 24 l/s (d) 12 l/s

22. An unconfined aquifer of porosity 35%, permeability 40 m/day, and specific yield of 0.15 has an area of 100 km^2. If the water table falls uniformly throughout the aquifer area by 0.2 m during a drought, the volume of water lost from storage is (IES, 2014)

(a) 1.5 million m^3 (b) 3.0 million m^3 (c) 7.0 million m^3 (d) 8.0 million m^3

23. Groundwater flows through an aquifer with a cross-sectional area of 1.0 × 104 m^2 and a length of 1500 m. Hydraulic heads are 300 m and 250 m at the ground water entry and exit points in the aquifer, respectively. Groundwater discharge into stream is at the rate of 1500 m^3/day. The corresponding hydraulic conductivity of the aquifer will be (IES, 2014)
(a) 3.5 m/day (b) 4.5 m/day (c) 5.0 m/day (d) 5.5 m/day

24. When 3.5 million m^3 of water was pumped out from an unconfined aquifer of 6.3 km^2 area extent, the water table was observed to go down by 2.5 m. The specific yield of the aquifer which is best approximated is (IES, 2014)
(a) 32% (b) 28% (c) 25% (d) 22%

25. The flow net constructed for the dam is shown in the figure below. Taking the coefficient of permeability as 3.8×10^{-6} m/s, the quantity of flow (cm^3/s) under the dam per metre of dam is (GATE, 2014)

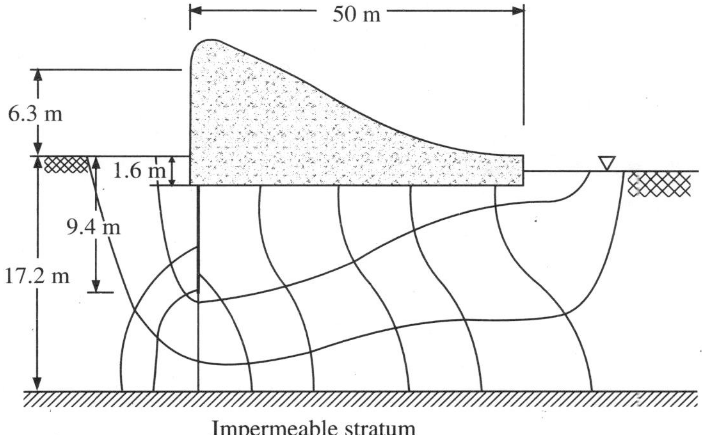

26. The relationship between porosity η, specific yield S_y and specific retention S_r of an unconfined aquifer is (GATE, 2015)
(a) $S_r + S_y = \eta$ (b) $S_y + \eta = S_r$ (c) $S_r + \eta = S_y$ (d) $S_y + S_r + \eta = 1$

27. Water table of an aquifer drops by 100 cm over an area of 1000 km^2. The porosity and specific retention of the aquifer material are 25% and 5%, respectively. The amount of water (expressed in km^3) drained out from the area is .... (GATE, 2016)

28. A tracer takes 100 days to travel from Well-1 to Well-2 which are 100 m apart. The elevation of water surface in Well-2 is 3 m below that in Well-1. Assuming porosity equal to 15%, the coecient of permeability (expressed in m/day) is (GATE, 2016)
(a) 0.30 (b) 0.45 (c) 1.00 (d) 5.00

Answers

1. (d) **2.** (a) **3.** (d) **4.** (a) **5.** (b) **6.** (d) **7.** (d) **8.** (a) **9.** (b) **10.** (b)
11. (d) **12.** (d) **13.** (b) **14.** (d) **15.** (c) **16.** (b) **17.** (c) **18.** (c) **19.** (*) **20.** (b)
21. (c) **22.** (b) **23.** (b) **24.** (d) **26.** (b) **27.** 0.19:0.21 **28.** (c)

Explanations

1. $Q = nv_a A = KiA$

   where n is porosity, v_a is actual flow velocity of groundwater, A is area of the aquifer opening into the wells, K is permeability of soil and i is the hydraulic gradient at the surface of well.

4. Transmissibility, T is calculated as

$$T = \Sigma kd = (4 \times 30) + (2 \times 10) + (6 \times 20)$$
$$= 120 + 20 + 120 = 260 \text{ m}^2/\text{day}$$

11. Yield of a well is given by:

    $Q = KiA$

    Here, i is the hydraulic gradient at the surface of well. If hydraulic conductivity remains constant, then the discharge is proportional to the area of an aquifer opening into the well. So, the discharge should change from 10 m^3/s to 15 m^3/s when aquifer area is changed from 50 m^2 to 75 m^2.

13. Initially, the infiltration rate will be more than the hydraulic conductivity of the soil because soil absorbs more water when it is not saturated. As the soil becomes saturated after long time, it will absorb water at the rate of 0.2 cm/h.

16. Using relationship,

$$\eta = S_y + S_r$$
$$S_y = 0.25$$

    Aquifer volume change = 150 × (23 – 20) = 450 ha-m

    Water storage change = 450 × 0.25 = 112.5 ha-m

17. Volume of water stored = 0.35 × 3 × 1 = 1.05 m^3

19. Specific yield + Specific retention = Porosity

    Therefore, volume of water extracted or change in storage,

$$\Delta s = S_y \times V = 0.2 \times 25 \times 1$$
$$= 5 \text{ ha-m}$$

20. Using Sichardt formula,

    Radius of influence, $R = 3000\, S\sqrt{k}$

$$R \propto S$$

    where S is drawn down

22. The water table falls 0.2 m. Soil mass under water table is fully saturated.

    Hence water loss yielded water

$$0.15 \times 0.2 \times 100 \times 10^6 = 3 \text{ million m}^3$$

23. $Q = KiA \;\Rightarrow\; 1500 = K \times \dfrac{50}{1500} \times 1 \times 10^4$

    Therefore, K = 4/5 m/day

24. Using volumetric approach, specific yield = $\dfrac{3.5\times10^{6}}{6.3\times10^{6}\times2.5}$ = 22%

25. N_f = Number of flow channels = 3

N_d = Number of equipotential drops = 10

$K = 3.8 \times 10^{-6}$ m/s

$H = 6.3$ m

Now, $KH\dfrac{N_f}{N_d} = 3.8\times10^{-6}\times6.3\times\dfrac{3}{10} = 7.8\times10^{-6}\ \text{m}^3/\text{s/m} = 7.182\ \text{cm}^3/\text{s/m}$

IES CONVENTIONAL QUESTIONS

PROBLEM 9.1 What is water bearing stratum? On what basis, groundwater flows in it? What is the difference between specific yield and specific retention. (IES, 2012)

Solution Please see the text.

Theoretical Questions

1. What do you mean by aquifer? Discuss different types of aquifers.
2. Explain the following terms:
   (a) Porosity
   (b) Specific yield
   (c) Storage coefficient
   (d) Transmissivity
3. State the Darcy's law and discuss its applicability.
4. Define anisotropy and heterogeneity.
5. Briefly explain the following terms:
   (a) Radius of influence
   (b) Cone of depression
6. Derive the equation of steady flow towards the well in confined and unconfined aquifer along with assumptions.
7. What are the yield tests?

Unsolved Problems

Problem 1: Find out the hydraulic conductivity and intrinsic permeability of an aquifer if the following informations are obtained during a field experiment. Distance between two observation well = 65 m

Time taken by a tracer to reach from one observation well to another = 9 h
Difference in water table in the wells = 0.9 m
Porosity of the aquifer = 0.4
Viscosity of water = 0.9×10^{-3} Ns/m^2

Problem 2: Find out

(a) The hydraulic conductivity of an aquifer
(b) Pore velocity of water

if the following data are given.

Cross-sectional area of the aquifer=3.5×10^4 m^2
Length of the aquifer = 1650 m
Hydraulic head at entry = 375 m
Hydraulic head at exit = 280 m
Rate of groundwater discharge = 1900 m^3/day
Porosity of the soil = 0.5

Problem 3: The following information is given regarding a confined aquifer having three layers:

Hydraulic conductivity of top layer = 3.5 cm/s and thickness = 5.5 m
Hydraulic conductivity of middle layer = 0.003 cm/s and thickness = 4.5 m
Hydraulic conductivity of bottom layer = 0.025 cm/s and thickness = 5.2 m

Find our equivalent hydraulic conductivity of the confined aquifer along the

(i) Horizontal direction of the alignment of the layers
(ii) Vertical direction of the alignment of the layers.

Problem 4: Find out the storage coefficient of a confined aquifer if the following data are given.

Thickness of the aquifer = 45 m
Porosity of the aquifer = 0.55
Bulk modulus of elasticity of water = 2.2×105 N/cm^2
Bulk modulus of the soil = 5900 N/cm^2

Problem 5: Determine the intrinsic permeability of a soil sample if the following data are provided:

Hydraulic conductivity = K = 12.5 m/day
Viscosity of water = 0.01 g-cm/s
Density of water = 1 g/cc

Problem 6: Determine the amount of flow from canal A to canal B if they are separated by an aquifer and the following information is provided:

Depth of water flow in canal A and B = 13.2 m and 9.4 m, respectively
Width of the aquifer = 4 km
Hydraulic conductivity of the aquifer = 0.25 mm/s

Problem 7: Compute the average hydraulic conductivity of an unconfined aquifer if a well is fully penetrated in it and the following data are given:

Drawdown in two observation wells = 6.6 m and 1.5 m, respectively
Distance of the observation wells = 15 m and 86 m, respectively from the main well
Thickness of the aquifer = 25 m

Diameter of the well = 0.3 m
Rate of pumping from the well = 45 *l*/s
Also, find out the radius of influence (R) of the well

Problem 8: Calculate the steady-state discharge from a well, penetrated completely in an unconfined aquifer if the following information is given:
Thickness of the aquifer = 45 m
Hydraulic conductivity, K = 40 m/day
Radius of well = 0.65 m
Steady water head in the well = 35 m
Radius of influence of the well R = 450 m

Problem 9: Calculate the maximum possible water yield from a well, penetrated fully into a confined aquifer if the following information is given:
Maximum drawdown limit of the well (s_w) = 4 m
Thickness of the confined aquifer b = 25 m
Hydraulic conductivity of the aquifer $K = 7.6 \times 10^{-4}$ m/s
Diameter of the well = 0.7 m
Radius of influence R = 300 m

Problem 10: Calculate the following if the time taken for increasing of water level from 4 m to 2.25 m is 2 h (data are obtained in a recuperation test, conducted in an open well test):
(a) Specific yield of open well
(b) Yield from an open well of diameter 6.5 m under a depression head of 3.5 m
(c) Diameter of an open well producing yield of 20 l/s under depression head of 2.5 m

Further Reading

Das, Ghanshyam, *Hydrology and Soil Conservation Engineering Including Watershed Management*, PHI, Delhi, 2014.

Raghunath, H.M., *Hydrology—Principles, Analysis and Design*, New Age Publishers, New Delhi, 2014.

Suresh, R., *Watershed Hydrology*, Standard Publishers Distributors, New Delhi, 2015.

Deodhar, M.J., *Elementary Engineering Hydrology*, Pearson, New Delhi, 2013.

Yadupathi Putty, Mysooru R., *Principles of Hydrology*, I.K. International, New Delhi, 2013.

CHAPTER

10

Rainwater Harvesting

10.1 Introduction

Water is one of the basic human needs and is indispensable almost in every aspect of our life. Freshwater that is suitable for use accounts for only 2.5% of the Earth's water, and yet this available water is under tremendous pressure due to the growing populations, leading to more demands in order to meet human needs. Thus, water conservation has become the need of the day. Hence, necessary measures need to be taken to manage and conserve the available water, and these measures need to be done urgently on priority basis.

Rainwater harvesting is one of such important methods to conserve and manage water and can result in significant savings. The term 'water harvesting' was first used by Geddes in 1950, and he defined it as the collection and storage of any farm water, either runoff or creek flow, for irrigation use. Other experts' defined water harvesting as a method for inducing, collecting, storing and conserving runoff from different sources for various purposes. In general terms, water harvesting can be defined as a process of collecting, storing and conserving the runoff water resulting from rain in the soil profile as well as in the reservoirs both over the surface or under the surface. The rainwater collected can be used for different purposes such as drinking purpose, irrigation purpose, industrial and domestic uses and for groundwater recharge. Rainwater harvesting, as many other methods which are in use presently, is not unfamiliar. It was practised in the ancient period over the third century BC, where the agricultural communities in Balochistan (presently located in Pakistan, Afghanistan and Iran) and Kutch, India, used it for irrigation. However, the only thing is that it has never been performed systematically in all places. Due to high demands of water, the necessity has come to harvest the rainwater along with the roof water to solve the water problems all over not only in the arid but also in the humid region. The runoff from the rainwater should be arrested or trap using proper management and planning. The management of rainwater conservation by the government agencies is to be implemented in rural as well as urban areas.

10.2 Need for Rainwater Harvesting

India is one of the countries in the world which receive abundant amount of rainfall, but yet water scarcity continues to haunt various part of India. Rainfall distribution in India is uneven. As it is, Cherrapunji located in Meghalaya is known to receive the highest annual rainfall in the world, whereas the western part of Jaisalmer district located in Rajasthan is one of the driest parts of the country. As India is a monsoonic country, most of the rainfall occurs only during monsoon months (June to September) with high intensity, resulting in heavy runoff, but it remains dry during non-monsoon season. Hence, harvesting of rainwater during monsoon season, and its storage and reuse during non-monsoon period is a solution to solve the water crisis problem. With the increase in population, water demand increases and this demand is not only from the agriculture sector but also from the growing industries and domestic uses. As the urbanisation, industrialisation and population increase, water demand increases and groundwater level depletes. Water scarcity is found not only in arid region, but also in high rainfall area like Cherrapunji where people face drinking water problem before monsoon commence. Hence, the main problem of water in India is not because of less water availability but it is mainly due to mismanagement of water resources. Thus, it indicates the need of water harvesting measures to solve the water problem.

10.3 Advantages and Limitation of Rainwater Harvesting

The oceans, seas and bays are the major sources of water, but are not suitable for drinking or domestic uses. This results in limiting the water or shortage of water for drinking or other purposes. The surface water and groundwater are the two major sources to fulfil the demand for water, and the only way to refill and recharge them is by proper use of rainwater.

The *advantages* of rainwater harvesting include reduction in the demand for groundwater, fulfilment of the demand for water during dry season, and mitigation of the impact of drought. This system is flexible and adaptable. It also helps in improving well yield and reduces the cost of pumping, groundwater contamination as well as flood and soil erosion.

The *limitations* of rainwater harvesting include unpredictable rainfall patterns and storage limit. The system has high initial cost and requires regular maintenance.

10.4 Rainwater Harvesting Potential

Water harvesting is one of the most efficient and low-cost techniques to conserve rainwater, but yet water harvesting has not been adopted as a competitive method for water supplies, although around 3000 water harvesting systems have been installed around the world. Despite its slow acceptance, its potential for providing economic water is still tremendous.

If rainwater harvesting techniques are used, water supply can be based on precipitation rather than streamflow and groundwater, which can then preserve the groundwater and stream water resources. Arid areas such as Hawali and Jamaica are using the rainwater harvesting techniques developed by their researcher and provide more than 245000 l/day of water during a year of average rainfall in a 0.6 ha catchment area.

India is a developing country with highest growing rate of population and industrialisation. Due to these reasons, the demand for water also increases, making the water resources confronting and reckon with drawbacks and bounds. With the increase in population and the pace of development, water resources are exploited outrageously. Fresh water sources are being profoundly oppressed and the failure of monsoon makes the situation worse. As surface water sources decline to commence the ever-increasing demands, groundwater are tapped, often to unsustainable levels. Also, the quick state of urbanisation reduces the availability of open spaces for natural recharge of rainwater. If we solely redeem each drop of water and recharge the underground aquifer, we can protect ourselves from this problem of water scarcity. Thus, rainwater is the ultimate answer, and if collected and utilised properly, it has a tremendous potential to meet the water demands and to solve the water crisis.

EXAMPLE 10.1 Total area of a building rooftop is about 3000 m^2, having an average annual rainfall of about 1875 mm (Assume 12% losses during monsoon months). The total number of rainy days are 50. Two storage tanks of size 15 m length, 8 m wide and 4 m deep are constructed for rainwater harvesting. Calculate the number of days that would be sufficient during dry period with the amount of available water.

Solution Given,

$$\text{Total catchment area} = 3000 \text{ m}^2$$

$$\text{Average annual rainfall} = 1875 \text{ mm} = 1.875 \text{ m}$$

$$\text{Total losses} = 12\% = 0.12$$

$$\text{Total quantity of harvestable water} = 3000 \times 1.875 = 5625 \text{ m}^3$$

$$\text{Considering 12\% losses, total losses} = 5625 \times 0.12 = 675 \text{ m}^3$$

$$\text{Total available water for harvest during monsoon months} = 5624 - 675$$

$$= 4949 \text{ m}^3 \text{ for 50 day}$$

$$= 98.98 \text{ m}^3\text{/day}$$

$$\text{Assuming requirement of water per day} = 21 \text{ l/capita/day}$$

$$\text{Total water} = 21 \times 3000 = 63000 \text{ l/day}$$

$$= 63.0 \text{ m}^3\text{/day}$$

$$\text{Total volume of tanks} = 2(15 \times 8 \times 4)$$

$$= 960 \text{ m}^3$$

$$\text{Number of sufficient days} = \frac{960}{63} = 15.24 \text{ days}$$

This would be sufficient for 15.24 days in dry period.

EXAMPLE 10.2 The size of the catchment area is about 12000 m^2, with an average annual rainfall of about 1570 mm. Assume that only 50% of the rainwater is stored (due to losses), out of which only 10% can be utilised. Estimate the total available water for recharge. Also calculate the number of days for which this water would be sufficient.

Solution Given,

$$\text{Area of a catchment} = 12000 \text{ m}^2$$
$$\text{Average annual rainfall} = 1570 \text{ mm} = 1.57 \text{ m}$$
$$\text{Available water for harvesting} = 1.57 \times 0.5 = 0.785 \text{ m}$$
$$\text{Actual water that can be harvest} = 0.785.1 = 0.0785 \text{ m}$$
$$\text{Total available water for recharge} = 12000 \times 0.0785$$
$$= 942 \text{ m}^3 = 942000 \text{ l}$$

942000 l water is available for recharging the groundwater in a year.

Assume requirement of water is 15 l/capita/day

$$\text{Total requirement of water} = 15 \times 12000$$
$$= 180000 \text{ l/day}$$
$$= 180.0 \text{ m}^3\text{/day}$$
$$\text{Amount of water available in a year} = 942 \text{ m}^3$$
$$\text{Amount of water available per day} = \frac{942}{180} = 5.233 \text{ days}$$

This would be the sufficient for 5.233 days in dry period.

EXAMPLE 10.3 The catchment area of a house of 7 members is 200 m^2 receiving an average annual rainfall of about 830 mm. Suppose that only 60% of the rainwater is stored (losses due to evaporation and overflow). Calculate the amount of available rainwater which can be effectively harvested.

Solution Given,

$$\text{Area of rooftop} = 200 \text{ m}^2$$
$$\text{Average annual rainfall} = 830 \text{ mm} = 0.83 \text{ m}$$
$$\text{Total volume of water during rainfall} = 0.83 \times 200$$
$$= 166 \text{ m}^3 = 166000 \text{ l}$$
$$\text{Harvestable volume of rainfall} = 166000 \times 0.6 = 99600 \text{ l}$$

Assuming the requirement of water for domestic purposes is 15 l/capita/day.

$$\text{Total amount of water required per family (7 member)} = 15 \times 7 = 105 \text{ l/day}$$
$$= 38325 \text{ l/year}$$
$$\text{Total stored water per year} = 99600 - 38325$$
$$= 61275 \text{ l/year}$$

10.5 Components of Rainwater Harvesting Structure

A rainwater collection and storage system is composed of many components. All these components need to be put together to provide a good quality water. Rainwater collection system can be roughly separated into two types—those where the rainwater is collected from the building terrace or normal ground then stored in tanks or pits, which is then used for irrigation supply or domestic use (Figure 10.1) and those where the water is collected and used for groundwater recharge through recharge pits, borewells and dug wells (Figure 10.2).

Figure 10.1 Rainwater harvesting for direct uses.

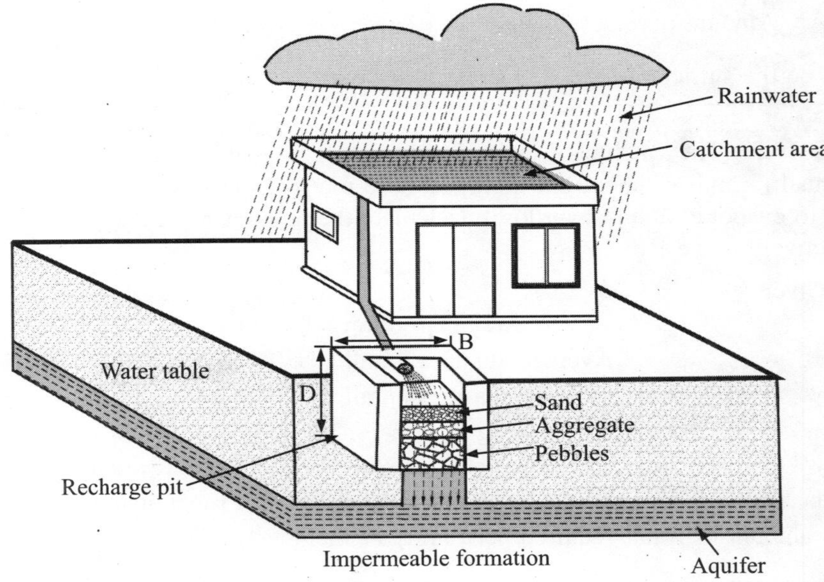

Figure 10.2 Rainwater harvesting for groundwater recharge.

The components of a rainwater harvesting are illustrated below:

10.5.1 Catchment Area

The catchment of rainwater harvesting system is the surface, which directly collects the rainfall and delivers the water to the system. It may be a concreted area like terrace roof, reinforced cement concrete (RCC) covered area or an unpaved area like open ground and grassland. The collected amount of water in the catchment depends on the rainfall intensity, rainfall duration, roofing material, etc.

10.5.2 Conveyance System

Conveyance system transports the collected water from the catchment to the harvesting system. This system may indclude drains that carry water or pipelines made with polyvinyl chloride (PVC) and iron, percolation pit, etc.

10.5.3 Filtering Device

Filters are used to remove suspended pollutants from rainwater collected from catchment roofs. The filtering media consist of sand (coarse) and gravel layers to remove pollutants from water before it enters into the storage tanks or devices. These filters are of various types and may be classified as follows:

1. Sand filter: It is a commonly used filter in which normal sand is used as a filter media. It is simple to construct and low in cost. This type of filter is used for the treatment of water in urban areas for the removal of turbidity, microorganisms and colours. The top layer of a simple sand, filter is of sand followed by the layer of aggregates and pebbles, as shown in Figure 10.3.

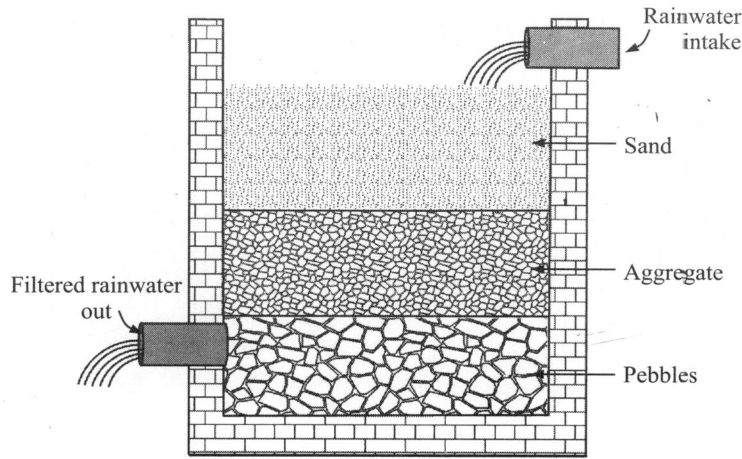

Figure 10.3 Sand bed filter.

2. Dewas filter: The Dewas filter was designed in a place called Dewas in Madhya Pradesh, where the district administration initiated a scheme for groundwater recharging. The filter consists of polyvinyl chloride (PVC) pipe having 14 cm diameter and 120 cm length. It is divided into three chambers for pollutants' purification. The first purification chamber consists of finer (2–6 mm) pebbles, the second consists of course (6–12 mm) size pebbles and the third one consists of more course (12–20 mm) pebbles, as shown in Figure 10.4. It also consists of a mesh at the end of the third chamber, which is used for cleaning of water after passing through the three chambers.

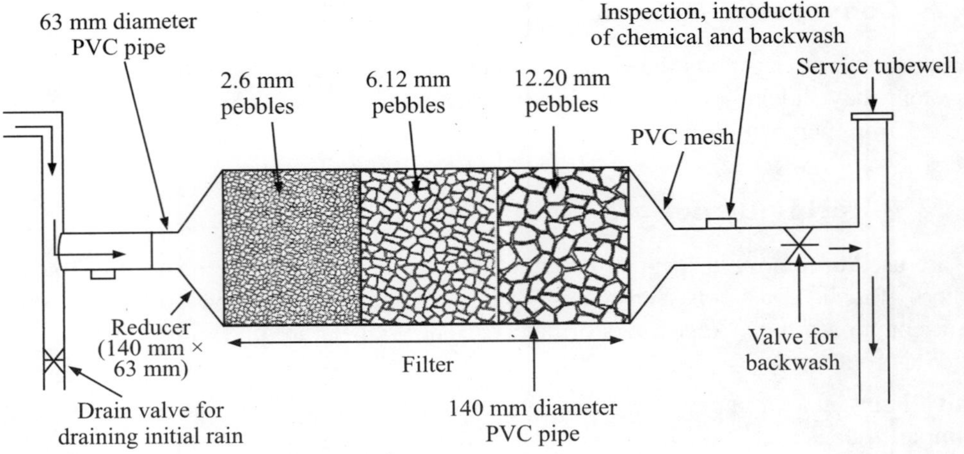

Figure 10.4 Dewas sand filter.

3. Filter for large rooftop area: This filter is used for filtering the rainwater from large roof areas. This system is designed with three chambers filled with different materials in which the innermost chamber is filled with pebbles, the middle one is filled with aggregate and outermost chamber is filled with sand, as shown in Figure 10.5. This system was designed by R. Jayakumar. So this is also known as Jayakumar filter.

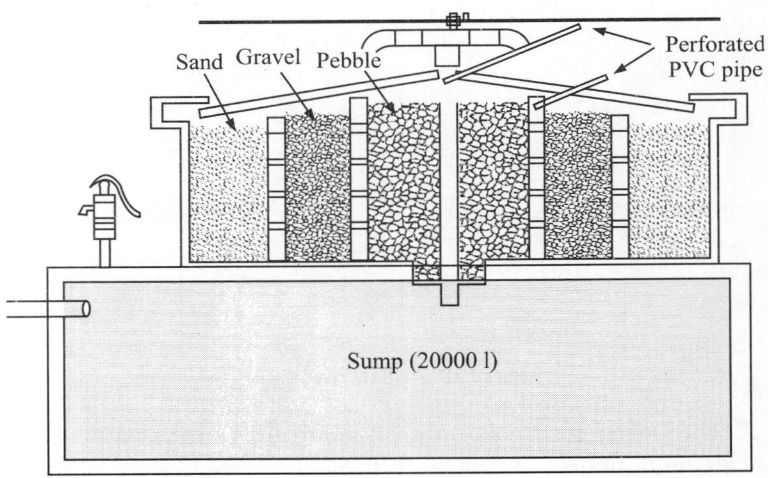

Figure 10.5 Jayakumar filter.

4. Horizontal roughing filter: It is also called *slow sand filter*, which is used to remove pollutants from surface water. This filter was designed for coastal area in Odisha. The dimension of the filter is 1 m^2 cross section wide and 8 m in length. It consists of three chambers filled with different materials of sizes ranging from 4 mm to 25 mm of fine sand, course sand and broken bricks as shown in Figure 10.6.

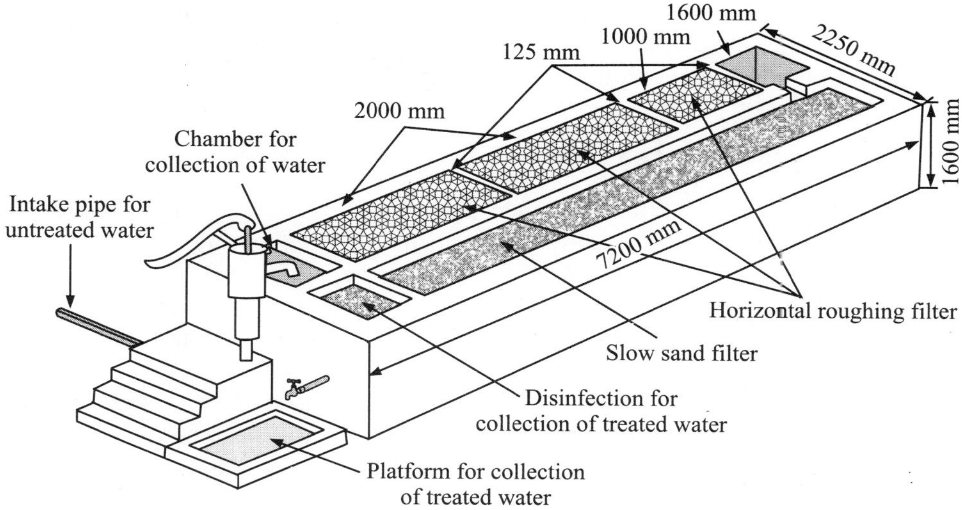

Figure 10.6 Horizontal roughing filter.

5. Storage device: This device is used for storing and collecting rainwater and is of various types like tank, storage pits, ponds made of concrete, iron container, PVC container, mud made pits, etc.

10.6 Factors Influencing Rainwater Harvesting

The important factors that influence the rainwater practices are as follows:

1. **Rainfall pattern:** Rainfall is an erratic process. The rainfall pattern is not uniformly distributed throughout the year or over the catchment area.
2. **Rainfall intensity:** In India, there is large variation in rainfall intensities from about 100 mm (Rajasthan) to about 11000 mm (Cherrapunji).
3. **Topography of the area:** Topography of the area determines the size of the basin; a flatter area will have a larger basin.
4. **Soil characteristics of catchment area:** It also varies with different soil characteristics such as whether soil is permeable or impermeable so as to enhance groundwater recharge.

10.7 Technology for Rainwater Harvesting

Rainwater harvesting is the simplest method that requires minimum knowledge and offers many benefits. Rainwater collected can be conserved by the following three main approaches:

1. Storage of water in storage tanks for direct use
2. Storage of water for groundwater recharge using rooftops catchment
3. Storages of water for groundwater recharge using runoff water.

There are different methods for rainwater harvesting both for urban and rural areas. These methods can be roughly divided into two parts—rooftops harvesting (mostly suitable for urban areas), and groundwater recharge through runoff (mostly applicable in rural areas).

10.7.1 Rooftop Rainwater Harvesting

Rooftop rainwater harvesting is the technique through which rainwater is captured from the roof catchments and stored in reservoirs. The main objective of rooftop rainwater harvesting is to make water available for future use. In most of the urban areas like Mumbai, Delhi, Bangalore where most of the areas are paved, the main way to conserve or prevent rainwater is to collect the rainwater from the building terraces and store them using different components. The rainwater should be then transferred into the recharge pits or storage tanks using pipes for direct usage or for recharging the groundwater aquifers. Few examples of rooftop harvesting are Deogiri Fort. Hilltop Fort and Meenakshi temple. The different ways where rooftop water can be stored for direct usage or groundwater recharging are described below:

1. Underground storage structures or recharge well: This method is generally used for domestic purposes where runoff water is collected from rooftops and stored into the shallow pits or wells. Dimensions of these structures depend on the available open space and generally range from 9 to 10 ft depth and 6 to 8 ft wide. It consists of two layers. The first layer is of boulders/stone layer and over the boulders a sand layer is filled which is used for purification, as shown in Figure 10.7. The capacity of these structures are about 25000 l of water.

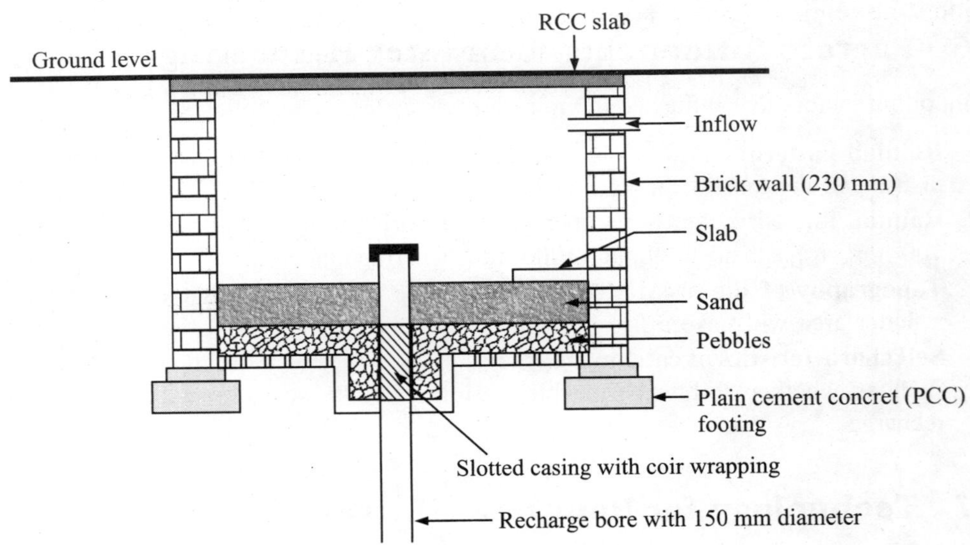

Figure 10.7 Recharge well or underground storage structure.

2. A Soak pit/recharge pit structure: An 8 ft deep and 5 ft wide soak pit, which is filled with stones, aggregate, boulders, is shown in Figure 10.8. The water collected from the rooftops and terraces is filled into the soak pit. The capacity of these structures are about 25000 l of water. The upper layer of this pit is covered with perforated cover and it should be lined with stone or brick walls.

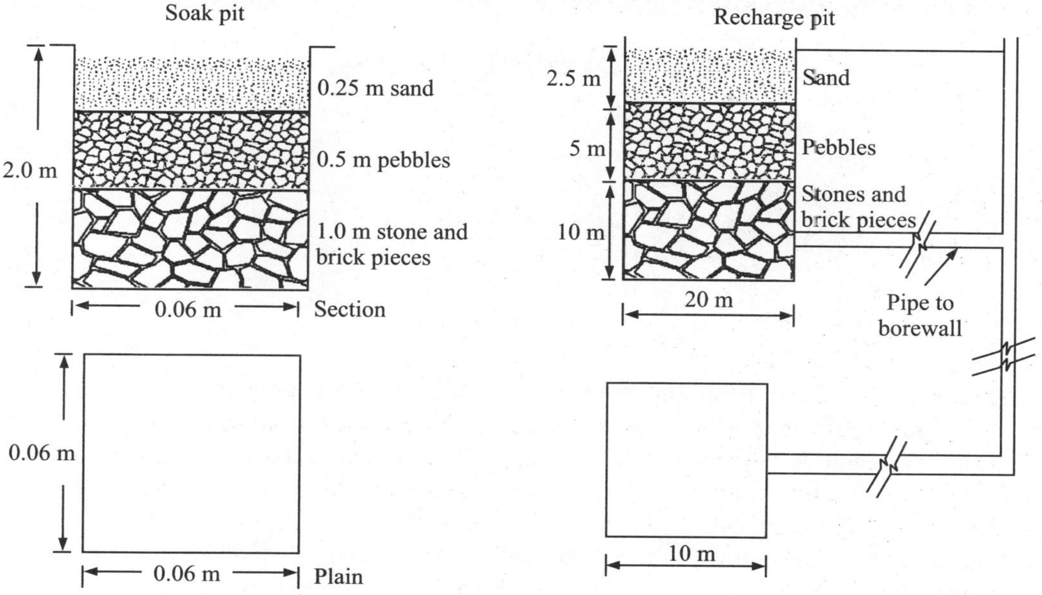

Figure 10.8 Soak pit or recharge pit.

3. Percolation pit: A *percolation pit* is a borehole which is of 30 cm diameter in the ground and extents to a depth of 3 to 10 m below. It is the easiest and efficient way of recharging the rainwater from rooftop. It is generally filled with stones, brick pieces and sand, which are then covered with concrete boundary as shown in Figure 10.9.

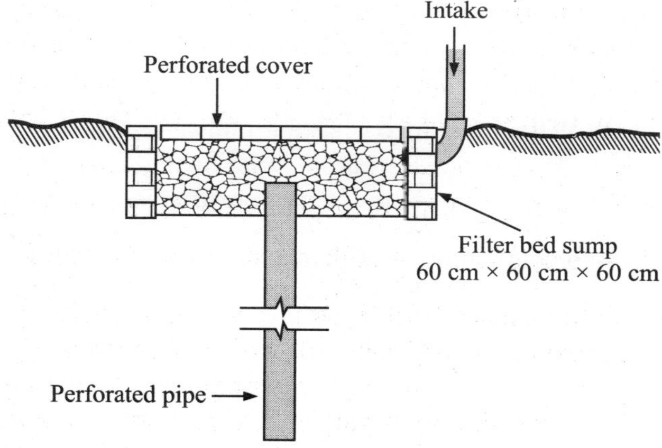

Figure 10.9 Percolation pit.

4. Recharging through tube well: In this case, rainwater collected from rooftop is collected in the storage tank nearby the tube well instead of direct supply of water into the well in order to avoid probabilities of contamination of groundwater, as shown in Figure 10.10. This storage tank is shallower than the water table. Storage tank which supplies water to the tube well consists of a filter having three chambers made of pebbles, aggregate and sand. The tube also consists of a PVC-casing pipe to avoid entering of soil particles.

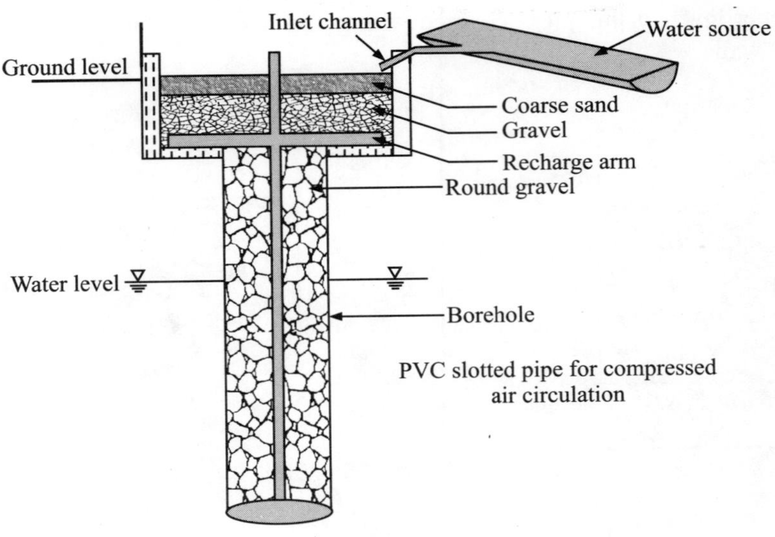

Figure 10.10 Recharging through tube well.

10.7.2 Rainwater Harvesting: Through Collection of Runoff Water

This method is generally applicable for the groundwater recharging in rural areas. In rural zones, large open space is available for different recharging methods. These methods are as follows:

1. Spreading method: As the name suggests the rainwater in this method is spread over the permeable surfaces of open lands and pits, where it directly infiltrates into the shallow aquifers, as shown in Figure 10.11. These structures are constructed on the sloping land to store all the water around the farm land. Low earth dykes (called *percolation bunds*) are constructed to spread and temporarily store the rainwater. The stored water then slowly percolates and joins the nearby aquifer. Recharging rate of this method depends on the depth of water stored, and also, on the permeability of the spread area.

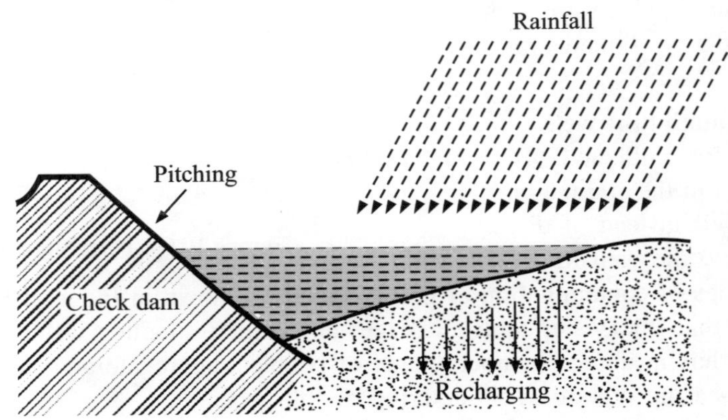

Figure 10.11 Recharging through spreading method.

2. Recharging well method: In this method, water is injected into the boreholes called *recharge wells*. Depending on the local surface conditions, the water is fed into the recharge wells either by gravity or by pumping through pipes. Dimension of the well is generally around 50 cm in diameter and a PVC pipe of about 20 cm in diameter as shown in Figure 10.7. The space between pipe and bore is filled with gravels to avoid suspended particles entering into the recharge well. Recharging rate of groundwater using this method is relatively higher. This method is widely used in Israel.

3. Open rainwater harvesting: In this method, polythene sheets and galvanised sheets placed across the open air far away from trees are used to collect direct rainfall. The rainwater collected is cleaner than the runoff water. However, polythene bags are costly and vulnerable to damage under heavy wind condition.

4. Induced infiltration method: This method is generally used for groundwater recharging, whereby an open well is constructed near a river bank. The percolating water is collected in the well with the help of radial collectors and is then discharged into a lower level aquifer. These types of wells are mostly found in France.

Summary

- Rainwater harvesting is a simple method of collecting water when it rains and using it later when the need arises. It is very economical and can be implemented both in rural and urban areas.
- Rainwater collection system can be roughly separated into two types—those where the rainwater is collected from the building terrace or normal ground and then stored in tanks or pits, which is then used for irrigation supply or domestic use, and those where the water is collected and used for groundwater recharge through recharge pits, borewells and dug wells.
- The main components of a rainwater harvesting system are catchment area, conveyance system, filtering devices and storing devices.
- The important factors that may influence the rainwater harvesting are rainfall pattern, rainfall intensity, topography of the area and soil characteristics.
- If the techniques for rainwater harvesting are implemented, water supply can be based on precipitation rather than streamflow and groundwater, which can then preserve the groundwater and stream water resources.
- With groundwater being exploited, surface water fails to meet the demands, rainwater harvesting is a potential solution to solve the water problem.

Theoretical Questions

1. Define and discuss the need of rainwater harvesting.
2. Discuss the advantages and limitations of rainwater harvesting.
3. What are the components of rainwater harvesting structures?
4. Discuss the various filter devices.

5. Describe the factors influencing rainwater harvesting.
6. Describe briefly the technology for rainwater harvesting.

Further Reading

Davie, Tim, *Fundamentals of Hydrology*, 2nd ed., Routledge, Taylor and Francis Group, London and New York, 2008.

Deshmukh, S.H. and Magar, R.B., 'The Scope of Rainwater Harvesting in Urban Areas', National Seminar on Rainwater Harvesting and Water Management, pp. 45–55, 2006.

Dhoble, R.M. and Bhole, A.G., 'Review of Rainwater Harvesting in India', National Seminar on Rainwater Harvesting and Water Management, pp. 159–170, 2006.

Kirloskar, S.G., 'Rainwater Harvesting and Water Management', National Seminar on Rainwater Harvesting and Water Management, pp. 65–69, 2006.

Parghane, R.K., Kulkarni, S.P., and Dhawale, A.W., 'Rainwater Harvesting and Recharging Groundwater', National Seminar on Rainwater Harvesting and Water Management, pp. 31–36, 2006.

Raghunath, H.M., *Hydrology—Principles, Analysis and Design*, New Age Publishers, New Delhi, 2014.

Singh, P.K., Singh, B., and Tewary, B.K., 'Rooftop Rainwater Harvesting for Artificial Recharge to Groundwater: An Urgent Need of Present Century', National Seminar on Rainwater Harvesting and Water Management, pp. 78–82, 2006.

Sinha, S.K., 'Rainwater Harvesting', National Seminar on Rainwater Harvesting and Water Management, pp. 146–158, 2006.

Sivanappan, R.K., 'Rainwater Harvesting, Conservation and Management Strategies for Urban and Rural Sectors', National Seminar on Rainwater Harvesting and Water Management, pp. 1–5. 2006.

Yadupathi Putty, Mysooru R., *Principles of Hydrology*, I.K. International, New Delhi, 2013.

CHAPTER

11

Climate Change and Water Resources Management

11.1 Introduction

Water is a necessary element for every human civilisation, living organisms and natural habitat. It is used for various purposes such as drinking, washing, agricultural productions, transportation, industrial development, restoration, animal husbandry, hydropower generation, manufacturing and commercial uses. The hydrological cycle shows the circulation of water in its different forms over the Earth system, i.e., atmosphere, lithosphere and sea. The world has modest potential in the availability of fresh water resources. In the Earth system, approximately 97.5% water is defined as salty seawater, and thus, it is non-usable for irrigation and drinking purposes. The remaining 2.5% water is present as fresh water reserves, as deep and ice-covered in Antarctica and as ice sheets/snow/glaciers in Greenland. Thus, only about 0.26% fresh water reserves are available in the form of rivers, lakes and groundwater, which are readily usable for mankind and can be used for various purposes. Fresh water resources have numerous benefits, but sometimes, the problems can also be occurred by their excess, overconsumption and quality deterioration. As per the incremental growth in urbanisation due to increasing population and number of cities, especially in the developing countries, the unwanted growth of organisations and infrastructure developmental plans are creating pressures in the context of water scarcity. India, as a developing country, has specific attention with regard to its imminent water resources. It has been analysed that India will undergo continue rapid population growth, and in future, it may get susceptible to the climate change.

Indian water resources establish one of their dynamic assets. The precipitation in India shows very high spatial and temporal variability and absurdity of the situation is from the north-eastern parts of India (highest rainfall) to Rajasthan state of India (lowest rainfall).

As per the mean scenario, around 4000 km^3 annual precipitation falls over the Indian sub-continent. However, the distribution of precipitation is highly variable and it shows significant spatio-temporal variability in the precipitation amount, and thus, the availability of water also fluctuates in time and space throughout the country. Precipitation can be categorised in the form of rainfall, snowpack, hail and other procedures, by which atmospheric water falls on the ground surface. The precipitation pattern over the Indian sub-continent, can be forecasted easily by Indian monsoon conditions. About 80% accumulated precipitation amount over the country brings through the south-west (SW) monsoon, which fulfils the needs of fresh water availability for drinking and irrigation purposes. However, if significant climatic changes occur over the Indian region, then mainly the SW monsoon can be critical and it shows a substantial impact on agriculture yields, water resources management and also on the overall economy of the country.

11.2 Indian Water Resources Systems

The major and perennial river basins of India have been considered as the fundamental hydrological units in the planning, management and enlargement of the water resources. Indian water resources system has been divided into six major hydrological regions, viz. water resources region (drainage flowing into Bay of Bengal such as Ganges, Indus basin, drainage flowing into Arabian Sea, Brahmaputra basin, minor rivers draining into other basins/countries such as Myanmar, Bangladesh and China, and Island drainage) primarily based on the drainage of rivers to outlet (Watershed Atlas of India). The watershed or hydrological boundary concepts can be easily accounted for the water resource development programmes. The classification of major river basins of Indian rivers, in a systematic manner, was initially presented by Dr. A.N. Khosla, CWPC, India under the direction of in 1949 and it was further followed by the CWC, New Delhi, India in their River basin Atlas of India (2012).

In India, a number of large river basins (around 12) are present having their basin area near to 20000 km^2 and above. The whole basin area corresponding to all major river basins is estimated around 25.3 lakh km^2. The major river basin has been defined as the Ganga-Brahmaputra-Meghna, which corresponds to the largest basin area of around 11.0 lakh km^2 (approximately 43% and area for all major river basins of the country). Another major river basins such as Indus, Mahanadi, Godavari and Krishna have more than 1.0 lakh km^2 basin area. Apart from these major river basins, around 46 medium-sized river basins having their basin between 2000 km^2 and 20000 km^2 also exist. The medium-sized river basins have about 2.50 lakh km^2 basin area. The medium-sized river and major river basins are the inter-state river basins, which correspond to about 81% geographical area of the country (River Basin Atlas of India). India has the maximum number of perennial channels in the world. However, the inland-based river basins and their streamflow depends on the Indian monsoon conditions.

11.3 Drought

11.3.1 Introduction

Drought is a natural disaster which generally has a large areal extent, and sometimes, may extend to continental scale. Drought is an environmental condition of lack in moisture supply over an extended period of time such as a season or a year. It arises due to subnormal precipitation over a large area for long period of time. Unlike other hydrological extreme phenomena like floods which are brief and short-lived, droughts are more gradual phenomenon. It slowly extends over an area and tightens its grip with time. In some extreme cases, it can last for many years, causing devastating impacts on water supplies and agriculture. Droughts occur in all climatic zones, in areas of high as well as low rainfalls.

11.3.2 Definition

There is no single definition which is used universally. Droughts have been defined differently by various organisations around the globe.

1. The World Meteorological Organisation has defined drought as a sustained, extended deficiency in precipitation.
2. According to the UN Convention to Combat Drought and Desertification, drought means the naturally occurring phenomenon that exists when precipitation has been significantly below normal recorded levels, causing serious hydrological imbalance that adversely affects land resource production system.
3. According to the Food and Agriculture Organisation (FAO) of the United Nations, drought hazard is the percentage of years when crops fails from the lack of moisture.
4. According to the Encyclopedia of Climate and Weather, the drought can be defined as an extended period—a season, a year or several years—of deficient rainfall relative to the statistical multi-year mean for a region.
5. As per the Indian Meteorological Department (IMD), drought is a condition when annual rainfall is less than 75% of the normal rainfall.

As per the IMD, a severe drought is the one when deficiency in rainfall is more 50% of the normal rainfall. An area is considered as drought-affected area if the drought has occurred in 20% years of the total number of years considered. Generally, rainfall for minimum 60 years is taken for identification of drought-affected areas. If the drought has occurred in more than 40% of the total years considered then that area is called *chronic drought area*.

11.3.3 Classification of Droughts

Droughts are classified as meteorological (based on meteorology, e.g. amount of rainfall), hydrological (based on hydrology, e.g. flow in steams, reservoirs, lakes) or agricultural (based on agricultural conditions, e.g. soil moisture) of the affected regions by the National Commission on Agriculture (1976). Apart from these three commonly used types of drought, authors have been using a fourth type of class based on the socio-economic impacts of the droughts. Figure 11.1 shows different classes of droughts.

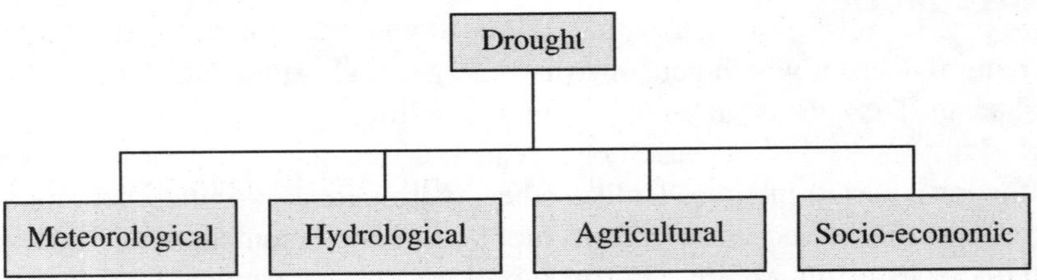

Figure 11.1 Different classes of droughts.

Meteorological drought

Meteorological drought is defined based on the amount of rainfall received by an area over a period of time. The situation of rainfall deficit more than 25% of the normal rainfall is called *meteorological drought*. Drought-affected areas are subclassified based on the rainfall deficit as per the Indian Meteorological Department. If the percent rainfall deficit lies between 26% and 50%, then the drought is classified as *moderate*. If the percent rainfall deficit is more than 50%, then the drought is classified as *severe*.

Hydrological drought

Hydrological drought is defined based on the scarcity in surface or subsurface water supply (i.e., streamflow, groundwater flow, reservoir water levels). The reason behind the depletion of surface and subsurface water resources is rainfall deficiency over a prolonged period of time. This kind of drought is generally studied on watershed or river basin scale. Streamflow has been extensively used by authors for hydrological drought studies. Hydrological droughts are usually out of phase with (or lag) the occurrence of meteorological and agricultural droughts.

Hydrological drought has the following four components:

1. **Magnitude:** It signifies the amount of deficiency in water supplies.
2. **Duration:** It is the time period between the beginning and the end of the drought.
3. **Severity:** It is the cumulative amount of deficiency in water supplies.
4. **Frequency of occurrence**

Agricultural drought

Agricultural drought occurs due to reduction in the soil moisture. The available soil moisture is not able to meet crop water demand of the plants under such conditions. Water demand of plants is dependent on prevailing weather conditions, biological characteristics of specific plant and stage of growth, and the physical and biological properties of soil. Hence, agricultural drought is region-, crop- and soil-specific. This type of drought also occurs due to prolonged deficiency in the rainfall amount.

Aridity index (AI) is generally used for quantification of the agricultural drought. It is defined as the ratio of annual water deficiency to the annual water required, and it is expressed in percentage. An aridity index is defined as:

$$AI = \frac{\text{PET} - \text{AET}}{\text{PET}} \times 100$$

where, PET is potential evapotranspiration and AET is actual evapotranspiration.

Classification of agricultural drought based on AI anomaly, which is the deviation of AI from normal value, is given in Table 11.1.

Table 11.1 Intensity of Agricultural Drought

AI anomaly	*Drought severity*
≤ 0	Non-arid
1–25	Mild arid
26–50	Moderate arid
> 50	Severe arid

Socio-economic drought

It is associated with the shortage of supply of some economic goods like drinking water, food grains, fish and hydroelectric power to fulfil the demand. With the unceasing increase in population and industrialisation, demand for economic goods is also increasing. When a shortfall in rainfall or surface and subsurface water supply causes lack of supplies to meet the demand of economic goods, the condition is called *socio-economic drought.*

11.3.4 Drought Indices

Drought indices are quantitative measures which are used to characterise the intensity of drought. Data for variables like precipitation and evaporation are converted into a single numeric value. Various commonly used drought indices for different classes of droughts are given in Table 11.2.

Table 11.2 Various Drought Indices Used

Drought class	*Indices used*
Meteorological drought	- Discrete and cumulative precipitation anomalies - Rainfall deciles - Palmer drought severity index (PDSI) - Drought area index (DAI) - Rainfall anomaly index (RAI) - Standardised precipitation index (SPI)
Hydrological drought	- Total water deficit - Cumulative streamflow anomaly - Palmer hydrological drought severity index (PHDI) - Surface water supply index (SWSI)
Agricultural drought	- Crop moisture index (CMI) - Palmer moisture anomaly index (Z index) - Computed soil moisture

11.4 Flood

11.4.1 Introduction

A flood is an excess of water (or mud) on land that is normally dry and is a situation wherein the inundation is caused by high flow or overflow of water in an established watercourse such as a river, stream or drainage ditch; or ponding of water at or near the point where the rain fell. A flood can strike anywhere without and warning and occurs when a large volume of rain falls within a short time.

Floods are relatively high flows of runoff resulted from high rainfall or due to melting of snow, which may overtop the natural or the artificial banks in a reach of a channel. Along a river, floods occur at different sections at different discharge. Amount of discharge that may cause flood at one reach may not overtop the banks at some other reaches.

11.4.2 Causes of Flood

The factors that can cause flood are as follows:

1. A high intensity rainfall over the catchment can result in huge discharge in the streams which may result into flood. Rainfall for long duration can also cause large accumulated flow in the streams.
2. A heavy landslide in the river may block the river flow and cause accumulation of water on the upstream. This may cause flood in the upstream area due to increase in water level.
3. Earthquakes may cause rise in the bed of river, which causes water to overtop the banks of the river banks.
4. A huge flow may be caused in the downstream of a dam due to sudden breaking of a dam. A similar phenomenon, called *glacial lake outburst flood (GLOF)*, is the release of huge amount of water due to bursting of glacial lakes.
5. A rapid melting of snow or ice can also result in the increase in discharge in the streams.

11.4.3 Classification of Flood

Floods are generally classified as follows:

1. Design flood: It is the amount of discharge that is considered for the design of hydraulic structure, such as bridges, flood banks, etc. It depends on the level of safety desirable to a particular structure. It can be standard project flood (SPF), maximum probable flood (MPF) or flood of any desired recurrence interval. Design flood is decided based on the nature of structure.

2. Standard project flood (STF): It is the flood which is likely to occur from the most severe combination of hydrological and meteorological characteristics of the watershed, but not due to extremely rare combinations. Its magnitude is approximately 80% of the maximum probable flood (MPF).

3. Maximum probable flood (MPF): The MPF is the largest flood that could conceivably occur at a particular location, usually estimated from probable maximum precipitation or snowmelt, coupled with the worst flood-producing catchment conditions. It is the extremely rare and catastrophic flood, and generally, it is not physically or economically possible to provide complete protection against this event. It results from the severemost combinations of critical hydrological and meteorological characteristics of the watershed.

4. Peak flood: It is the maximum intensity of the flood discharge which happens in the stream during flood duration. In other words, it is the maximum instantaneous flow in the river.

5. Maximum observed flood: It is the highest flood that occurs during a specified period of time. The observed time period can range from a week to years. It is defined for the time for which flood data are available.

6. Maximum flood: It is the largest flood which has ever occurred in the past. Data of this type of flood may not be available.

7. Annual flood: It is the highest flood which is observed in a water year or a hydrologic year. A *water year* is defined as the duration of 12 months of a year selected for maintaining or presenting records of flows.

8. Ordinary flood: It is the flood which occurs more than once during the project period.

9. Flash flood: It is the abrupt rise in the discharge in a stream with relatively high peaks. It may be caused by heavy rain associated with a severe thunderstorm, hurricane, tropical storm, or melted water from ice or snow flowing over ice sheets or snowfields. Flash floods may occur after the collapse of a natural ice or debris dam, or a human structure such as a man-made dam.

Flooding is also divided into two categories—first based on location and second based on duration.

1. Flooding based on location: In this category flood occurs at different locations such as

(a) *Coastal flood:* If flood occurs in coastal areas it is called *coastal flood.*
(b) *Arroyos flood:* If flood occurs in arroyos rivers which are mostly dry, it is called *arroyos flood.*
(c) *Urban flood:* If flood occurs in urban areas, it is called *urban flood.*
(d) *River flood:* If flood occurs in river site, it is called *river flood.*

2. Flooding based on location: In this category, flood occurs for different time durations such as

(a) *Slow-onset flood:* The flood occurs for longer time period, it may be one or two weeks, or even months.
(b) *Rapid-onset flood:* The flood occurs for one or two days.
(c) *Flash flood:* As the name suggests, the flood occurs instantly, within few minutes or few hours after heavy rainfall, reservoir break and glacial lake outburst, etc.

11.4.4 Factors Affecting Flood

The factors which affect flood can be categorised into two classes—Meteorological factors and physiological factors.

Meteorological factors

These are atmospheric parameters which affect floods. These climatic factors affect the rainfall happening in the catchment. The rainfall water, after losses, flows over the land and becomes streamflow. Some of the atmospheric factors are as follows:

1. Intensity, duration and coverage of precipitation: More the intensity, more the duration and more the coverage of the precipitation, more will be the discharge in the streams.

2. Movement of storm: If the storm movers along the direction of the flow, the time of concentration will be less and the peak discharge will be high. On the other hand, if the storm moves against the flow direction, the time of concentration as well as the peak discharge will be less.

3. Climate change: With the increase in the carbon emission and global warming, the climate is changing around the globe. Researchers have observed change in rainfall pattern in various parts of the world. Flood frequency and peak flood discharge are expected to increase due to the changing climate.

Physiological factors

These factors include the physical characteristics of the watershed, which affect the flow of precipitation over the land and accumulation of this water into the streams. These factors include the following:

1. Size and shape of the watershed: The shape and size of the watershed affect floods. Time of concentration for a fan-shaped catchment is less and hence, the peak discharge will be more. On the other hand, in case of an elongated-shaped catchment, the time of concentration will be more.
2. Vegetation in and around the river
3. Hydraulic structures present in the river
4. Deforestation and watershed management
5. Drainage pattern and density.

11.5 Climate Change

As per the exact definition of the climate change given by the Intergovernmental Panel on Climate Change (IPCC) (2007), *Climate change* refers to a change in the state of the climate that can be identified (e.g., by using statistical tests) by changes in the mean and/or the variability of its properties, and that persists for an extended period, typically decades or longer. Due to the adverse impacts and negative consequences of the climate system, people livings, species (flora and fauna) diversities, ecosystem behaviour, environmental functions, conservation services, and natural resources, infrastructure, or economic, social, or cultural assets in places and settings could be affected unpleasantly. Many climate change studies revealed that over the last few decades, changes in climate have caused impacts on natural and human systems on all continents and across the oceans.

The mean global average Earth's surface and ocean's surface temperature calculated by a linear trend illustrates an increase in temperature of 0.85 (0.65 to 1.06)°C for a historical time duration from 1880 to 2012 (IPCC). The total increments, between the mean of the 1850–1900 period and the mean of the 2003–2012 time duration is 0.78 (0.72 to 0.85)°C, based on the single longest dataset available. The atmospheric radiative forcing (RF) is recorded around 43% higher than the records from 2005 reported in the IPCC Assessment Report 4 (AR4). This is causing due to the continuous growth in greenhouse gas concentration, and it indicates a frailer net cooling effect. In comparison the larger scales, on regional scales, the confidence in model ability to simulate surface temperature is lesser.

11.6 National Demand and Water Resources Availability

Due to agricultural growth, urbanisation, growing population, fast industrialisation and economic development, the requirement of water has already been grown up and will be more serious in the coming years, especially in India. The land-use changeability, alterations in cropping pattern and overconsumption of water due to increasing demand in irrigation and drainage sectors are seriously affecting the hydrological cycle in many regions of India. To see the growth of the future national requirements of water, a broad assessment of water resources, as per their availability and requirement, is needed. Due to the negative effects of the changing climatic conditions, water availability is critical for sustainable development and significant national and regional long-term development strategies. The decreasing availability of fresh water reserves or water scarcity is a key factor driving the demand for water worldwide, which is directly affected by the factors such as population progression, urbanisation and industrial development. In India, the demand for water is expected to be grown up by 20%, and the industrial requirements are projected to be doubled as 24.1 trillion litres in the contemporary to 46 trillion litres.

In India, the domestic requirement of water is expected to be grown up by 40% from 41 to 55 trillion litres, whereas in the irrigation sector, it will be required only 14% more 10 years, and thus, 592 trillion litres up from 517 trillion litres at present. The per capita obtainability of water has come done in past and is possible to come down further with the additional demand due to increasing populations. Expressively came down and is possible to come down additional with the demand and increasing population. According to the CWC and Water Resources Ministry of India, the per capita water accessibility in India between the years 2025 and 2050 is predicted to come down by approximately 36% and 60% correspondingly, in comparison to the 2001 levels.

11.7 Climate Change Impacts on Water Resources

In the last few years, majority of studies revealed observed changes in the air temperature, precipitation intensity and volume, evaporation rate and extreme climate events since the 20th century and it is also projected that these climatic severities will be more vulnerable and frequent in 21st century. In the future time (21st century) domain, it has been projected that the carbon dioxide concentration in the atmosphere will be increased up to about 400 ppm and more. As per the Intergovernmental Panel on Climate Change (IPCC-2007), the global average annual temperature of the Earth has increased by 1.2°C due to anthropogenic activities. Due to increment in the extreme events, the amount of overland flow and streamflow runoff will be increased in the 21st century and the global mean sea level will rise by 15 cm to 30 cm.

Past few studies have revealed that there has been 40% decrements in the past 45–50 years, notified in Arctic sea's ice thickness in late summer to early autumn. As per the IPCC-2007, the frequency of severe extreme events in the context of water resources such as floods and flash floods, especially in large and Himalayan river basins, have increased during the 20th century and will be more disastrous in the 21st century. Due to the global climate change, the spatial variability in precipitation pattern has been increased, and thus, in India, major portions are facing severe droughts and many places are facing floods.

The main perennial rivers of India such as Ganga, Brahmaputra and Indus have originated from Himalayan continents and these rivers maintain their flow throughout the year. The water melting from temporary snow covers and glaciers/ice sheets and the baseflow contributes to the discharge in spring season and regulates the natural flow. It has been noticed that the current climatic uncertainties, occurring mainly due to increasing global annual temperature, are influencing the hydrological system and disrupting the environmental flow of these perennial river channels. The mean intensities of mountain glaciations vary from 3.4% to 1.3% (e.g., 3.4% for Indus, 3.2% for Ganges and 1.3% for Brahmaputra). However, their tributaries show maximum intensity of glaciations for Indus river system (2.5 to 10.8%), for Ganges, it varies from 0.4% to 10%, and for Brahmaputra river system it varies from 0.4% to 4%. From the above statistics, it is concluded that the precipitation events are more substantial in the eastern region, whereas, the melt water contributions are more predominant and more significant in the western and central Himalayan regions.

The impact of global changing climate is very often over the entire Himalayan regions, and thus, periodic and chaotic variations on a vast range of spatial and temporal scales are also observed. At present, due to climate change, the climatic uncertainties are effectively accelerating the hydrological cycle, and thus, the precipitation pattern and its intensity are highly variable in nature as their magnitude and timing of runoff are shifting. At present, approximately 40% to 45% of the mean annual precipitation, including rainfall and snowfall in the Himalayan rivers, is wasting to the sea by natural discharge process. There are many indicators of climate change such as surface and air temperature, atmospheric water vapour, precipitation, extreme events, glacier retreating, and melting of land ice noticed in the Himalayan regions. Due to the increasing trends of air temperature across the globe (IPCC 2013), Himalayan glaciers are retreating with faster rate than their normal rate. In the recent years, glacial retreat is a major concern that has been noticed over the entire Himalayan regions.

A few studies focused on the Himalayas show Himalayan glaciers are under serious threat due to the changing climate. On a regional scale, snow cover and valley glaciers are the important sources for the local water availability, streamflows and groundwater recharge over the entire Himalayan continents. The precipitation variability has also been increased over the Himalayas (IPCC-2013). Several climate change issues have been noticed over the eastern and western Himalaya in India. Kedarnath disaster experienced in 2013 in the Uttarakhand state of India, which is a part of western Himalaya, itself explained about the lack of current research findings and its understanding. In Indian, the groundwater irrigation contributes around 70–80% of the value of irrigated production. A few studies has also revealed the negative impacts of climate change on the Indian groundwater reserves. In many parts of India such as Haryana and Punjab, the groundwater level is significantly decreased due to overdrafting because of less rainfall. Therefore, better management and mitigation strategies are necessary to minimise these threats. Here, some observations have been analysed in the 21st century scenarios using the Fifth Generation Coupled Model Intercomparison Project (CMIP5) simulations, reported in the IPCC AR.5 as below:

1. Due to increasing the surface temperature, the vapour carrying capacity is affecting, and thus, the variability and the ratio of precipitation to snowfall and rainfall are increasing non-uniformly.
2. Warming rate is affecting the snow-melt and glacier-mass budgets.
3. The precipitation as snowfall is decreasing and the extent of snow cover is reducing by earlier melting.
4. Wet regions and seasons are becoming wetter and dry regions are becoming drier.
5. Global average precipitation has increased in the warmer world and the variability is also increasing from region to region.

11.8 Climate Change Assessment and Forecasting

The IPCC AR. provides comprehensive assessment of climate change issues, their impacts and response strategies through the evaluation of scientific approaches. The first IPCC AR was produced in the year 1990 to analyse the scientific understandings and related uncertainties underlying assessment findings. The IPCC AR5 explores the literature on the scientific, technological, economic and social aspects of mitigation of climate change. The long-term scenarios measured in the IPCC AR5 were produced principally by large-scale coupled models or integrated models. Integrated models deliver measureable, long-term projections of many of the most important characteristics of mitigation pathways. They account for many of the most important interactions between the natural systems and numerous relevant human systems.

The Fifth Generation Coupled Model Intercomparison Project (CMIP5) has been built after the accomplishments of earlier phases of CMIP3. The CMIP5 models include two types of experiments—(a) long-term (century time scale) and (b) near-term scenarios (10–30 years). Both these experiments are coupled with ocean-atmosphere GCMs (General Circulation Models). The CMIP5 scenarios of climate change are determined by emission scenarios reliable with the RCP (Representative Concentration Pathways) experiments, which explore an estimate of the RF (Radiative Forcing) in the 21st century. The RCP8.5 has been categorised as the high emission scenarios experiment, and as per RCP8.5, radiative forcing will be increased in the 21st century. Likewise, the RCP2.6 and the RCP4.5 are the intermediate scenarios, in which the RF will be reached a maximum near the middle of the 21st century before decreasing. The model responses are examined by the detection and attribution simulations of these model experiments to subsets of historical forcing, which include well-mixed greenhouse gases and ozone only, natural forcing only, aerosol forcing only, and anthropogenic forcing only.

However, with respect to the adverse effects of climate change, a long-term planning and conservation-management practices are needed to save our water resources which is extremely uncertain. For the betterment of the water resource system, a supple group of ideas, which are capable to minimise the impacts of climate change and can be applied effectively, could be beneficial because there can be allowed several policies to evolve progressively. Many adaptive methods, which can demonstrate principally functioning, including conservation tillage, rainwater harvesting, plantation of trees, protection vegetation cover in a high degree of slopes, mini-terracing for soil and moisture preservation, better grassland management, water reuse, distillation, and more efficient soil and irrigation water management, have been used as the water resources management practices.

A number of statistical downscaling methods are utilised by the various researchers around the world all these methods use the basic concept of founding quantitative relations between large-scale GCM predictors and the point source observational predictands or observed datasets. The statistical downscaling demonstrates a commonly used approach to downscale high-resolution or in situ climate projections. To examine the impact of global climate change on the local climate, a new dynamical downscaling methodology has been utilised worldwide. The most straightforward approach of interpolation or downscaling of GCMs is to work on the linear regression methods or more sophisticated methods between regional scale grid points nearest to the location to suppose the regional scale using statistical downscaling models (SDSM). The SDSM, developed by Wilby et al. (2006), is an integration of stochastic weather generator and numerous linear regression mathematical functions. The SDSM has been evaluated by many researchers across the world mainly in the downscaling of temperature, precipitation, evapotranspiration etc. It has three sub-models such as monthly, seasonal, and annual to develop, though many authors concluded that the monthly sub-model can be utilised perfectly and more accurately to develop the SDSM for downscaling local climate variables.

Water resources planning and management applications have been demonstrated by various authors and hydrologists using the framework constructed at the larger scale by the actions of hydrological and hydrodynamic modelling tools and techniques. Many researchers found that the accurate prediction of the modelled scenarios is difficult when a watershed/catchment involves snow and glacier-induced hydrological processes additionally. Deterministic hydrological models, which have been evaluated over mountainous catchments for climate change projections such as precipitation-runoff modelling system (PRMS), Topnet generated as per the TopModel (Beven et al., 1995) and soil water assessment tool (SWAT) (Neistch et al., 2011; Arnold et al., 1998), were shown their utilities over complex high-hilly terrains. Combined deterministic and probabilistic methods for steep slopes and flood plain-based catchments have been evaluated.

Summary

This chapter explores the current and futuristic needs of water resources of India. As per the IPCC, the atmospheric warming will be increased in the 21st century, and thus, the severity of extreme events such as droughts and floods will also be frequent in the future time. Apart from the climate change issues, agricultural expansions and deforestation, due to increasing rate of population, will increase the severity on fresh water reserves such as groundwater reserves, and fresh water lakes and ponds. All these factors will increase the load on the availability water resources of in the coming future. Thus, a proper framework, utilising advance water resources conservation and management practices and tools and techniques, is needed for the sustainable management and development of the water resources systems. A basin scale to micro watershed scale implementation of watershed management practices are necessarily required to protect the current availability of fresh water resources. In India, various water resources projects have been established and also proposed by various central governmental agencies such as Central Water Commission (CWC), Central Groundwater Management Board (CGWB) and other national and state water departmental agencies, which are working for the management of water resources of India.

Theoretical Questions

1. Discuss about Indian water resources system.
2. Define drought. What are the types of drought?
3. What are the causes of flood?
4. Define climate change and discuss the impact of climate change on water resources.

Further Reading

IPCC, 2007: Summary for Policymakers, In: Climate Change, *The Physical Science Basis, Contribution of Working Group I to the Fourth Assessment Report of the Intergovernmental Panel on Climate Change* [Solomon, S., D. Quin, M. Manning, Z. Chen, M. Marquis, K.B. Averyt, M. Tignor and H.L. Miller (eds.)], Cambridge University Press, Cambridge, United Kingdom and New York, USA.

IPCC, 2013: Climate Change 2013: The Physical Science Basis, Contribution of Working Group I to the Fifth Assessment Report of the Intergovernmental Panel on Climate Change [Stocker, T.F., D. Quin, G.K. Plattner, M. Tignor, S.K. Allen, J. Boschung, A. Navels, Y. Xia, V. Bex and P.M. Midgley (eds.)], Cambridge University Press, Cambridge, United Kingdom and New York, USA, pp. 1535, the Twelfth Session of Working Group I (WGI-12) Stockholm, Sweden.

Raghunath, H.M., *Hydrology—Principles, Analysis and Design*, New Age Publishers, New Delhi, 2014.

'River Basin Atlas of India', Central Water Commission, New Delhi and National Remote Sensing Center (NRSC), ISRO, Jodhpur, India, 2012.

Yadupathi Putty, Mysooru R., *Principles of Hydrology*, I.K. International, New Delhi, 2013

Appendix

Normal Distribution z-variate

$$\phi(z) = \int_{z}^{-\infty} \frac{1}{\sqrt{2\pi}} e^{-u^2/2du}$$

1	0	0.01	0.02	0.03	0.04	0.05	0.06	0.07	0.08	0.09
0	0.5	0.50399	0.50798	0.51197	0.51595	0.51994	0.52392	0.5279	0.53188	0.53586
0.1	0.53983	0.5438	0.54776	0.55172	0.55567	0.55962	0.56356	0.56749	0.57142	0.57535
0.2	0.57926	0.58317	0.58706	0.59095	0.59483	0.59871	0.60257	0.60642	0.61026	0.61409
0.3	0.61791	0.62172	0.62552	0.6293	0.63307	0.63683	0.64058	0.64431	0.64803	0.65173
0.4	0.65542	0.6591	0.66276	0.6664	0.67003	0.67364	0.67724	0.68082	0.68439	0.68793
0.5	0.69146	0.69497	0.69847	0.70194	0.7054	0.70884	0.71226	0.71566	0.71904	0.7224
0.6	0.72575	0.72907	0.73237	0.73565	0.73891	0.74215	0.74537	0.74857	0.75175	0.7549
0.7	0.75804	0.76115	0.76424	0.7673	0.77035	0.77337	0.77637	0.77935	0.7823	0.78524
0.8	0.78814	0.79103	0.79389	0.79673	0.79955	0.80234	0.80511	0.80785	0.81057	0.81327
0.9	0.81594	0.81859	0.82121	0.82381	0.82639	0.82894	0.83147	0.83398	0.83646	0.83891
1	0.84134	0.84375	0.84614	0.84849	0.85083	0.85314	0.85543	0.85769	0.85993	0.86214
1.1	0.86433	0.8665	0.86864	0.87076	0.87286	0.87493	0.87698	0.879	0.881	0.88298
1.2	0.88493	0.88686	0.88877	0.89065	0.89251	0.89435	0.89617	0.89796	0.89973	0.90147
1.3	0.9032	0.9049	0.90658	0.90824	0.90988	0.91149	0.91309	0.91466	0.91621	0.91774
1.4	0.91924	0.92073	0.9222	0.92364	0.92507	0.92647	0.92785	0.92922	0.93056	0.,93189
1.5	0.93319	0.93448	0.93574	0.93699	0.93822	0.93943	0.94062	0.94179	0.94295	0.94408
1.6	0.9452	0.9463	0.94738	0.94845	0.9495	0.95053	0.95154	0.95254	0.95352	0.95449

(Contd.)

1	0	0.01	0.02	0.03	0.04	0.05	0.06	0.07	0.08	0.09
1.7	0.95543	0.95637	0.95728	0.95818	0.95907	0.95994	0.9608	0.96164	0.96246	0.96327
1.8	0.96407	0.96485	0.96562	0.96638	0.96712	0.96784	0.96856	0.96926	0.96995	0.97062
1.9	0.97128	0.97193	0.97257	0.9732	0.97381	0.97441	0.975	0.97558	0.97615	0.9767
2	0.97725	0.97778	0.97831	0.97882	0.97932	0.97982	0.9803	0.98077	0.98124	0.98169
2.1	0.98214	0.98257	0.983	0.98341	0.98382	0.98422	0.98461	0.985	0.98537	0.98574
2.2	0.9861	0.98645	0.98679	0.98713	0.98745	0.98778	0.98809	0.9884	0.9887	0.98899
2.3	0.98928	0.98956	0.98983	0.9901	0.99036	0.99061	0.99086	0.99111	0.99134	0.99158
2.4	0.9918	0.99202	0.99224	0.99245	0.99266	0.99286	0.99305	0.99324	0.99343	0.99361
2.5	0.99379	0.99396	0.99413	0.9943	0.99446	0.99461	0.99477	0.99492	0.99506	0.9952
2.6	0.99534	0.99547	0.9956	0.99573	0.99585	0.99598	0.99609	0.99621	0.99632	0.99643
2.7	0.99653	0.99664	0.99674	0.99683	0.99693	0.99702	0.99711	0.9972	0.99728	0.99736
2.8	0.99744	0.99752	0.9976	0.99767	0.99774	0.99781	0.99788	0.99795	0.99801	0.99807
2.9	0.99813	0.99819	0.99825	0.99831	0.99836	0.99841	0.99846	0.99851	0.99856	0.99861
3	0.99865	0.99869	0.99874	0.99878	0.99882	0.99886	0.99889	0.99893	0.99896	0.999
3.1	0.99903	0.99906	0.9991	0.99913	0.99916	0.99918	0.99921	0.99924	0.99926	0.99929
3.2	0.99931	0.99934	0.99936	0.99938	0.9994	0.99942	0.99944	0.99946	0.99948	0.9995
3.3	0.99952	0.99953	0.99955	0.99957	0.99958	0.9996	0.99961	0.99962	0.99964	0.99965
3.4	0.99966	0.99968	0.99969	0.9997	0.99971	0.99972	0.99973	0.99974	0.99975	0.99976
3.5	0.99977	0.99978	0.99978	0.99979	0.9998	0.99981	0.99981	0.99982	0.99983	0.99983
3.6	0.99984	0.99985	0.99985	0.99986	0.99986	0.99987	0.99987	0.99988	0.99988	0.99989
3.7	0.99989	0.9999	0.9999	0.9999	0.99991	0.99991	0.99992	0.99992	0.99992	0.99992
3.8	0.99993	0.99993	0.99993	0.99994	0.99994	0.99994	0.99994	0.99995	0.99995	0.99995
3.9	0.99995	0.99995	0.99996	0.99996	0.99996	0.99996	0.99996	0.99996	0.99997	0.99997

Index